C.B. Allan O. Hume

The Nests And Eggs Of Indian Birds Vol. I

C.B. Allan O. Hume

The Nests And Eggs Of Indian Birds Vol. I

ISBN/EAN: 9783742839558

Manufactured in Europe, USA, Canada, Australia, Japa

Cover: Foto ©berggeist007 / pixelio.de

Manufactured and distributed by brebook publishing software
(www.brebook.com)

C.B. Allan O. Hume

The Nests And Eggs Of Indian Birds Vol. I

WOODBURY COMPY

ALLAN OCTAVIAN HUME

THE

NESTS AND EGGS

OF

INDIAN BIRDS.

BY

ALLAN O. HUME, C.B.

SECOND EDITION.

EDITED BY

EUGENE WILLIAM OATES,

AUTHOR OF 'A HANDBOOK TO THE BIRDS OF BRITISH BURMAH' AND OF
THE BIRDS IN 'THE FAUNA OF BRITISH INDIA.'

VOL. I.

WITH FOUR PORTRAITS.

LONDON:

R. H. PORTER.
18 PRINCES STREET, CAVENDISH SQUARE, W.
1889.

ALERE FLAMMAM.

PRINTED BY TAYLOR AND FRANCIS,
RED LION COURT, FLEET STREET.

AUTHOR'S PREFACE.

I HAVE long regretted my inability to issue a revised edition of 'Nests and Eggs.' For many years after the first Rough Draft appeared, I went on laboriously accumulating materials for a re-issue, but subsequently circumstances prevented my undertaking the work. Now, fortunately, my friend Mr. Eugene Oates has taken the matter up, and much as I may personally regret having to hand over to another a task, the performance of which I should so much have enjoyed, it is some consolation to feel that the readers, at any rate, of this work will have no cause for regret, but rather of rejoicing that the work has passed into younger and stronger hands.

One thing seems necessary to explain. The present Edition does not include quite all the materials I had accumulated for this work. Many years ago, during my absence from Simla, a servant broke into my museum and stole thence several cwts. of manuscript, which he sold as waste paper. This manuscript included more or less complete life-histories of some 700 species of birds, and also a certain number of detailed accounts of nidification. All small notes on slips of paper were left, but almost every article written on full-sized foolscap sheets was abstracted. It was not for many months that the theft was discovered, and then very little of the MSS. could be recovered.

It thus happens that in the cases of some of the most interesting species, of which I had worked up all the notes into a connected whole, nothing, or, as in the case of *Argya subrufa*, only a single isolated note, appears in the text. It is to be greatly regretted, for my work was imperfect enough as it was; and this 'Selection from the Records,' that my Philistine-servant saw fit to permit himself, has rendered it a great deal more imperfect still; but neither Mr. Oates nor myself can be justly blamed for this.

In conclusion, I have only to say that if this compilation should find favour in any man's sight he must thank Mr. Oates for it, since not only has he undergone the labour of arranging my materials and seeing the whole work through the press—not only has he, I believe, added himself considerably to those materials—but it is solely owing to him that the work appears *at all*, as I know no one else to whom I could have entrusted the arduous and, I fear, thankless duty that he has so generously undertaken.

ALLAN HUME.

Rothney Castle, Simla,
October 19th, 1889.

EDITOR'S NOTE.

Mr. Hume has sufficiently explained the circumstances under which this edition of his popular work has been brought about. I have merely to add that, as I was engaged on a work on the Birds of India, I thought it would be easier for me than for anyone else to assist Mr. Hume. I was also in England, and knew that my labour would be very much lightened by passing the work through the press in this country. Another reason, perhaps the most important, was the fear that, as Mr. Hume had given up entirely and absolutely the study of birds, the valuable material he had taken such pains to accumulate for this edition might be irretrievably lost or further injured by lapse of time unless early steps were taken to utilize it.

A few words of explanation appear necessary on the subject of the arrangement of this edition. Mr. Hume is in no way responsible for this arrangement nor for the nomenclature employed. He may possibly disapprove of both. He, however, gave me his manuscript unreservedly, and left me free to deal with it as I thought best, and I have to thank him for reposing this confidence in me. Left thus to my own devices, I have considered it expedient to conform in all respects to the arrangement of my work on the Birds, which I am writing, side by side, with this work. The classification I have elaborated for my purpose is totally different to that employed by Jerdon and familiar to Indian ornithologists; but a departure from Jerdon's arrangement was merely a question of time, and no better opportunity than the present for readjusting the classification of Indian birds appeared likely to present itself. I have therefore adopted a new system, which I have fully set forth in my other work.

I take this opportunity to present the readers of Mr. Hume's work with portraits of Mr. Hume himself, of Mr. Brian Hodgson, the late Dr. Jerdon, and the late Colonel Tickell.

EUGENE W. OATES.

ERRATA.

Page 103. *After* **Drymocataphus tickelli** *insert* (Blyth).

Page 216. *For* Bhringa tenuirostris *read* B. tectirostris.

Page 223. *For* **Pnoepyga albiventris** (Hodgs.), *read* **Pnoepyga sq mata** (Gould).

Page 311. *After* **Lanius vittatus** *insert* Valenc.

WOODBURY COMPY.

THOMAS CAVERHILL JERDON.

WOODBURY COMPY.

BRIAN HOUGHTON HODGSON.

WOODBURY COMP?

SAMUEL RICHARD TICKELL.

THE

NESTS AND EGGS

OF

INDIAN BIRDS.

Order PASSERES.

Family CORVIDÆ.

Subfamily CORVINÆ.

1. Corvus corax, Linn. *The Raven.*

Corvus corax, *Linn., Jerd. B. Ind.* ii, p. 293.
Corvus lawrencii, *Hume; Hume, Rough Draft N. & E.* no. 657.

I separated the Punjab Raven under the name of *Corvus law-rencei* ('Lahore to Yarkand,' p. 83), and I then stated, what I wish now to repeat, that if we are prepared to consider *C. corax, C. littoralis, C. thibetanus,* and *C. japonensis* all as one and the same species, then *C. lawrencei* too must be suppressed; but if any of these are retained as distinct, then so must *C. lawrencei* be*.

The Punjab Raven breeds throughout the Punjab (except perhaps in the Dehra Ghazee Khan District), in Bhawulpoor, Bikaneer, and the northern portions of Jeypoor and Jodhpoor, extending rarely as far south as Sambhur. To Sindh it is merely a seasonal

---

* I think it impossible to separate the Punjab Raven from the Ravens of Europe and other parts of the world, and I have therefore merged it into *C. corax.*—ED.

visitant, and I could not learn that they breed there, nor have I ever known of one breeding anywhere east of the Jumna. Even in the Delhi Division of the Punjab they breed sparingly, and one must go further north and west to find many nests.

The breeding-season lasts from early in December to quite the end of March; but this varies a little according to season and locality, though the majority of birds always, I think, lay in January.

The nest is generally placed in single trees of no great size, standing in fields or open jungle. The thorny Acacias are often selected, but I have seen them on Sisoo and other trees.

The nest, placed in a stout fork as a rule, is a large, strong, compact, stick structure, very like a Rook's nest at home, and like these is used year after year, whether by the same birds or others of the same species I cannot say. Of course they never breed in company: I *never* found two of their nests within 100 yards of each other, and, as a rule, they will not be found within a quarter of a mile of each other.

Five is, I think, the regular complement of eggs; very often I have only found four fully incubated eggs, and on two or three occasions six have, I know, been taken in one nest, though I never myself met with so many.

I find the following old note of the first nest of this species that I ever took :—

"At Hansie, in Skinner's Beerh, December 19, 1867, we found our first Raven's nest. It was in a solitary Keekur tree, which originally of no great size had had all but two upright branches lopped away. Between these two branches was a large compact stick nest fully 10 inches deep and 18 inches in diameter, and not more than 20 feet from the ground. It contained five slightly incubated eggs, which the old birds evinced the greatest objection to part with, not only flying at the head of the man who removed them, but some little time after they had been removed similarly attacking the man who ascended the tree to look at the nest. After the eggs were gone, they sat themselves on a small branch above the nest side by side, croaking most ominously, and shaking their heads at each other in the most amusing manner, every now and then alternately descending to the nest and scrutinizing every portion of the cavity with their heads on one side as if to make sure that the eggs were really gone."

Mr. W. Theobald makes the following note of this bird's nidification in the neighbourhood of Pind Dadan Khan and Katas in the Salt Range :—

"Lay in January and February; eggs, four only; shape, ovato-pyriform; size, 1·7 by 1·3; colour, dirty sap green, blotched with blackish brown; also pale green spotted with greenish brown and neutral; nest of sticks difficult to get at, placed in well-selected trees or holes in cliffs."

I have not verified the fact of their breeding in holes in cliffs, but it is very possible that they do. All I found near Pind Dadan Khan and in the Salt Range were doubtless in trees, but I explored a very limited portion of these hills.

Colonel C. H. T. Marshall, writing from Bhawulpoor on the 17th February, says : " I succeeded yesterday in getting four eggs of the Punjab Raven. The eggs were hard-set and very difficult to clean."

From Sambhur Mr. R. M. Adam tells us :—" This Raven is pretty common during the cold weather, but pairs are seen about here throughout the year. They are very fond of attaching themselves to the camps of the numerous parties of Banjaras who visit the lake.

" I obtained a nest at the end of January which contained three eggs, and a fourth was found in the parent bird. The nest was about 15 feet from the ground in a Kaggera tree (*Acacia leucophloea*) which stood on a bare sandy waste with no other tree within half a mile in any direction."

The eggs of the Punjab bird are, as might be expected, much the same as those of the European Raven. In shape they are moderately broad ovals, a good deal pointed towards the small end, but, as in the Oriole, greatly elongated varieties are very common, and short globular ones almost unknown. The texture of the egg is close and hard, but they usually exhibit little or no gloss. In the colour of the ground, as well as in the colour, extent, and character of the markings, the eggs vary surprisingly. The ground-colour is in some a clear pale greenish blue ; in others pale blue ; in others a dingy olive ; and in others again a pale stone-colour. The markings are blackish brown, sepia and olive-brown, and rather pale inky purple. Some have the markings small, sharply defined, and thinly sprinkled ; others are extensively blotched and streakily clouded ; others are freckled or smeared over the entire surface, so as to leave but little, if any, of the ground-colour visible. Often several styles of marking and shades of colouring are combined in the same egg. Almost each nest of eggs exhibits some peculiarity, and varieties are endless. With sixty or seventy eggs before one, it is easy to pick out in almost every case all the eggs that belong to the same nest, and this is a peculiarity that I have observed in the eggs of many members of this family. All the eggs out of the same nest usually closely resemble each other, while almost *any* two eggs out of different nests are markedly dissimilar.

They vary from 1·72 to 2·25 in length, and from 1·2 to 1·37 in width ; but the average of seventy-two eggs measured is 1·94 by 1·31.

Mandelli's men found four eggs of the larger Sikhim bird in Native Sikhim, high up towards the snows, where they were shooting Blood-Pheasants.

These eggs are long ovals, considerably pointed towards one end ; the shell is strong and firm, and has scarcely any gloss. The ground-colour is pale bluish green, and the eggs are smudged and clouded all over with pale sepia ; on the top of the eggs there are a few small spots and streaks of deep brownish black. They were found on the 5th March, and vary in length from 1·83 to 1·96, in breadth from 1·18 to 1·25.

3. Corvus corone, Linn. *The Carrion-Crow.*

Corvus corone, *Linn., Jerd. B. Ind.* ii, p. 295; *Hume, Rough Draft N. & E.* no. 659 *.

The only Indian eggs of the Carrion-Crow which I have seen, and one of which, with the parent bird, I owe to Mr. Brooks, were taken by the latter gentleman on the 30th May at Sonamerg, Cashmere.

The eggs were broad ovals, somewhat compressed towards one end, and of the regular Corvine type—a pretty pale green ground, blotched, smeared, streaked, spotted, and clouded, nowhere very profusely but most densely about the large end, with a greenish or olive-brown and pale sepia. The brown is a brighter and greener, or duller and more olive, lighter or darker, in different eggs, and even in different parts of the same egg. The shell is fine and close, but has only a faint gloss.

The eggs only varied from 1·67 to 1·68 in length, and from 1·14 to 1·18 in breadth.

Whether this bird breeds regularly or only as a straggler in Cashmere we do not know; it is always overlooked and passed by as a "Common Crow." Future visitors to Cashmere should try and clear up both the identity of the bird and all particulars about its nidification.

4. Corvus macrorhynchus, Wagler. *The Jungle-Crow.*

Corvus culminatus, *Sykes, Jerd. B. Ind.* ii, p. 295.
Corvus levaillantii, *Less., Hume, Rough Draft N. & E.* no. 660.

The Jungle-Crow (under which head I include † *C. culminatus,* Sykes, *C. intermedius,* Adams, *C. andamanensis,* Tytler, and each and all of the races that occur within our limits) breeds almost everywhere in India, alike in the low country and in the hills both of Southern and Northern India, to an elevation of fully 8000 feet.

March to May is, I consider, the normal breeding-season; in the plains the majority lay in April, rarely later, and in the hills in May; but in the plains a few birds lay also in February.

The nest is placed as a rule on good-sized trees and pretty near their summits. In the plains mangos and tamarinds seem to be preferred, but I have found the nests on many different kinds of trees. The nest is large, circular, and composed of moderate-sized twigs; sometimes it is thick, massive, and compact; sometimes loose and straggling; always with a considerable depression in the centre, which is smoothly lined with large quantities of horsehair,

* Mr. Hume, at one time separated the Indian Carrion-Crow from *Corvus corone* under the name *C. pseudo-corone.* In his 'Catalogue' he re-unites them. I quite agree with him that the two birds are inseparable.—ED.
† See 'Stray Feathers,' vol. ii. 1874, p. 243, and 'Lahore to Yarkand,' p. 85.

or other stiff hair, grass, grass-roots, cocoanut-fibre, &c. In the hills they use *any* animal's hair or fur, if the latter is pretty stiff. They do not, according to my experience, affect luxuries in the way of soft down; it is always something moderately stiff, of the coir or horsehair type; nothing soft and fluffy. Coarse human hair, such as some of our native fellow-subjects can boast of, is often taken, when it can be got, in lieu of horsehair.

They lay four or five eggs. I have quite as often found the latter as the former number. I have never myself seen six eggs in one nest, but I have heard, on good authority, of six eggs being found.

Captain Unwin writes: "I found a nest of the Bow-billed Corby in the Agrore Valley, containing four eggs, on the 30th April. It was placed in a Cheer tree about 40 feet from the ground, and was made of sticks and lined with dry grass and hair."

Mr. W. Theobald makes the following remarks on the breeding of this bird in the Valley of Cashmere :—

"Lays in the third week of April. Eggs four in number, ovato-pyriform, measuring from 1·6 to 1·7 in length and from 1·2 to 1·25 in breadth. Colour green spotted with brown; valley generally. Nest placed in Chinar and difficult trees."

Captain Hutton tells us that the Corby " occurs at Mussoorie throughout the year, and is very destructive to young fowls and pigeons; it breeds in May and June, and selects a tall tree, near a house or village, on which to build its nest, which is composed externally of dried sticks and twigs, and lined with grass and hair, which latter material it will pick from the backs of horses and cows, or from skins of animals laid out to dry. I have had skins of the Surrow (*Nœmorhœdus thar*) nearly destroyed by their depredations. The eggs are three or four in number."

From the plains I have very few notes. I transcribe a few of my own.

"On the 11th March, near Oreyah, I found a nest of a Corby— a good large stick nest, built with tamarind twigs, and placed fully 40 feet from the ground in the fork of a mango-tree standing by itself. The nest measured quite 18 inches in diameter and five in thickness. It was a nearly flat platform with a central depression 8 inches in diameter, and not more than 2 deep, but there was a solid pad of horsehair more than an inch thick below this. I took the mass out; it must have weighed half a pound. Four eggs much incubated.

"*Etawah, 14th March.*—Another nest at the top of one of the huge tamarind-trees behind the Asthul: could not get up to it. A boy brought the nest down; it was not above a foot across, and perhaps 3 inches deep; cavity about 6 inches in diameter, thickly lined with grass-roots, inside which again was a coating of horsehair perhaps a rupee in thickness; nest swarming with vermin. Eggs five, quite fresh; four eggs normal; one quite round, a pure pale slightly greenish blue, with only a few very minute spots and specks of brown having a tendency to form a feeble zone round the large end. Measures only 1·25 by 1·2. Neither in shape, size, nor colour is it like a Corby's egg; but it is not a Koel's, or that of any of our parasitic Cuckoos, and I have seen at home similar pale eggs of the Rook, Hooded Crow, Carrion-Crow, and Raven.

"*Bareilly, May 10th.*—Three fresh eggs in large nest on a mango-tree. Nest as usual, but lined with an immense quantity of horsehair. We brought this home and weighed it; it weighed six ounces, and horsehair is very light."

Major C. T. Bingham writes :—

"This Crow, so common at Allahabad, is very scarce here at Delhi. In fact I have only seen one pair.

"At Allahabad it lays in February and March. I have, however, only found one nest, a rather loose structure of twigs and a few thick branches with rather a deep depression in the centre. It was placed on the very crown of a high toddy palm (*Borassus flabelliformis*), and was unlined save for a wad of human hair, on which the eggs, two in number, lay; these I found hard-set (on the 13th March); in colour they were a pale greenish blue, boldly blotched, spotted, and speckled with brown."

Colonel Butler has furnished me with the following note on the breeding of the Jungle-Crow :—

"Belgaum, 12th March, 1880.—A nest containing four fresh eggs. It consisted of a loose structure of sticks lined with hair and leaves, and was placed at the top of and in the centre of a green-foliaged tree in a well-concealed situation about 30 feet from the ground. 18th March: Two nests, each containing three slightly incubated eggs; one of the nests was quite low down in the centre of an 'arbor vitæ' about 12 feet from the ground. 31st March: Another nest containing four slightly incubated eggs. Some of the latter nests were very solidly built, and not so well concealed. 11th April: Two more nests, containing five incubated and three slightly incubated eggs respectively; and on the 14th April a nest containing four slightly incubated eggs. These birds, when the eggs are at all incubated, often sit very close, especially if the nest is in an open situation, and in many instances I have thrown several stones at the nest, and made as much row as I could below without driving the old bird off, and I have seen my nest-seeker within a few yards of the nest after climbing the tree before the old bird flew off. On the 26th of April I found two more nests, one containing four young birds just hatched, the other three fresh eggs. On the 27th another nest containing three fresh eggs, and on the 28th a nest of three fresh eggs. On the 5th May two more nests containing four fresh and four incubated eggs respectively."

"In the Nilghiris," writes Mr. Davison, "the Corby builds a coarse nest of twigs, lined with cocoanut-fibre or dry grass high up in some densely-foliaged tree. The eggs are usually four, often five, in number. The birds lay in April and May."

Miss Cockburn again says :—"They build like all Crows on large trees merely by laying a few sticks together on some strong branch, generally very high up in the tree. I do not remember ever seeing more than one nest on a tree at a time, so that they differ very much from the Rook in that respect. They lay four eggs of a bluish green, with dusky blotches and spots, and nothing can exceed the care and attention they bestow on their young. Even when the latter are able to leave their nests and take long flights, the parent birds will accompany them as if to prevent their getting into mischief. The nests are found in April and May."

Mr. J. Darling, jun., writes from the Nilghiris :—"I have found the nest of this Crow pretty nearly all over the Nilghiris. The usual number of eggs laid is four, but on one occasion, near the Quinine Laboratory in the Government Gardens at Ooty, I procured six from one nest. The breeding-season is from March to May, but I have taken eggs as early as the 12th February."

From Ceylon, we hear from Mr. Layard that "about the villages the Carrion-Crow builds its nest in the cocoanut-trees. In the jungles it selects a tall tree, amid the upper branches of which it fixes a framework of sticks, and on this constructs a nest of twigs and grasses. The eggs, from three to five, are usually of a dull greenish-brown colour, thickly mottled with brown, these markings being most prevalent at the small end. They are usually laid in January and February."

Mr. J. R. Cripps informs us that in Eastern Bengal it is "common and a permanent resident. Occasionally found in the clumps of jungle that are found about the country, which the next species never affects. Breeds in the cold weather. I had noticed a pair building on a Casuarina tree in my garden, about 50 feet off the ground, and on the 18th December, 1877, I took two perfectly fresh eggs from it ; and again on the 9th January, 1878, I found two callow young in this same nest, the birds never having deserted it. The lining used for this nest was principally jute-fibre—any tree is selected to build on; the nests are placed from 15 to 50 feet off the ground. Some nests are very well concealed, whereas others are quite exposed. On the 15th January I found a nest about 15 feet up a small kudum tree, standing in a large plain, and which had a lining of hair from the tail-tufts of cows. There was one fresh egg, and a week later I got another fresh egg from this very nest. From two to four eggs are in each nest."

Mr. Oates writes from Pegu :—"These birds all begin to build about the same time, and I have taken numerous nests at the end of January. At the end of February most nests contain young birds."

Mr. W. Theobald gives the following notes on the nidification of this bird in Tenasserim and near Deoghur :—

" Lays in the third week of February and fourth week of March : eggs ovato-pyriform ; size 1·66 by 1·15; colour, dull sap-green much blotched with brown ; nest carefully placed in tall trees."

The eggs, though smaller, closely resemble, as might have been expected, those of the Raven, but they are, I think, typically somewhat broader and shorter. Almost every variety, as far as coloration goes, to be found amongst those of the Raven, are found amongst the eggs of the present species, and *vice versâ*; and for a description of these it is only necessary to refer to the account of the former species; but I may notice that amongst the eggs of *C. macrorhynchus* I have not yet noticed any so boldly blotched as is occasionally the case with some of the eggs of the Raven, which remind one not a little, so far as the character of the markings go, of eggs of *Œdicnemus crepitans* and *Esacus recurvirostris.* Like those of the Raven the eggs exhibit little gloss, though here and there a fairly glossy egg is met with. Eggs from various parts of the Himalayas, of the plains of Upper India, of the hills and plains of Southern India, do not differ in any respect. *Inter se* the eggs from each locality differ surprisingly in size, in tone of colour, and in character of markings; but when you compare a dozen or twenty from each locality, you find that these differences are purely individual and in no degree referable to locality.

There are just as big eggs and just as small ones from Simla and Kotegurh, from Cashmere, from Etawah, Bareilly, Futtehgurh, from Kotagherry, and Conoor : all that one can possibly say is that perhaps the Plains birds do on the *average* lay a *shade larger* eggs than the Himalayan or Nilghiri ones.

Taking the eggs as a whole, I think that in size and shape they are about intermediate between the eggs of the European Carrion-Crow and Rook. But they vary, as I said, astonishingly in size, from 1·5 to 1·95 in length, and in breadth from 1·12 to 1·22, and I have one perfectly spherical egg, a deformity of course, which measures 1·25 by 1·2.

The average of thirty Himalayan eggs is 1·73 by 1·18, of twenty Plains eggs 1·74 by 1·2, and of fifteen Nilghiri eggs 1·7 by 1·18. I would venture to predict that with fifty of each, there would not be a hundredth of an inch between their averages.

7. Corvus splendens, Vieill.   *The Indian House-Crow.*

Corvus splendens, *Vieill. Jerd. B. Ind.* ii, p. 298.
Corvus impudicus, *Hodgs., Hume, Rough Draft N. & E.* no. 603.

Throughout India and Upper Burma the Common Crow resides and breeds, not ascending the hills either in Southern or Northern India to any great elevation, but breeding up to 4000 feet in the Himalayas.

The breeding-season *par excellence* is June and July, but occasional nests will be found earlier even in Upper India, and in Southern and Eastern India a great number lay in May. The nests are commonly placed in trees without much regard to size or kind, though densely foliaged ones are preferred, and I have just as often found several in the same tree as single ones. At times they will build in nooks of ruins or large deserted buildings, where these are in well inhabited localities, but out of many thousands I have only seen three or four nests in such abnormal positions.

The nest is placed in some fork, and is usually a ragged stick platform, with a central depression lined with grass-roots; but they are not particular as to material; I have found wool, rags, grass, and all kinds of vegetable fibre, and Mr. Blyth mentions that he has " seen several nests composed more or less, and two almost exclusively, of the wires taken from soda-water bottles, which had been purloined from the heaps of these wires commonly set aside by the native servants until they amount to a saleable quantity." Four is the normal number of eggs laid, but I often have found five, and on two occasions six. It is in this bird's nest that the Koel chiefly lays.

Writing of Nepal, Dr. Scully remarks :—" In the valley it lays in May and June ; some twenty nests were once examined on the 23rd June, and half the number then contained young birds."

Major Bingham says :—" Very common, of course, both at Allahabad and at Delhi, and breeds in June, July, and beginning of August. At Allahabad it is much persecuted by the Koel (*Eudynamys orientalis*), every fourth or fifth nest that I found in some topes of mango-trees having one or two of the Koel's eggs."

Colonel Butler informs me that in Karachi it " begins to lay in the mangrove bushes in the harbour as early as the end of May ;" and that it " breeds in the neighbourhood of Deesa in June, July, and August, commencing to build in the last week of May."

Later, he writes :—" Belgaum, 15th May, 1879. Found numerous nests in the native infantry lines in low trees, containing fresh and incubated eggs and young birds of all sizes. In the same locality, on the 30th March, 1880, I found a nest containing four young birds able to fly; the eggs must therefore have been laid quite as early as the middle of February, if not earlier."

Mr. G. W. Vidal writes :—" The Common Crow appears to have two broods in the year in our district (Ratnagiri), the first in April and May, and the second in November and December. In these four months I have found nests, eggs, and young birds in several different places in the district, and as yet at no other times. It is extremely improbable that there should be one breeding-season lasting from April to December, and I think I may state with certainty that the Crows *do not* breed at Ratnagiri during the months of heaviest rainfall, viz. July, August, and September. As their breeding in November and December appears to be exceptional, I subjoin a record of the few nests I examined.

" Nov. 22, 1878. Ratnagiri :
" One nest with 3 young birds.
"   "   " 1 fresh egg.
" Nov. 23, 1878. Ratnagiri :
" One nest with 1 fresh egg.
"   "   " 1 fresh egg.
" Dec. 4, 1878. Saugmeshwar.—One nest with 3 eggs hard-set ; another nest probably containing young birds, but the Crows pecked so viciously at the man who was climbing the tree, that he got frightened and came down again without reaching the nest.

Crows with sticks and feathers in their mouths are flying about all day.

"Dec. 5, 1878. Aroli.—Found a nest with a Crow sitting in it; no one to climb the tree."

Mr. Benjamin Aitken has favoured me with the following interesting note :—" I send you an account of a nest of the Common Crow, found in October, 1874, in the town of Madras. My attention was first directed to the remarkable pair of Crows to which the nest belonged, in the end of July, when they were determinedly and industriously attempting to fix a nest on the top ledge of a pillar in the verandah of the ' Madras Mail ' office. The ledge was so narrow that one would have thought the Sparrow alone of all known birds would have selected it for a site; and even the Sparrow only under the condition of a writing or toilet-table being underneath to catch the lime, sticks, straws, rags, feathers, and other innumerable materials that commonly strew the ground below a Sparrow's nest. I was told that the Crows had been at their task for two months before I saw them, and I then watched them till nearly the end of October. The celebrated spider that taught King Bruce a lesson in patience was eager and fitful compared with this pair of Crows. I kept no account of the number of times their structure was blown down, only to be immediately begun again ; but as there was a good deal of rain and wind at that season, in addition to the regular sea-breeze, it was a common thing for the sticks to be cleared off day after day. But perseverance will often achieve seeming impossibilities, and, moreover, the Crows worked more indefatigably as the season went on, and used to run up their nest with great rapidity (no doubt, also, they improved by their practice); so that several times the structure was completed, or nearly completed, before being swept to the ground, though how it remained in its place for a moment seems a mystery; and twice I saw a broken egg among the scattered *débris*. At length, about the middle of September, the Crows determined to try the pillar at the other end of the verandah. By this time, of course, all the Crows in Madras had long brought up their broods and sent them adrift ; and what they thought to see an eccentric pair of their own species forsaking society, and *building* in September, may be imagined. The new site selected differed in no respect from the old one, and was no less exposed to the wind ; but the birds had grown expert at building ' castles in the air,' and now met with fewer mishaps. In the first week of October the hen bird was sitting regularly, so on the 8th of the month I sent a man up by a ladder, and he held up four eggs for me to look at. It fairly seemed after this that patience was to have its reward, but on the night of the 20th there came a storm of wind and rain, and when I went to the office in the morning, the nest was lying on the ground, with two young Crows in it, with the feathers just beginning to appear. The other two, I suppose, had fallen over into the street. And thus ended one of the most persevering attempts on record to overcome a difficulty insurmountable from the first. The old birds thought it time now to stop operations, and frequented the office no more.

" I am told by a gentleman in the ' Mail ' office that the Crows have built in that verandah regularly for five or six years past, but nobody seems to have watched the nests. I am, therefore, hopeful that the attempt will be repeated this year, in which case I will keep a diary of all that takes place."

He writes subsequently :—" I sent you a long story in my last batch of notes about two eccentric Crows that succeeded in building a nest upon the narrow ledge of a pillar in the verandah of my office, several months after all well-conducted Crows had sent out their progeny to battle with the world. I mentioned to you that they were said to build in that unnatural place every year, and I said that I would watch them this year.

" Well, would you believe it ? on the 26th July, when every other Crow's nest in Madras had hard-set eggs, or newly-hatched young ones, these two indefatigable birds set methodically to work to construct a nest on the south pillar—the one where all their earlier efforts were made last year, but not the one on which they succeeded in fixing their nest. They worked all the 26th and 27th, putting up sticks as fast as they fell down, and then desisted till the 4th August, when they began operations on the opposite (north) pillar with redoubled energy. Meeting with no better success they left off operations after a couple of days' fruitless labour. Yesterday (after a delay of five weeks) they set to work on the south pillar again and succeeded in raising a great pile, which, however, was ignominiously blown down in the afternoon. Today they are continuing their work indefatigably."

Mr. J. R. Cripps has the following note in his list of birds of Furreedpore, Eastern Bengal :—" Very common, and a permanent resident, affecting the haunts of man. They build and lay in May. The Koel lays its eggs in this bird's nest. In April, 1876, I saw two nests in the compound of the house in which I lived at Howrah, which were made *entirely* of galvanized wire, the thickest piece of which was as thick as a slate pencil. How the birds managed to bend these thick pieces of wire was a marvel to us ; not a stick was incorporated with the wires, and the lining of the nest (which was of the ordinary size) was jute and a few feathers. The railway goods-yard, which was alongside the house, supplied the wire, of which there was ever so much lying about there."

Typically the eggs may, I think, be said to be rather broad ovals, a good deal pointed towards the small end ; but really the eggs vary so much in shape that, even with nearly two hundred before me, it is difficult to decide what is really the most typical form. Pyriform, elongated, and globular varieties are common ; long Cormorant-shaped eggs and perfect ovals are not uncommon. As regards the colour of the ground, and colour, character, and extent of marking, all that I have above said of the Raven's eggs applies to those of this species, but varieties occur amongst those of the latter which I have not observed in those of the former. In some the

ground is a very pale pure bluish green; in others it is dingier and
greener. All are blotched, speckled, and streaked more or less
with somewhat pale sepia markings; but in some the spots and
specks are a darker brown and, as a rule, well defined, and there
is very little streaking, while in others the brown is pale and
muddy, the markings ill-defined, and nearly the whole surface of
the egg is freckled over with smudgy streaks. Sometimes the
markings are most numerous at the large end, sometimes at the
small; no two eggs are exactly alike, and yet they have so strong
a family resemblance that there is no possibility of mistaking them.
Generally the markings as a whole are less bold, and the general
colour of a large body of them laid together is bluer and brighter
than that of a similar drawer-full of Ravens' eggs. As a whole,
too, they are more glossy. I have one egg before me bright blue
and almost as glossy as a Mynah's, thickly blotched and speckled
at the broad end, and thinly spotted elsewhere with olive-green,
blackish-brown, and pale purple. Another egg, a pale pure blue,
is spotless, except at the large end, where there is a conspicuous
cap of olive-brown and olive-green spots and speckles, and there
are numerous other abnormal varieties which I have not observed
amongst the Ravens.

On the whole the eggs do *not* vary much in size; out of one
hundred and ninety-seven, one hundred and ninety-five varied
between 1·28 and 1·65 in length, and 0·98 and 1·15 in breadth.
One egg measures only 1·2 in length, and one is only 0·96 in
breadth; but the average of the whole is 1·44 by 1·06.

8. Corvus insolens, Hume.   *The Burmese House-Crow.*

Corvus insolens, *Hume*; *Hume, Cat.* no. 663 bis.

The Burmese House-Crow breeds pretty well over the whole of
Burma.

Mr. Oates, writing from Pegu, says:—" Nesting operations are
commenced about the 20th March. The nest and eggs require no
separate description, for both appear to be similar to those of
C. splendens."

When large series of the eggs of both these species are compared, those of the Burmese Crow strike one as *averaging* somewhat brighter coloured, otherwise they are precisely alike and need
no separate description.

9. Corvus monedula, Linn.   *The Jackdaw.*

Colæus monedula (*Linn.*), *Jerd. B. Ind.* ii, p. 302.
Corvus monedula, *Linn., Hume, Rough Draft N. & E.* no. 665.

I only know positively of Jackdaws breeding in one district
within our limits, viz. Cashmere; but I have seen it in the hills in
summer, as far east as the Valley of the Beas, and it must breed
everywhere in suitable localities between the two.

In the cold season of course the Jackdaw descends into the plains of the North-west Punjaub, is very numerous near the foot of the hills, and has been found in cis-Indus as far east as Umballa, and south at Ferozpoor, Jhelum, and Kalabagh. In Trans-Indus it extends unto the Dehra Ghazi Khan district.

I have never taken its eggs myself.

Mr. W. Theobald makes the following remarks on its nidification in the Valley of Cashmere :—

" Lays in the first week of May ; eggs four, five, and six in number, ovato-pyriform and long ovato-pyriform, measuring from 1·26, 1·45, to 1·60 in length, and from 0·9 to 1·00 in breadth ; colour pale, clear bluish green, dotted and spotted with brownish black; valley generally; in holes of rocks, beneath roofs, and in tall trees."

Dr. Jerdon says :—" It builds in Cashmere in old ruined palaces, holes in rocks, beneath roofs of houses, and also in tall trees, laying four to six eggs, pale bluish green, dotted and spotted with brownish black."

Mr. Brookes writes :—" The Jackdaw breeds in Cashmere in all suitable places : holes in old Chinar (Plane) trees, and in house-walls, under the eaves of houses, &c. I did not note the materials of the nests, but these will be the same as in England."

The eggs of this species are typically rather elongated ovals, somewhat compressed towards one end. The shell is fine, but has only a faint gloss. The ground-colour is a pale greenish white, but in some eggs there is very little green, while in a very few the ground is quite a bright green. The markings, sometimes very fine and close, sometimes rather bold and thinly set, consist of specks or spots of deep blackish brown, olive-brown, and pale inky purple. In most eggs all these colours are represented, but in some eggs the olive-, in others the blackish brown is almost entirely wanting. In some eggs the markings are very dense towards the large end, in others they are pretty uniformly distributed over the whole surface ; in some they are very minute and speckly, in others they average the tenth of an inch in diameter.

The eggs that I possess vary from 1·34 to 1·52 in length, and from 0·93 to 1·02 in breadth ; but the average of sixteen eggs was 1·4 by 0·98.

10. Pica rustica (Scop.). *The Magpie.*

Pica bactriana, *Bp.*, *Hume*, *Rough Draft N. & E.* no. 668 bis.

The Magpie breeds, we know, in Afghanistan, and also throughout Ladak from the Zojee-la Pass right up to the Pangong Lake, but it breeds so early that one is never in time for the eggs. The passes are not open until long after they are hatched.

Captain Hutton says this bird "is found all the year round from Quettah to Girishk, and is very common. They breed in

March, and the young are fledged by the end of April. The nest is like that of the European bird, and all the manners of the Afghan Magpie are precisely the same. They may be seen at all seasons."

From Afghanistan, Lieut. H. E. Barnes writes :—

" The Magpie is not uncommon in the hills wherever there are trees, but it seldom descends to the plains. They commence breeding in March, in which month and April I have examined scores of nests, which in every case were built in the ' Wun,' a species of *Pistacia*—the only tree found hereabouts. A stout fork near the top is usually selected.

" The nest is shallow and cup-shaped, with a superstructure of twigs, forming a canopy over the egg-cavity. The eggs, generally five in number, are of the usual corvine green, blotched, spotted, and streaked, as a rule, most densely about the large end with umber mingled with sepia-brown. The average of thirty eggs is 1·25 by ·97."

Colonel Biddulph writes in ' The Ibis ' that in Gilgit he took a nest with five eggs, hard set, in a mulberry-tree at Nonval (5600 feet) on the 9th May. Also another nest with three fresh eggs at Dayour (5200 feet) on the 25th May.

The eggs are typically rather elongated ovals, rather pointed towards the small end, but shorter and broader varieties, and occasionally ones with a pyriform tendency, occur. The ground is a greenish or brownish white. In some eggs it has none, in others a slight gloss. Everywhere the eggs are finely and streakly freckled with a brown that varies from olive almost to sepia; about the large end the markings are almost always most dense, forming there a more or less noticeable, but quite irregular and undefined cap or zone. In one or two eggs dull purplish-brown clouds or blotches underlie and intermingle with this cap, and occasionally a small spot of this same tint may be noticed elsewhere when the egg is closely examined.

12. Urocissa occipitalis (Bl.). *The Red-billed Blue Magpie.*

Urocissa sinensis (*Linn.*), *Jerd. B. Ind.* ii, p. 309.
Urocissa occipitalis (*Bl.*), *Hume, Rough Draft N. & E.* no. 671.

I have never myself found the nest of the Red-billed Blue Magpie ; although it does breed sparingly as far east as Simla and Kotegurh, it is not till you cross the Jumna that it is abundant. East of the Jumna, about Mussoorie, Teeree, Gurhwal, Kumaon, and in Nepal, it is common.

From Mussoorie Captain Hutton tells us that " this species occurs at Mussoorie throughout the year. It breeds at an elevation of 5000 feet in May and June, making a loose nest of twigs externally and lined with roots. The nest is built on trees, sometimes high up, at others about 8 or 10 feet from the ground. The eggs are from three to five, of a dull greenish ash-grey, blotched and speckled with brown dashes confluent at the larger end, the ends nearly equal in size. It is very terrene in its habits, feeding almost entirely on the ground."

Colonel G. F. L. Marshall remarks :—

"The Red-billed Blue Magpie is, as far as I know, an early breeder at Naini Tal; common as the bird is I have only found one nest and that on the 24th April; it was a shallow slenderly built structure of fine roots, chiefly of maiden-hair fern, in a rough outer casing of twigs, placed on a horizontal bough overhanging a nullah about fifteen feet from the ground. The tree had moderately dense foliage, and was about twenty-five feet high in a small clump on a hillside covered with low scrub at 5000 feet elevation above the sea. Around the nest several small boughs and twigs grew out, and being very slight in structure it was not easy to see. The old bird sat very close. There were six eggs in the nest about half-incubated : in two of them the markings were densest at the small end. The egg-cavity was 6 inches in diameter by about $1\frac{1}{4}$ deep. On the 5th June I saw old birds accompanied by young ones able to fly, but without the long tails."

The eggs of this species much resemble those of the European Magpie, but are considerably smaller. They are broad, rather perfect ovals, somewhat elongated and pointed in many specimens. They exhibit but little gloss. The ground-colour varies much, but in all the examples that I possess, which I owe to Captain Hutton's kindness, it is either of a yellowish-cream, pale café au lait or buff colour, or pale dull greenish. The ground is profusely blotched, spotted, and streaked (the general character of the markings being striations parallel to the major axis), with various shades of reddish and yellowish brown and pale inky purple. The markings vary much in intensity as well as in frequency, some being so closely set as to hide the greater part of the ground-colour; but in the majority of the eggs they are more or less confluent at the large end, where they form a comparatively dark, irregular blotchy zone.

The eggs vary from 1·25 to 1·4 in length, and from 0·89 to 0·96 in breadth; but the average of 11 eggs is 1·33 by 0·93.

Major Bingham, referring to the Burmese Magpie, which has been separated under by the name of *U. magnirostris*, says :—

"This species I have only found common in the Thoungyeen Valley. Elsewhere it seemed to me scarce. Below I give a note about its breeding.

"I have found three nests of this handsome Magpie—two on the bank of the Meplay choung on the 14th April, 1879, and 5th March, 1880, respectively, and one near Meeawuddy on the Thoungyeen river on the 19th March, 1880.

"The first contained three, the second four, and the third two eggs.

"These are all of the same type, dead white, with pale claret-coloured dashes and spots rather washed-out looking, and lying chiefly at the large end. One egg has the spots thicker at the small end. They are moderately broad ovals, and vary from 1·19 to 1·35 in length, and from 0·93 to 1·08 in breadth.

"The nests were all alike, thick solid structures of twigs and branches, lined with finer twigs about 8 or 9 inches in diameter, and placed invariably at the top of tall straight saplings of teak, pynkado (*Xylia dolabriformis*), and other trees at a height of about 15 feet from the ground."

All the eggs of the Burmese bird that I have seen, nine taken by Major Bingham, were of one and the same type. The eggs broad ovals, in most cases pointed towards the small end. The shell fine, but as a rule with scarcely any perceptible gloss. The ground-colour a delicate creamy white. The markings moderate-sized blotches, spots, streaks, and specks, as a rule comparatively dense about one, generally the large, end, where only as a rule any at all considerable sized blotches occur, elsewhere more or less sparsely set, and generally of a speckly character. The markings are of two colours: brown, varying in shade in different eggs, olive-yellowish, chocolate, and a grey, equally varying in different eggs from pale purple to pale sepia. None of my eggs of the Himalayan bird (I have unfortunately but few of these) correspond at all closely with these.

13. Urocissa flavirostris (Bl.). *The Yellow-billed Blue Magpie.*

Urocissa flavirostris (*Bl.*), *Jerd. B. Ind.* ii, p. 310; *Hume, Rough Draft N. & E.* no. 672.

The Yellow-billed Blue Magpie breeds throughout the lower ranges of the Himalayes in well-wooded localities from Hazara to Bhootan, and very likely further east still, from April to August, mostly however, I think, laying in May. The nest, which is rather coarse and large, made of sticks and lined with fine grass or grass-roots, is, so far as my experience goes, commonly placed in a fork near the top of some moderate-sized but densely foliaged tree.

I have never found a nest at a lower elevation than about 5000 feet; as a rule they are a good deal higher up.

They lay from four to six eggs, but the usual number is five.

Colonel C. H. T. Marshall writes:—"The Yellow-billed Blue Magpie breeds commonly about Murree. I have never seen the bird below 6000 feet in the breeding-season. They do not commence laying till May, and I have taken eggs nearly fresh as late as the 15th August. I do not think the bird breeds twice, as the earliest eggs taken were found on the 10th May.

"They build in hill oaks as a rule, the height of the nest from the ground varying much, some being as low as 10 feet, others nearer 30 feet. The hen bird sits close, and sometimes (when the nest is high up) does not even leave the nest when the tree is struck below. The nest is a rough structure built close to the trunk, externally consisting of twigs and roots and lined with fibres. The egg-cavity is circular and shallow, not at all neatly lined. The outer part of the nest is large compared to what I should call the true nest, and consists of a heap of twigs, &c., like what is gathered together for the platform of a Crow's nest.

" The eggs, which are four in number, vary in length from 1·45 to 1·25, and in breadth from 0·9 to 0·75. The ordinary type is an egg a good deal pointed at the thinner end. The ground-colour is greenish white, blotched and freckled with ruddy brown, with a ring at the larger end of confluent spots. The young birds are of a very dull colour until after the first month. The normal number of eggs laid appears to be four."

Captain Cock wrote to me :—" *U. flavirostris* is common at Dhurmsala, but the nest is rather difficult to find. I have only taken six in three years. It is usually placed amongst the branches of the hill oak, where it has been polled, and the thickly growing shoots afford a good cover ; but sometimes it is on the top of a small slender sapling. The nest is a good-sized structure of sticks with a rather deep cup lined with dried roots ; in fact, it is very much like the nest of *Garrulus lanceolatus*, only larger and much deeper. They generally lay four eggs, which differ much in colour and markings."

Dr. Jerdon says :—" I had the nest and eggs brought me once. The nest was made of sticks and roots. The eggs, three in number, were of a greenish-fawn colour very faintly blotched with brown."

The eggs are of the ordinary Indian Magpie type, scarcely, if at all, smaller than those of *U. occipitalis*, and larger than the average of eggs of either *Dendrocitta rufa* or *D. himalayensis*. Doubtless all kinds of varieties occur, as the eggs of this family are very variable ; but I have only seen two types—in the one the ground is a pale dingy yellowish stone-colour, profusely streaked, blotched, and mottled with a somewhat pale brown, more or less olivaceous in some eggs, the markings even in this type being generally densest towards the large end, where they form an irregular mottled cap : in the other type the ground is a very pale greenish-drab colour ; there is a dense confluent raw-sienna-coloured zone round the large end, and only a few spots and specks of the same colour scattered about the rest of the egg. All kinds of intermediate varieties occur. The texture of the shell is fine and compact, and the eggs are mostly more or less glossy.

The eggs vary from 1·22 to 1·48 in length, and from 0·8 to 0·96 in breadth ; but the average of twenty-seven eggs is 1·3 by 0·92.

14. **Cissa chinensis** (Bodd.). *The Green Magpie.*

Cissa sinensis (*Briss.*), *Jerd. B. Ind.* ii, p. 312.
Cissa speciosa (*Shaw*), *Hume, Rough Draft N. & E.* no. 673.

According to Mr. Hodgson's notes the Green Magpie breeds in Nepal in the lower valleys and in the Terai from April to July. The nest is built in clumps of bamboos and is large and cup-shaped, composed of sticks and leaves, coated externally with bamboo-leaves and vegetable fibres, and lined inside with fine roots. It lays four eggs, one of which is figured as a broad oval, a good deal pointed towards one end, with a pale stone-coloured ground freckled and mottled all over with sepia-brown, and measuring 1·27 by 0·89.

Mr. Oates writes :—" In the Pegu Hills on the 19th April I found the nest of the Green Magpie, and shot the female off it.

" The nest was placed in a small tree, about 20 feet from the ground, in a nullah and well exposed to view. The nest was neatly built, exteriorly of leaves and coarse roots, and finished off interiorly with finer fibres and roots ; depth about 2 inches ; inside diameter 6 inches. Contained three eggs nearly hatched; all got broken ; I have the fragments of one. The ground-colour is greenish white, much spotted and freckled with pale yellowish-brown spots and dashes, more so at the larger end than elsewhere."

Sundry fragments that reached me, kindly sent to me by Mr. Oates, had a dull white ground, very thickly freckled and mottled all over, as far as I could judge, with dull, pale, yellowish brown and purplish grey, the former preponderating greatly. As to size and shape, this deponent sayeth nought.

Major Bingham writes from Tenasserim :—" On the 18th April I found a nest of this most lovely bird placed at a height of 5 feet from the ground in the fork of a bamboo-bush. It was a broad, massive, and rather shallow cup of twigs, roots, and bamboo-leaves outside, and lined with finer roots. It contained three eggs of a pale greenish stone-colour, thickly and very minutely speckled with brown, which tend to coalesce and form a cap at the larger end. I shot the female as she flew off the nest."

Major Bingham subsequently found another nest in Tenasserim, about which he says :—

" Crossing the Wananatchoung, a little tributary of the Thoun-gyeen, by the highroad leading from Meeawuddy to the sources of the Thoungyeen, I found in a small thorny tree on the 8th April a nest of the above bird—a great, firmly-built but shallow saucer of twigs, 6 feet or so above the ground, and lined with fine black roots. It contained three fresh eggs of a dingy greyish white, thickly speckled chiefly at the large end, where it forms a cap, with light purplish brown. The eggs measure 1·25 × 0·89, 1·18 × 0·92, and 1·20 × 0·90."

Mr. James Inglis notes from Cachar :—" This Jay is rather rare ; it frequents low quiet jungle. In April last a Kuki brought me three young ones he had taken from a nest in a clump of tree-jungle; he said the nest was some 20 feet from the ground and made of bamboo-leaves and grass." ·

A nest of this species taken below Yendong in Native Sikhim, on the 28th April, contained four fresh eggs. It was placed on the branches of a medium-sized tree at a height of about 12 feet from the ground ; it was a large oval saucer, 8 inches by 6, and about 2·5 in depth, composed mainly of dry bamboo-leaves, bound firmly together with fine stems of creepers, and was lined with moderately fine roots; the cavity was 5 inches by 4, and about 1 in depth.

The eggs received from Major Bingham, as also others received from Sikhim, where they were procured by Mr. Mandelli on the 21st and 28th of April, are rather broad ovals, somewhat pointed towards the small end. The shell is fine, but has only a little gloss. The ground-colour is white or slightly greyish white, and they are uniformly freckled all over with very pale yellowish and greyish brown. The frecklings are always somewhat densest at the large end, where in some eggs they form a dull brown cap or zone. In some eggs the markings are everywhere denser, in some sparser, so that some eggs look yellower or browner, and others paler.

The eggs are altogether of the *Garruline* type, not of that of the *Dendrocitta* or *Urocissa* type. I have eggs of *G. lanceolatus*, that but for being smaller precisely match some of the *Cissa* eggs. Jerdon is, I think, certainly wrong in placing *Cissa* between *Urocissa* and *Dendrocitta*, the eggs of which two last are of the same and quite a distinct type*.

The eggs vary from 1·15 to 1·26 in length, and from 0·9 to 0·95 in breadth, but the average of eight is 1·21 by 0·92.

15. Cissa ornata (Wagler). *The Ceylonese Magpie.*

Cissa ornata (*Wagl.*), *Hume, Cat.* no. 673 bis.

Colonel Legge writes in his ' Birds of Ceylon ':—" This bird breeds during the cool season. I found its nest in the Kandapolla jungles in January; it was situated in a fork of the top branch of a tall sapling, about 45 feet in height, and was a tolerably bulky structure, externally made of small sticks, in the centre of which was a deep cup 5 inches in diameter by 2½ in depth, made entirely of fine roots; there was but one egg in the nest, which unfortunately got broken in being lowered to the ground. It was ovate and slightly pyriform, of a faded bluish-green ground thickly spotted all over with very light umber-brown over larger spots of bluish-grey. It measured 0·98 inch in diameter by *about* 1·3 in length."

16. Dendrocitta rufa (Scop.). *The Indian Tree-pie.*

Dendrocitta rufa (*Scop.*), *Jerd. B. Ind.* ii, p. 314; *Hume, Rough Notes N. & E.* no. 674.

The Indian Tree-pie breeds throughout the continent of India, alike in the plains and in the hills, up to an elevation of 6000 or 7000 feet.

---

* I am responsible, and not Mr. Hume, for calling this bird a Magpie. Jerdon calls it a Jay, but places it among the Magpies, which is, I consider, its proper position, notwithstanding the colour of its eggs.—ED.

I personally have found the nest with eggs in May, June, July, and during the first week of August, in various districts in the North-West Provinces, and have had them sent me from Saugor (taken in July) and from Hansi (taken in April, May, and June); but perhaps because the bird is so common scarcely any one has sent me notes about its nidification, and I hardly know whether in other parts of India and Burma its breeding-season is the same as with us.

The nest is always placed in trees, generally in a fork, near the top of good large ones; babool and mango are very commonly chosen in the North-West Provinces, though I have also found it on neem and sisso trees. It is usually built with dry twigs as a foundation, very commonly thorny and prickly twigs being used, on which the true nest, composed of fine twigs and lined with grass-roots, is constructed. The nests vary much: some are large and loosely put together, say, fully 9 inches in diameter and 6 inches in height externally; some are smaller and more densely built, and perhaps not above 7 inches in diameter and 4 inches in depth. The egg-cavity is usually about 5 inches in diameter and 2 inches in depth, but they vary very much both in size and materials; and I see that I note of one nest taken at Agra on the 3rd August—"A very shallow saucer some 6 inches in diameter, and with a central depression not above 1½ inch in depth. It was composed *exclusively* of roots; externally somewhat coarse, internally of somewhat finer ones. It was very loosely put together."

Five is the full complement of eggs, but it is very common to find only four fully incubated ones.

Mr. W. Blewitt writes that he "found several nests in the latter half of April, May, and the early part of June in the neighbourhood of Hansie.

"Four was the greatest number of eggs I found in any nest.

"The nests were placed in neem, keekur, and shishum trees, at heights of from 10 to 17 feet from the ground, and were densely built of twigs mostly of the keekur and shishum, and more or less thickly lined with fine straw and leaves. They varied from 6 to 8 inches in diameter and from 2 to 3 inches in depth."

Mr. A. Anderson writes:—"The Indian Magpie lays from April to July, and I have once actually seen a pair building in February. Their eggs are of two very distinct types,—the one which, according to my experience, is the ordinary one, is covered all over with reddish-brown spots or rather blotches, chiefly towards the big end, on a pale greenish-white ground, and is rather a handsome egg; the other is a pale green egg with *faint brown* markings, which are confined almost entirely to the obtuse end. I have another clutch of eggs taken at Budaon in 1865, which presents an intermediate variety between the above two extremes; these are profusely blotched with russet-brown on a dirty-white ground.

"The second and third nests above referred to contained five eggs; but the usual complement is not more than four. On the 2nd August, 1872, I made the following note relative to the breeding of this bird :—The bird flew off immediately we approached the tree, and never appeared again. The nest viewed from below looked larger; this is owing to dry *babool* twigs or rather small branches (some of them having thorns from an inch to 2 inches long!) having been used as a foundation, and actually encircling the nest, no doubt by way of protection against vermin; some of these thorny twigs were a foot long, and they had to be removed piecemeal before the nest proper could be got at. The egg-cavity is deep, measuring 5 inches in depth by 4 in breadth inside measurement; it is well lined with khus grass."

Major Bingham says :—

"Common as is this bird I have only found one nest, and that was at Allahabad on the 9th July, and contained one half-fledged young one and an addled egg. The nest, which was placed at the very top of a large mango-tree, was constructed of branches and twigs of the same lined with fine grass-roots. The egg is a yellowish white, thickly speckled, chiefly at the large end, with rusty. Length 1·10 by 0·82 in breadth."

Colonel Butler tells us that it "breeds in Sind, in the hot weather. Mr. Doig took a nest containing three fresh eggs on the 1st May, 1878. The eggs, which seem to me to be remarkably small for the size of the bird, are of the first type mentioned in Rough Draft of 'Nests and Eggs,' p. 422."

Lieut. H. E. Barnes says in his 'Birds of Bombay:'—"In Sind they breed during May and June, always choosing babool trees, placing the nest in a stoutish fork near the top; they are composed at the bottom of thorny twigs, which form a sort of foundation upon which the true nest is built; the latter consists of fine twigs lined with grass-roots; the nest is frequently of large size."

Mr. G. W. Vidal, writing of the South Konkan, says :—"Common about all well-wooded villages from coast to Ghâts. Breeds in April."

With regard to Cachar Mr. Inglis writes:—"This Magpie is very common in all the neighbouring villages, but I have not often seen it in the jungles. It remains all the year and breeds during April and May."

The eggs are typically somewhat elongated ovals, a good deal pointed towards the small end. They vary extraordinarily in colour and character, as well as extent of markings, but, as remarked when speaking of the Raven, all the eggs out of the same nest closely resemble each other, while the eggs of different nests are almost invariably markedly distinct. There are, however, two leading types—the one in which the markings are bright red, brownish red, or pale pinkish purple; and the other in which they are olive-brown and pale purplish brown. In the first type the ground-colour is either pale salmon, or else very pale greenish white, and the markings are either bold blotches, more or less confluent at the large end, where they are far most numerous, and only a few specks and spots towards the smaller end, or they are spots and small blotches thickly distributed over the whole surface, or they are streaky smudges forming a mottled ill-defined cap at the large end, and running down thence in streaks and spots longitudinally; in the other type the ground-colour is greenish white or pale yellowish stone-colour, and the character of the markings varies as in the preceding type. Besides these there are a few eggs with a dingy greyish-white ground, with very faint, cloudy, ill-defined spots of pale yellowish brown pretty uniformly distributed over the whole surface. In nine eggs out of ten the markings are most dense at the large end, where they form irregular, more or less imperfect caps or zones. A few of the eggs are slightly glossy.

Of the salmon-pink type some specimens in their coloration resemble eggs of *Dicrurus longicaudatus* and some of our Goatsuckers, while of those with the greenish-white ground-colour some strongly recall the eggs of *Lanius lahtora*.

In length the eggs vary from 1·0 to 1·3, and in breadth from 0·78 to 0·95; but the average of forty-four eggs is 1·17 by 0·87.

17. Dendrocitta leucogastra, Gould.   *The Southern Tree-pie.*

Dendrocitta leucogastra, *Gould, Jerd. B. Ind.* ii, p. 317; *Hume, Rough Draft N. & E.* no. 678.

From Travancore Mr. Bourdillon has kindly sent me an egg and the following note on the nidification of the Southern Tree-pie :—

"Three eggs, very hard-set, of an ashy-white colour, marked with ashy and greenish-brown blotches, 1·12 long and 0·87 broad, were taken on 9th March, 1873, from a nest in a bush 8 or 10 feet from the ground. The nest of twigs was built after the style of the English Magpie's nest, minus the dome. It consisted of a large platform 6 inches deep and 8 or 10 inches broad, supporting a nest 1½ inch deep and 3½ inches broad. The bird is not at all uncommon on the Assamboo Hills between the elevations of 1500 and 3000 feet above the sea, seeming to prefer the smaller jungle and more open parts of the heavy forest."

Later he writes :—" On the 8th April I found another nest containing three half-fledged Magpies (*D. leucogastra*). The nest was entirely composed of twigs, roughly but securely put together; interior diameter 3 inches and depth 2 inches, though there was a good-sized base or platform, say, 5 inches in diameter. The nest was situated on the top fork of a sapling about 12 feet from the ground. I tried to rear the young birds, but they all died within a week."

The egg is very like that of our other Indian Tree-pies. It is in shape a broad and regular oval, only slightly compressed towards one end. The shell is fine and compact and is moderately glossy. The ground is a creamy stone-colour. It is profusely blotched and streaked with a somewhat pale yellowish brown, these markings being most numerous and darkest in a broad, irregular, imperfect zone round the large end, and it exhibits further a number of pale inky-purple clouds and blotches, which seem to underlie the brown markings, and which are chiefly confined to the broader half of the egg. The latter measures 1·13 by 0·86.

18. **Dendrocitta himalayensis**, Bl. *The Himalayan Tree-pie.*

Dendrocitta sinensis (*Lath.*), *Jerd. B. Ind.* ii, p. 316.
Dendrocitta himalayensis, *Bl., Hume, Rough Draft N. & E.* no. 676.

Common as is the Himalayan Tree-pie throughout the lower ranges of those mountains from which it derives its name, I personally have never taken a nest.

It breeds, I know, at elevations of from 2000 to 6000 feet, during the latter half of May, June, July, and probably the first half of August.

A nest in my museum taken by Mr. Gammie in Sikhim, at an elevation of about 2500 feet, out of a small tree, on the 30th of July, contained two fresh eggs. It was a very shallow cup, composed entirely of fine stems, apparently of some kind of creeper, strongly but not at all compactly interwoven ; in fact, though the nest holds together firmly, you can see through it everywhere. It is about 6 inches in external diameter, and has an egg-cavity of about 4 inches wide and 1·5 deep. It has no pretence for lining of any kind.

Of another nest which he took Mr. Gammie says :—" I found a nest containing three fresh eggs in a bush, at a height of about 10 feet from the ground. The nest was a very loose, shallow, saucer-like affair, some 6 or 7 inches in diameter and an inch or so in thickness, composed entirely of the dry stems and tendrils of creepers. This was at Labdah, in Sikhim, at an elevation of about 3000 feet, and the date the 14th May, 1873." Later he writes :—

" This Magpie breeds in the Darjeeling District in May, June, and July, most commonly at elevations between 2000 and 4000 feet. It affects clear cultivated tracts interspersed with a few standing shrubs and bamboos, in which it builds. The nest is generally placed from 6 to 12 feet from the ground in the inner part of the shrubs, and is made of pieces of creeper stems intermixed with a few small twigs loosely put together without any lining. There is scarcely any cup, merely a depression towards the centre for the eggs to rest in. Internally it measures about 4·8 in breadth by 1·5 in depth. The eggs are three or four in number.

" This is a very common and abundant bird between 2000 and 4000 feet, but is rarely found far from cultivated fields. It seems to be exceedingly fond of chestnuts, and, in autumn, when they are ripe, lives almost entirely on them; but at other times is a great pest in the grain-fields, devouring large quantities of the grain and being held in detestation by the natives in consequence. Jerdon says ' it usually feeds on trees,' but I have seen it quite as frequently feeding on the ground as on trees."

Mr. Hodgson has two notes on the nidification of this species in Nepal :—" *May 18th.*—Nest, two eggs and two young ; nest on the fork of a small tree, saucer-shaped, made of slender twigs twisted circularly and without lining ; cavity 3·5 in diameter by 0·5 deep;

eggs yellowish white, blotched with pale olive chiefly at the larger end; young just born.

"*Jaha Powah, 6th June.*—Female and nest in forest on a largish tree placed on the fork of a branch; a mere bunch of sticks like a Crow's nest; three eggs, short and thick, fawny white blotched with fawn-brown chiefly at the thick end."

Dr. Jerdon says :—"I have had the nest and eggs brought me at Darjeeling frequently. The nest is made of sticks and roots, and the eggs, three or four in number, are of a pale dull greenish-fawn colour, with a few pale reddish-brown spots and blotches, sometimes very indistinct."

Captain Hutton tells us that this species "occurs abundantly at Mussoorie, at about 5000 feet elevation, during summer, and more sparingly at greater elevations. In the winter it leaves the mountains for the Dhoon.

"It breeds in May, on the 27th of which month I took a nest with three eggs and another with three young ones. The nest is like that of *Urocissa occipitalis*, being composed externally of twigs and lined with finer materials, according to the situation; one nest, taken in a deep glen by the side of a stream, was lined with the long fibrous leaves of the Mare's tail (*Equisetum*) which grew abundantly by the water's edge; another, taken much higher on the hillside and away from the water, was lined with tendrils and fine roots. The nest is placed rather low, generally about 8 or 10 feet from the ground, sometimes at the extremity of a horizontal branch, sometimes in the forks of young bushy oaks. The eggs somewhat resemble those of *U. occipitalis,* but are paler and less spotted, being of a dull greenish ash with brown blotches and spots, somewhat thickly clustered at the larger end."

Mr. J. R. Cripps says :—"On the 15th June, 1880, I found a nest [in the Dibrugarh District] with three fresh eggs. It was fixed in the middle branches of a sapling, about ten feet off the ground, in dense forest, and was built of twigs, presenting a fragile appearance; the egg-cavity was 4½ inches [in diameter] and 1 inch deep, and lined with fine twigs and grass-roots."

Captain Wardlaw Ramsay writes :—"I obtained two eggs of this species at an elevation of 4200 feet in the Karen hills east of Toungngoo on the 16th April, 1875."

Taking the eggs as a body they are rather regular, somewhat elongated ovals, but broader and again more pointed varieties occur. The ground-colour varies a great deal: in a few it is nearly pure white, generally it has a dull greenish or yellowish-brown tinge, in some it is creamy, in some it has a decided pinky tinge. The markings are large irregular blotches and streaks, almost always most dense at the large end, where they are often more or less confluent, forming an irregular mottled cap, and not unfrequently very thinly set over the rest of the surface of the egg. In one egg, however, the zone is about the thick end, and there are scarcely any markings elsewhere. As a rule the markings are of an olive-brown of one shade or another; but when the ground is at all pinkish then the markings are more or less of a reddish brown. Besides these primary markings, all the eggs exhibit a greater or smaller number of faint lilac or purple spots or blotches, which chiefly occur where the other markings are most dense. In length they vary from 1·06 to 1·22, and in breadth from 0·8 to 1·0, but the average of 34 eggs is 1·14 by 0·85.

21. Crypsirhina varians (Lath.). *The Black Racket-tailed Magpie.*

Crypsirhina varians (*Lath.*), *Hume, Cat.* no. 678 quat.

This Magpie is very common in Lower Pegu, where Mr. Oates found many nests. He says :—

" This bird appears to lay from the 1st of June to the 15th of July; most of my nests were taken in the latter month. It selects either one of the outer branches of a very leafy thorny bush, or perhaps more commonly a branch of a bamboo, at heights varying from 5 to 20 feet.

" The nest is composed of fine dead twigs firmly woven together. The interior is lined with twisted tendrils of convolvulus and other creepers. The uniformity with which this latter material is used in all nests is remarkable. The inside diameter is 5 inches, and the depth only 1, thus making the structure very flat. The exterior dimensions are not so definite, for the twigs and creepers stick out in all directions ; but making all allowances, the outside diameter may be put down at 7 or 8 inches, and the total depth at 1½ inches.

" The eggs are usually three in number, but occasionally only two well incubated eggs may be found. In a nest from which two fresh eggs had been taken, a third was found a few days later.

" The eggs measure from 1·09 to ·88 in length, and from ·76 to ·68 in breadth. The average of 22 eggs is ·98 by ·72."

In shape the eggs are typically moderately broad, rather regular ovals, but some are distinctly compressed towards the small end, some are slightly pyriform, some even pointed, though in the great majority of cases the egg is pretty obtuse at the small end ; the shell is compact and tolerably fine, and has a faint gloss. The ground-colour seems to be invariably a pale yellowish stone-colour. The markings vary a good deal: in some they are more speckly, in others more streaky, but taking them as a whole they are intermediate between those of *Dendrocitta* and those of *Garrulus*, neither so bold and streaky as the former, nor so speckly as the latter. The markings are a yellowish olive-brown ; they consist of spots, specks, small streaky blotches and frecklings ; they are always pretty densely set over the whole surface of the egg, but they are always most dense in a zone or sometimes a cap at the large end, where they are often, to a great extent, confluent. In some eggs small dingy brownish-purple spots and little blotches are intermingled in the zone. The eggs differ in general appearance a good deal, because in some almost all the markings are fine grained and freckly, and in such eggs but little of the ground-colour is visible, while in other eggs the markings are bolder (in comparison, for they are never really bold) and thinner set, and leave a good deal of the ground-colour visible.

23. Platysmurus leucopterus (Temm.). *The White-winged Jay.*

Platysmurus leucopterus (*Temm.*), *Hume, Cat.* no. 678 quint.

Mr. W. Davison writes :—

"I found a nest of this bird on the 8th of April at the hot springs at Ulu Langat. The nest was built on the frond of a *Calamus*, the end of which rested in the fork of a small sapling. The nest was a great coarse structure like a Crow's, but even more coarsely and irregularly built, and with the egg-cavity shallower. It was composed externally of small branches and twigs, and loosely lined with coarse fibres and strips of bark. It contained two young birds about a couple of days old. The nest was placed about 6 feet from the ground. The surrounding jungle was moderately thick, with a good deal of undergrowth."

24. Garrulus lanceolatus, Vigors. *The Black-throated Jay.*

Garrulus lanceolatus, *Vig.*, *Jerd. B. Ind.* ii, p. 308; *Hume, Rough Draft N. & E.* no. 670.

The Black-throated Jay breeds throughout the Himalayas, at elevations of from 4000 to 8000 feet, from the Valley of Nepal to Murree.

They lay from the middle of April until the middle of June.

They build on trees or thick bushes, never at any great height from the ground, and often within reach of the hand. They always, I think, choose a densely foliaged tree, and place the nest sometimes in a main fork and sometimes on some horizontal bough supported by one or more upright shoots.

All the nests I have seen were moderately shallow cups, built with slender twigs and sticks, some 6 inches in external diameter, and from less than 3 inches to nearly 4 inches in height, with a nest-cavity some 4 inches across and 2 inches deep, lined with grass and moss-roots. Once only I found a nest almost entirely composed of grass, and with no lining but fine grass-stems.

The eggs vary from four to six, but this latter number is rarely met with.

Colonel C. H. T. Marshall writes :—"This is one of the commonest birds about Murree; we always found it well to the front during our rambles, chattering about in the trees. They breed from the middle of April till the end of June. We have taken their eggs between the 20th April and the 16th June. They keep above 5000 feet. I never observed any in the lower ranges. The nest is not a difficult one to find, being large and of loose coustruction; from 15 to 30 feet up a medium-sized tree close to the trunk or sometimes in a large fork. They never seem to build in the spruce firs which abound about Murree. They are by no means shy birds, and hop about the trees close by while their nest is being examined. Five is the ordinary number of eggs, which differ very much in appearance and size: the longest I have measures 1·25 and the shortest 1·1. Some are paler, some darker; some are of a uniform pale greenish-ash colour with a darker ring, while others are thickly speckled and freckled with a darker shade of the same colour. Some lack the odd ink-scratch which is so often to be seen on the larger end, and is the most peculiar feature of the egg, while a few have it at the thinner end.

"I should describe the average type as a long egg for its breadth; ground-colour greenish ashy with very thick sprinklings of spots of a darker and more greenish shade of the same colour, a ring of a darker dull olive round the large end, on which are one or two lines that look like a haphazard scratch from a fine steel pen."

From Dhurmsala Captain Cock wrote to me that this was "a most common bird at Dhurmsala; appears in large flocks during the winter, and often mixes with *Garrulus bispecularis* and *Urocissa flavirostris*. Pairs off about the end of April, when nidification begins. Builds a rather rough nest of sticks, generally placed on a tall sapling oak near the top; sometimes among the thicker branches of a pollard oak: outer nest small twigs roughly put together; inner nest dry roots and fibres, rather deep cup-shaped. Eggs number from four to five and vary in shape. I have found them sometimes nearly round, but more generally the usual shape. They vary in their colour, too, some being much lighter than others, but most of them have a few hair-like streaks on the larger end."

From Mussoorie Captain Hutton tells us that "the Black-throated Jay breeds in May and June, placing the nest sometimes on the branch of a tall oak tree (*Quercus incana*), at other times in a thick bush. It is composed of a foundation of twigs, and lined with fine roots of grass &c. mixed with the long black fibres of ferns and mosses, which hang upon the forest trees, and have much the appearance of black horse-hair. The nest is cup-shaped, rather shallow, loosely put together, circular, and about 4½ inches in diameter. The eggs are sometimes three, sometimes four in number, of a greenish stone-grey, freckled, chiefly at the larger end, with dusky and a few black hair-like streaks, which are not always present; they vary also in the amount of dusky freckling at the larger end. The nestling bird is devoid of the lanceolate markings on the throat."

From Nynee Tal Colonel G. F. L. Marshall writes:—"The Black-throated Jay builds a very small cup-shaped nest of black hair-like creepers and roots, intertwined and placed in a rough irregular casing of twigs. A nest found on the 2nd June containing three hard-set eggs was placed conspicuously on the top of a young oak sapling about 7 feet high, standing alone in an open glade, in the forest on Aya Pata, which is about 7000 feet above the sea. Another nest, found at an elevation of about 4500 feet on the 9th June, contained two eggs ; it was placed about 10 feet from the ground in a small tree in a hedgerow amongst cultivated fields."

Mr. Hodgson notes from Jaha Powah :—" Found five nests of this species between 18th and 30th May. Builds near the tops of moderate-sized trees in open districts, making a very shallow nest of thin elastic grasses sparingly used and without lining. The nest is placed on some horizontal branch against some upright twig, or at some horizontal fork. It is nearly round and has a diameter of about 6 inches. They lay three or four eggs of a sordid vernal green clouded with obscure brown."

The eggs are somewhat lengthened ovals, very much smaller than, though so far as coloration goes very similar to, those of *G. glandarius*. The ground-colour in some is a brown stone colour, in others pale greenish white, and intermediate shades occur, and they are very minutely and feebly freckled and mottled over the whole surface with a somewhat pale sepia-brown. This mottling differs much in intensity ; in some few eggs indeed it is absolutely wanting, while in others, though feeble elsewhere, it forms a distinct, though undefined, brownish cap or zone at the large end. The eggs generally have little or no gloss. It is not uncommon to find a few hair-like dark brown lines, more or less zigzag, about the larger end.

In length they vary from 1·03 to 1·23, and in breadth from 0·78 to 0·88; but the average of twenty-four eggs is 1·12 by 0·85.

25. Garrulus leucotis, Hume. *The Burmese Jay.*

Garrulus leucotis, *Hume, Hume, Cat.* no. 669 bis.

The nest of this Jay has not yet been found, but Capt. Bingham writes :—

" Like Mr. Davison I have found this very handsome Jay affecting only the dry *Dillenia* and pine-forests so common in the Thoungyeen valley. I have seen it feeding on the ground in such places with *Gecinus nigrigenys*, *Upupa longirostris*, and other birds. I shot one specimen, a female, in April, near the Meplay river, that must have had a nest somewhere, which, however, I failed to find, for she had a full-formed but shell-less egg inside her."

26. Garrulus bispecularis, Vigors. *The Himalayan Jay.*

Garrulus bispecularis, *Vig., Jerd. B. Ind.* ii, p. 307 ; *Hume, Rough Draft N. & E.* no. 669.

The Himalayan Jay breeds pretty well throughout the lower ranges of the Himalayas. It is nowhere, that I have seen, numerically very abundant, but it is to be met with everywhere. It lays in March and April, and, though I have never taken the nest myself, I have now repeatedly had it sent me. It builds at moderate heights, rarely above 25 feet from the ground, in trees or thick shrubs, at elevations of from 3000 to 7000 feet. The nest is a moderate-sized one, 6 to 8 inches in external diameter, composed of fine twigs and grass, and lined with finer grass and roots.

The nest is usually placed in a fork.

The eggs are four to six in number.

Mr. Hodgson notes that he " found a nest " of this species " on the 20th April, in the forest of Shewpoori, at an elevation of 7000 feet. The nest was placed in the midst of a large tree in a fork. The nest was very shallow, but regularly formed and compact. It was composed of long seeding grasses wound round and round, and lined with finer and more elastic grass-stems. The nest measured about $6\frac{1}{2}$ inches in diameter, but the cavity was only about half an inch deep."

Colonel C. H. T. Marshall remarks :—" I only took one authenticated set of eggs of this species (I found several with young), as it is an early breeder—I say authenticated eggs, because I *think* we may have attributed some to *Garrulus lanceolatus*, as the nests and eggs are very similar, and having a large number of the eggs of the latter, I took some from my shikaree without verifying them.

" The nest I took on the 6th May, 1873, at Murree, was at an elevation, I should say, of between 6500 and 7000 feet (as it was near the top of the hill), in the forest. The tree selected was a horse-chestnut, about 25 feet high. The nest was near the top, which is the case with nearly all the Crows' and Magpies' nests that I have taken. It was of loose construction, made of twigs and fibres, and contained five partially incubated eggs.

" The eggs are similar to those of *G. lanceolatus*. I have carefully compared the five of the species which I am now describing with twenty of the other, and find that the following differences exist. The egg of *G. bispecularis* is more obtuse and broader, there is a brighter gloss on it, and the speckling is more marked; but with a large series of each I think the only perceptible difference would be its greater breadth, which makes the egg look larger than that of the Black-throated Jay. My four eggs measure 1·15 by 0·85 each.

" This species only breeds once in a year, and from my observations lays in April, all the young being hatched by the 15th May. Captain Cock and myself carefully hunted up all the forests round Murree, where the birds were constantly to be seen, commencing our work after the 10th May, and we found nothing but young ones."

Colonel G. F. L. Marshall writes :—" I have found nests of this species for the first time this year; the first on the 22nd of May, by which time, as all recorded evidence shows it to be an early breeder, I had given up all hopes of getting eggs. The first nest contained two fresh eggs; it was on a horizontal limb of a large oak, at a bifurcation about eight feet from the trunk and about the same from the ground. The nest was more substantial than that of *G. lanceolatus*, much more moss having been used in the outer casing, but the lining was similar; it was a misshapen nest, and appeared, in the distance, like an old deserted one; the bird was sitting at the time; I took one egg, hoping more would be laid, but the other was deserted and destroyed by vermin. Another nest I found on the 2nd June; it contained three eggs just so much incubated that it is probable no more would be laid; this nest was much neater in construction and better concealed than the former one; it was in a rhododendron tree, in a bend about ten feet from the ground, between two branches upwards of a foot each in diameter, and covered with moss and dead fern; the tree grew out of a precipitous bank just below a road, and though the nest was on the level of the edge it was almost impossible to detect it; it was a very compact thick cup of roots covered with moss outside. The eggs were larger, more elongated, and much more richly coloured than in the first nest. Both nests were at about 7000 feet elevation, and in both instances the bird sat very close."

The eggs of this species are, as might be expected, very similar to those of *G. lanceolatus*, but they are perhaps slightly larger, and the markings somewhat coarser. The eggs are rather broad ovals, a good deal pointed towards one end. The ground-colour is pale greenish white, and they are pretty finely freckled and speckled (most densely so towards the large end, where the markings are almost confluent) with dull, rather pale, olive-brown, amongst which a little speckling and clouding of pale greyish purple is observable. The eggs are decidedly smaller than those of the English Jay, and few of the specimens I have exhibit any of those black hair-like lines often noticeable in both the English Jay and *G. lanceolatus*.

In length the eggs that I have measured varied from 1·1 to 1·21, and in breadth they only varied from 0·84 to 0·87.

27. Nucifraga hemispila, Vigors. *The Himalayan Nutcracker.*

Nucifraga hemispila, *Vig., Jerd. B. Ind.* ii, p. 304; *Hume, Rough Draft N. & E.* no. 666.

The Himalayan Nutcracker is *very* common in the fir-clad hills north of Simla, where it particularly affects forests of the so-called pencil cedar, which is, I think, the *Pinus excelsa*. I have never been able to obtain the eggs, for they must lay in March or early in April; but I have found the nest near Fagoo early in May with nearly full-fledged young ones, and my people have taken them with young in April below the Jalouri Pass.

The tree where I found the nest is, or rather *was* (for the whole hill-slope has been denuded for potatoe cultivation), situated on a steeply sloping hill facing the south, at an elevation of about 6500 feet. The nest was about 50 feet from the ground, and placed on

GRACULUS.—PARUS. 31

two side branches just where, about 6 inches apart, they shot out of the trunk. The nest was just like a Crow's—a broad platform of sticks, but rather more neatly built, and with a number of green juniper twigs with a little moss and a good deal of grey lichen intermingled. The nest was about 11 inches across and nearly 4 inches in external height. There was a broad, shallow, central depression 5 or 6 inches in diameter and perhaps 2 inches in depth, of which an inch was filled in with a profuse lining of grass and fir-needles (the long ones of _Pinus longifolia_) and a little moss. This was found on the 11th May, and the young, four in number, were sufficiently advanced to hop out to the ends of the bough and half-fly half-tumble into the neighbouring trees, when my man with much difficulty got up to the nest.

29. **Graculus eremita** (Linn.). _The Red-billed Chough._

Fregilus himalayanus, _Gould, Jerd. B. I._ ii, p. 319.

Mr. Mandelli obtained three eggs of this species from Chumbi in Thibet; they were taken on the 8th of May from a nest under the eaves of a high wooden house.

Though larger than those of the European Chough, they resemble them so closely that there can be no doubt as to their authenticity.

In shape the eggs are moderately elongated ovals, very slightly compressed towards the small end. The shell is tolerably fine and has a slight gloss. The ground-colour is white with a faint creamy tinge, and the whole egg is profusely spotted and striated with a pale, somewhat yellowish brown and a very pale purplish grey. The markings are most dense at the large end, and there, too, the largest streaks of the grey occur.

One egg measures 1·74 by 1·2.

Subfamily PARINÆ.

31. **Parus atriceps**, Horsf. _The Indian Grey Tit._

Parus cinereus, _Vieill., Jerd. B. Ind._ ii, p. 278.
Parus cæsius, _Tick., Hume, Rough Draft N. & E._ no. 645.

The Indian Grey Tit breeds throughout the more wooded mountains of the Indian Empire, wherever these attain an altitude of 5000 feet, at elevations of from 4000 or 5000 to even (where the hills exceed this height) 9000 feet.

In the Himalayas the breeding-season extends from the end of March to the end of June, or even a little later, according to the season. They have two broods—the first clutch of eggs is generally laid in the last week of March or early in April; the second towards the end of May or during the first half of June.

In the Nilghiris they lay from February to May, and _probably_ a second time in September or October.

The nests are placed in holes in banks, in walls of buildings or of terraced fields, in outhouses of dwellings or deserted huts and houses, and in holes in trees, and very frequently in those cut in some previous year for their own nests by Barbets and Wood-peckers.

Occasionally it builds *on* a branch of a tree, and my friend Sir E. C. Buck, C.S., found a nest containing six half-set eggs thus situated on the 19th June at Gowra. It was on a "Banj" tree 10 feet from the ground.

The only nest that I have myself seen in such a situation was a pretty large pad of soft moss, slightly saucer-shaped, about 4 inches in diameter, with a slight depression on the upper surface, which was everywhere thinly coated with sheep's wool and the fine white silky hair of some animal. The nest is usually a shapeless mass of downy fur, cattle-hair, and even feathers and wool, but when on a branch is strengthened exteriorly with moss. Even when in holes, they sometimes round the nest into a more or less regular though shallow cup, and use a good deal of moss or a little grass, or grass-roots; but as a rule the hairs of soft and downy fur con-stitute the chief material, and this is picked out by the birds, I believe, from the dung of the various cats, polecats, and ferrets so common in all our hills.

I have never found more than six eggs, and often smaller numbers, more or less incubated.

Mr. Brooks tells us that the Indian Grey Tit is "common at Almorah. In April and May I found the nest two or three times in holes in terrace-walls. It was composed of grass-roots and feathers, and contained in each case nearly fully-grown young, five in number."

From Dhurmsala Captain Cock wrote:—"*Parus cinereus* built in the walls of Dr. C.'s stables this year. When I found the nest it contained young ones. I watched the parents flying in and out, but to make sure put my ear to the wall and could hear the young ones chirruping. The nest was found in the early part of May 1869."

Colonel Butler writes:—"Belgaum, 12th June, 1879. A nest built in a hollow bamboo which supported the roof of a house in the native infantry lines. I did not see the nest myself, as un-fortunately the old bird was captured on it, and the nest and eggs destroyed; however, the hen bird was brought to me alive by the man who caught her, and I saw at once, by the bare breast, that she had been sitting, and on making enquiries the above facts were elicited. The broken egg-shells were white thickly spotted with rusty red.

"Belgaum, 8th June, 1880.—A nest in a hole of a tree about 7 feet from the ground, containing five fresh eggs. The nest con-sisted of a dense pad of fur (goat-hair, cow-hair, human hair, and hare's fur mixed) with a few feathers intermixed, laid on the top of a small quantity of dry grass and moss, which formed the foundation."

Lieut. H. E. Barnes notes from Chaman in Afghanistan :—
" This Tit is very common, and remains with us all the year
round. I found a nest on the 10th April, built in a hole in a tree ;
it was composed entirely of sheep's wool, and contained three incubated eggs, white, with light red blotches, forming a zone at the
larger end. They measured ·69 by ·48."

Mr. Benjamin Aitkin says :—

" When I was in Poona, in the hot season of 1873, the Grey
Tits, which are very common there, became exceedingly busy about
the end of May, courting with all their spirit, and examining every
hole they could find. One was seen to disappear up the mouth of
a cannon at the arsenal. Finally, in July, two nests with young
birds were discovered, one by myself, and one by my brother. The
nests were in the roofs of houses, and were not easily accessible,
but the parent birds were watched assiduously carrying food to
the hungry brood, which kept up a screaming almost equal to that
of a nest of minahs. On the 27th July a young one was picked up
that had escaped too soon from a third nest. The Indian Grey
Tit does not occur in Bombay, and I never saw it in Berar."

Speaking of Southern India Mr. Davison remarks that " the
Grey Tit breeds in holes either of trees or banks ; when it builds in
trees it very often (whenever it can apparently) takes possession of
the deserted nest-hole of *Megalœma viridis* ; when in banks a rat-
hole is not uncommonly chosen. All the nests I have ever seen or
taken were composed in every single instance of fur obtained from
the dried droppings of wild cats."

From Kotagherry, Miss Cockburn sends the following interesting
note :—

" Their nests are found in deep holes in earth-banks, and sometimes in stone walls. Once a pair took possession of a bamboo in
one of our thatched out-houses—the safest place they could have
chosen, as no hand could get into the small hole by which they
entered. These Tits show great affection and care for their young.
While hatching their eggs, if a hand or stick is put into the nest
they rise with enlarged throats, and, hissing like a snake, peck at
it till it is withdrawn. On one occasion I told my horse-keeper to
put his hand into a hole into which I had seen one of these birds
enter. He did so, but soon drew it out with a scream, saying a
' snake had bit him.' I told him to try again, but with no better
success ; he would not attempt it the third time, so the nest was
left with the bold little proprietor, who no doubt rejoiced to find
she had succeeded in frightening away the unwelcome intruder.
The materials used by these birds for their nests consist of soft
hair, downy feathers, and moss, all of which they collect in large
quantities. They build in the mouths of February and March ;
but I once found a nest of young Indian Grey Tits so late as the
10th November. They lay six eggs, white with light red spots.
On one occasion I saw a nest in a bank by the side of the road ;
when the only young bird it contained was nearly fledged the road
had to be widened, and workmen were employed in cutting down

the bank. The poor parent birds appeared to be perfectly aware that their nest would soon be reached, and after trying in vain to persuade the young one to come out, they pushed it down into the road but could get it no further, though they did their utmost to take it out of the reach of danger. I placed it among the bushes above the road, and then the parents seemed to be immediately conscious of its safety."

Mr. H. R. P. Carter notes that he "found a nest of the Grey Tit at Coonoor, on the Nilgiris, on the 15th May. It was placed in a hole in a bank by the roadside. It was a flat pad, composed of the fur of the hill-hare, hairs of cattle, &c., and was fluffy and without consistence. It contained three half-set eggs."

Mr. J. Darling, Jun., says:—"I have found the nests at Ooty, Coonoor, Neddivattam, and Kartary, at all heights from 5000 to nearly 8000 feet above the sea, on various dates between 17th February and 10th May.

"It builds in banks, or holes in trees, at all heights from the ground, from 3 to 30 feet. It is fond of taking possession of the old nest-holes of the Green Woodpecker. The nest is built of fur or fur and moss, and always lined with fine fur, generally that of hares. Its shape depends upon that of the hole in which it is placed, but the egg-cavity or depression is about 3 inches in diameter and an inch in depth.

"It lays four, five, and sometimes six eggs, but I think more commonly only four."

Dr. Jerdon remarks:—"I once found its nest in a deserted bungalow at Kallia, in the corner of the house. It was made chiefly of the down of hares (*Lepus nigricollis*), mixed with feathers, and contained six eggs, white spotted with rusty red."

The eggs resemble in their general character those of many of our English Tits, and though, I think, typically slightly longer, they appear to me to be very close to those of *Parus palustris*. In shape they are a broad oval, but somewhat elongated and pointed towards the small end. The ground-colour is pinkish white, and round the large end there is a conspicuous, though irregular and imperfect, zone of red blotches, spots, and streaks. Spots and specks of the same colour, or occasionally of a pale purple, are scantily sprinkled over the rest of the surface of the egg, and are most numerous in the neighbourhood of the zone. The eggs have a faint gloss. Some eggs do not exhibit the zone above referred to, but even in these the markings are much more numerous and dense towards the large end.

In length the eggs vary from 0·65 to 0·78, and in breadth from 0·5 to 0·58; but the average of thirty-eight is 0·71 by 0·54, so that they are really, as indeed they look *as a body*, a shade shorter and decidedly broader than those of *P. monticola*.

34. **Parus monticola**, Vig. *The Green-backed Tit.*

Parus monticolus, *Vig., Jerd. B. Ind.* ii, p. 277; *Hume, Rough Draft N. & E.* no. 644.

The Green-backed Tit breeds through the Himalayas, at elevations of from 4000 to 7000 or 8000 feet.

The breeding-season lasts from March to June, and some birds at any rate must have two broods, since I found three fresh eggs in the wall of the Pownda dak bungalow about the 20th June. More eggs are, however, to be got in April than in any other month.

They build in holes, in trees, bamboos, walls, and even banks, but walls receive, I think, the preference.

The nests are loose dense masses of soft downy fur or feathers, with more or less moss, according to the situation.

The eggs vary from six to eight, and I have repeatedly found seven and eight young ones; but Captain Beavan has found only five of these latter, and although I consider from six to eight the normal complement, I believe they very often fail to complete the full number.

Captain Beavan says :—" At Simla, on May 4th, 1866, I found a nest of this species in the wall of one of my servant's houses. It contained five young ones, and was composed of fine grey pushm or wool resting on an understructure of moss."

At Murree Colonel C. H. T. Marshall notes that this species " breeds early in May in holes in walls and trees, laying white eggs covered with red spots."

Speaking of a nest he took at Dhurmsala, Captain Cock says :— " The nest was in a cavity of a rhododendron tree, and was a large mass of down of some animal; it looked like rabbit's fur, which of course it was not, but it was some dark, soft, dense fur. The nest contained seven eggs, and was found on the 28th April, 1869. The eggs were all fresh."

Mr. Gammie says :—" I got one nest of this Tit here on the 14th May in the Chinchona reserves (Sikhim), at an elevation of about 4500 feet. It was in partially cleared country, in a natural hole of a stump, about 5 feet from the ground. The nest was made of moss and lined with soft matted hair; but I pulled it out of the hole carelessly and cannot say whether it had originally any defined shape. It contained four hard-set eggs."

The eggs are very like those of *Parus atriceps;* but they are somewhat longer and more slender, and as a rule are rather more thickly and richly marked.

They are moderately broad ovals, sometimes almost perfectly symmetrical, at times slightly pointed towards one end, and almost entirely devoid of gloss. The ground is white, or occasionally a delicate pinkish white, in some richly and profusely spotted and blotched, in others more or less thickly speckled and spotted with darker or lighter shades of blood-, brick-, slightly purplish-, or

3*

brownish-red, as the case may be. The markings are much denser towards the large end, where in some eggs they form an imperfect and irregular cap. In size they vary from 0·68 to 0·76 in length, and from 0·49 to 0·54 in breadth; but the average of thirty-two eggs is 0·72 by 0·52 nearly.

35. Ægithaliscus erythrocephalus (Vig.). *The Red-headed Tit.*

Ægithaliscus erythrocephalus (*Vig.*), *Jerd. B. Ind.* ii, p. 270 ; *Hume, Rough Draft N. & E.* no. 634.

The Red-headed Tit breeds throughout the Himalayas from Murree to Bhootan, at elevations of from 6000 to 9000 or perhaps 10,000 feet.

They commence breeding very early. I have known nests to be taken quite at the beginning of March, and they continue laying till the end of May.

The nest is, I think, most commonly placed in low stunted hill-oak bushes, either suspended between several twigs, to all of which it is more or less attached, or wedged into a fork. *I have* found the nest in a deodar tree, *laid* on a horizontal bough. I have seen them in tufts of grass, in banks and other unusual situations ; but the great bulk build in low bushes, and of these the hill-oak is, I think, their favourite.

The nests closely resemble those of the Long-tailed Tit (*Acredula rosea*). They are large ovoidal masses of moss, lichen, and moss-roots, often tacked together a good deal outside with cotton-wool, down of different descriptions, and cobwebs. They average about 4½ inches in height or length, and about 3½ inches in diameter. The aperture is on one side near the top. The egg-cavity, which may average about 2¼ inches in diameter and about the same in depth below the lower edge of the aperture, is densely lined with very soft down or feathers.

They lay from six to eight eggs, but I once found only four eggs in a nest, and these fully incubated.

From Murree, Colonel C. H. T. Marshall notes that this species "builds a globular nest of moss and hair and feathers in thorny bushes. The eggs we found were pinkish white, with a ring of obsolete brown spots at the larger end. Size 0·55 by 0·43. Lays in May."

Captain Hutton tells us that the Red-cap Tit is "common at Mussoorie and in the hills generally, throughout the year. It breeds in April and May. The situation chosen is various, as one taken in the former month at Mussoorie, at 7000 feet elevation, was placed on the side of a bank among overhanging coarse grass, while another taken in the latter month, at 5000 feet, was built among some ivy twining round a tree, and at least 14 feet from the ground. The nest is in shape a round ball with a small lateral entrance, and is composed of green mosses warmly lined with feathers. The eggs are five in number, white with a pinkish tinge, and sparingly sprinkled with lilac spots or specks, and having a well-defined lilac ring at the larger end."

From Nynee Tal, Colonel G. F. L. Marshall writes :—"This species makes a beautifully neat nest of fine moss and lichens, globular, with side entrance, and thickly lined with soft feathers. A nest found on Cheena, above Nynee Tal, on the 24th May, 1873, at an elevation of about 7000 feet, was wedged into a fork at the end of a bough of a cypress tree, about 10 feet from the ground, the entrance turned inwards towards the trunk of the tree. It contained one tiny egg, white, with a dark cloudy zone round the larger end.

"About the 10th of May, at Naini Tal, I was watching one of these little birds, which kept hanging about a small rhododendron stump about 2 feet high, with very few leaves on it, but I could see no nest. A few days later I saw the bird carry a big caterpillar to the same stump and come away shortly without it; so I looked more closely and found the nest, containing nearly full-fledged young, so beautifully wedged into the stump that it appeared to be part of it, and nothing but the tiny circular entrance revealed that the nest was there. It was the best-concealed nest for that style of position that I have ever seen."

These tiny eggs, almost smaller than those of any European bird that I know, are broad ovals, sometimes almost globular, but generally somewhat compressed towards one end, so as to assume something of a pyriform shape. They are almost entirely glossless, have a pinkish or at times creamy-white ground, and exhibit a conspicuous reddish or purple zone towards the large end, composed of multitudes of minute spots almost confluent, and interspaced with a purplish cloud. Faint traces of similar excessively minute purple or red points extend more or less above and below the zone. The eggs vary from 0·53 to 0·58 in length, and from 0·43 to 0·46 in breadth; but the average of twenty-five is 0·56 nearly by 0·45 nearly.

41. Machlolophus spilonotus (Bl.). *The Black-spotted Yellow Tit.*

Machlolophus spilonotus (*Bl.*), *Jerd. B. Ind.* ii, p. 281.

Mr. Mandelli found a nest of this species at Lebong in Sikhim on the 15th June in a hole in a dead tree, about 5 feet from the ground. The nest was a mere pad of the soft fur of some animal, in which a little of the brown silky down from fern-stems and a little moss was intermingled. It contained three hard-set eggs.

One of these eggs is a very regular oval, scarcely, if at all, pointed towards the lesser end; the ground-colour is a pure dead white, and the markings, spots, and specks of pale reddish brown, and underlying spots of pale purple, are evenly scattered all over the egg; it measures 0·78 by 0·55.

42. Machlolophus xanthogenys (Vig.). *The Yellow-cheeked Tit.*

Machlolophus xanthogenys (*Vig.*), *Jerd. B. Ind.* ii, p. 279; *Hume, Rough Draft N. & E.* no. 647.

The Yellow-cheeked Tit is one of the commonest birds in the neighbourhood of Simla, yet curiously enough I have never found a nest.

I have had eggs and nest sent me, and I know it breeds throughout the Western Himalayas, at elevations of from 4000 to 7000 feet; and that it lays during April and May (and probably other months), making a soft pad-like nest, composed of hair and fur, in holes in trees and walls; but I can give no further particulars.

Captain Hutton tells us that it is "common in the hills throughout the year. It breeds in April, in which month a nest containing four fledged young ones was found at 5000 feet elevation; it was constructed of moss, hair, and feathers, and placed at the bottom of a deep hole in a stump at the foot of an oak tree."

Writing from Dhurmsala, Captain Cock says:—"Towards the end of April this bird made its nest in a hole of a tree just below the terrace of my house. Before the nest was quite finished a pair of *Passer cinnamomeus* bullied the old birds out of the place, which they deserted. After they had left it I cut the nest out and found it nearly ready to lay in, lined with soft goat-hair and that same dark fur noticed in the nest of *Parus monticola.*"

Later he wrote to me that this species "breeds up at Dhurmsala in April and May. It chooses an old cleft or natural cavity in a tree, usually the hill-oak, and makes a nest of wool and fur at the bottom of the cavity, upon which it lays five eggs much like the eggs of *Parus monticola.* Perhaps the blotches are a little larger, otherwise I can see no difference. I noticed on one occasion the male bird carry wool to the nest, which, when I cut it out the same day, I found contained hard-set eggs. I used to nail a sheepskin up in a hill-oak, and watch it with glasses, during April and May, and many a nest have I found by its help. *Parus atriceps, P. monticola, Machlolophus xanthogenys, Abrornis albisuperciliaris,* and many others used to visit it and pull off flocks of wool for their nests. Following up a little bird with wool in its bill through jungle requires sharp eyes and is no easy matter at first, but one soon becomes practised at it."

The eggs are regular, somewhat elongated ovals, in some cases slightly compressed towards one end. The ground is white or reddish white, and they are thickly speckled, spotted, and even blotched with brick-dust red; they have little or no gloss.

They vary in length from 0·7 to 0·78, and in breadth from 0·52 to 0·55; but I have only measured six eggs.

43. **Machlolophus haplonotus** (Bl.). *The Southern Yellow Tit.*

Machlolophus jerdoni (*Bl.*), *Jerd. B. Ind.* ii, p. 280.

Col. E. A. Butler writes :—" Belgaum, 12th Sept., 1879.—Found a nest of the Southern Yellow Tit in a hole of a small tree about 10 feet from the ground. My attention was first attracted to it by seeing the hen-bird with her wings spread and feathers erect angrily mobbing a palm-squirrel that had incautiously ascended the tree, and thinking there must be a nest close by, I watched the sequel, and in a few seconds the squirrel descended the tree and the Tit disappeared in a small hole about halfway up. I then put a net over the hole and tapped the bough to drive her out, but this was no easy matter, for although the nest was only about $\frac{3}{4}$ foot from the entrance, and I made as much noise as a thick stick could well make against a hollow bough, nothing would induce her to leave the nest until I had cut a large wedge out of the branch, with a saw and chisel, close to the nest, when she flew out into the net.

" The nest, which contained, to my great disappointment, five young birds about a week old, was very massively built, and completely choked up the hollow passage in which it was placed. The foundation consisted of a quantity of dry green moss, of the kind that natives bring in from the jungles in the rains, and sell for ornamenting flower vases, &c. Next came a thick layer of coir, mixed with a few dry skeleton-leaves and some short ends of old rope and a scrap or two of paper, and finally a substantial pad of blackish hair, principally human, but with cow- and horse-hair intermixed, forming a snug little bed for the young ones. The total depth of the nest exteriorly was at least 7 inches.

" The bough, about 8 inches in diameter, was partly rotten and hollow the whole way down, having a small hole at the side above by which the birds entered, and another rather larger about a foot below the nest all choked up with moss that had fallen from the base of the nest. It is strange that it should have escaped my eye previously, as the tree overhung my gateway, through which I passed constantly during the day. Immediately below the nest a large black board bearing my name was nailed to the tree.

" At Belgaum, on the 10th July, 1880, I observed a pair of Yellow Tits building in a crevice of a large banian tree about 9 feet from the ground. The two birds were flying to and from the nest in company, the hen carrying building-materials in her beak. I watched the nest constantly for several days, but never saw the birds near it again until the 18th inst., when the hen flew out of the hole as I passed the tree. I visited the spot on the 19th and 20th inst., tapping the tree loudly with a stick as I passed, but without any result, as the bird did not fly off the nest.

" On the 21st, thinking the nest must either be forsaken or contain eggs, I got up and looked into the hole, and to my surprise found the hen bird comfortably seated on the nest, notwithstanding the noise I had been making to try and put her off. As the crevice was too small to admit my hand, I commenced to enlarge the entrance with a chisel, the old bird sitting closer than ever the whole time. Finding all attempts to drive her off the eggs fruitless, I tried to poke her off with a piece of stick, whereupon she stuck her head into one of the far corners and sulked. I then inserted my hand with some difficulty and drew her gently out of the hole, but as soon as she caught sight of me, she commenced fighting in the most pugnacious manner, digging her claws and beak into my hand, and finally breaking loose, flying, not away as might have been expected, but straight back into the hole again, to commence sulking once more. Again I drew her out, keeping a firm hold of one leg until I got her well away from the hole, when I released her. I then extracted five fresh eggs from the hole by means of a small round net attached to the loop end of a short piece of wire. The nest was a simple pad of human and cows' hair, with a few horse-hairs interwoven, and one or two bits of snake's skin in the lining, having a thin layer of green moss and thin strips of inner bark below as a foundation—in fact a regular Tit's nest. The eggs, of the usual parine type, were considerably larger than the eggs of *P. atriceps*, broad ovals, slightly smaller at one end than the other, having a white ground spotted moderately thickly all over with reddish chestnut; no zone or cap, but in some eggs more freely marked at one end (either small or large end) than the other, some of the markings almost amounting to blotches and the spots as a rule rather large."

Messrs. Davidson and Wenden remark of this bird in the Deccan :—" Specimens of this Tit were procured at Lanoli in August and at Egutpoora in March. They certainly breed at these places, as in September, at the latter place, W. observed two parent birds with four young ones capable of flying out very short distances."

And Mr. Davidson further states that it is "common through-out the district of Western Kandeish. I saw a pair building in the hole of a large mango tree at Malpur in Pimpalnir in the end of May."

44. **Lophophanes melanolophus** (Vig.). *The Crested Black Tit.*

Lophophanes melanolophus (*Vig.*), *Jerd. B. Ind.* ii, p. 273; *Hume, Rough Draft N. & E.* no. 638.

The Crested Black Tit breeds throughout the Lower Himalayas west of Nepal, at elevations of from 6000 to 8000 feet.

The breeding-season lasts from March to June, but the majority have laid, I think, for the first hatch by the end of the first week in April, unless the season has been a very backward one. They usually rear two broods.

They build, so far as I know, always in holes, in trees, rocks, and walls, preferentially in the latter. Their nests involve gener-ally two different kinds of work—the working up of the true nests on which the eggs repose, and the preliminary closing in and making comfortable the cavity in which the former is placed. For this latter work they use almost exclusively moss. Sometimes very little filling-in is required ; sometimes the mass of moss used to level and close in an awkward-shaped recess is surprisingly great. A pair breed every year in a terrace-wall of my garden at Simla ; elevation about 7800 feet. One year they selected an opening a foot high and 6 inches wide, and they closed up the whole of this, leaving an entrance not 2 inches in diameter. Some years ago I disturbed them there, and found nearly half a cubic foot of dry green moss. Now they build in a cavity behind one of the stones, the entrance to which is barely an inch wide, and in this, as far as I can see, they have no moss at all.

The nests are nothing but larger or smaller pads of closely felted wool and fur ; sometimes a little moss, and sometimes a little vegetable down, is mingled in the moss, but the great body of the material is always wool and fur. They vary very much in size : you may meet with them fully 5 inches in diameter and 2 inches thick, comparatively loosely and coarsely massed together ; and you may meet with them shallow saucers 3 inches in diameter and barely half an inch in thickness anywhere, as closely felted as if manufactured by human agency.

Six to eight is considered the full complement of eggs, but the number is very variable, and I have taken three, four, and five well-incubated eggs.

Captain Beavan, to judge from his description, seems to have found a regular cup-shaped nest such as I have never seen. He says :—" At Simla, April 20th, 1866, I found a nest of this species with young ones in it in an old wall in the garden. I secured the old bird for identification, and then released her. The nest contained seven young ones, and was large in proportion. The outside and bottom consists of the softest moss, the nest being carefully built between two stones, about a foot inside the wall ; the rest of it is composed of the finest grey wool or fur. Diameter inside 2·5 ; outside about 5 inches. Depth inside nearly 3 inches ; outside 3·6."

Captain Cock told me that he " found several nests in May and June in Cashmere. The first nest I found was in a natural cavity high up in a tree, containing three eggs, which I unfortunately broke while taking them out of the nest. The interior of the cavity was thickly lined with fur from some small animal, such as a hare or rat. I found my second nest close to my tent in a cleft of a pine, quite low down, only 3 feet from the ground. I cut it out and it contained five eggs of the usual type—broad, blunt little eggs, white, with rusty blotches."

Colonel G. F. L. Marshall writes :—" I have only found two nests of this species in Naini Tal, both had young (two in one nest, in the other I could not count) on the 25th April ; they were at about 7000 feet elevation, built in holes in walls, the entrance in both cases being very small, having nothing to distinguish it from other tiny crevices, and nothing to lead any one to suppose that there was a nest inside. It was only by seeing the parent birds go in that the nest was discovered."

The eggs of this species are moderately broad ovals, with a very slight gloss. The ground-colour is a slightly pinkish white, and they are richly blotched and spotted, and more or less speckled (chiefly towards the larger end), with bright, somewhat brownish red.

The markings very commonly form a dense, almost confluent zone or cap about the large end, and they are generally more thinly scattered elsewhere, but the amount of the markings varies much in different eggs. In some, although they are thicker in the zone, they are still pretty thickly set over the entire surface, while in others they are almost confined to one end of the egg, generally the broad end.

These eggs vary much in size and in density of marking. The ordinary dimensions are about 0·61 by 0·47, but in a large series they vary in length from 0·57 to 0·72, and in breadth from 0·43 to 0·54. The very large eggs, however, indicated by these *maxima* are rare and abnormal.

47. Lophophanes rufinuchalis (Bl.). *The Simla Black Tit.*

Lophophanes rufonuchalis (*Bl.*), *Jerd. B. Ind.* ii, p. 274.

Mr. Brooks informs us that this Tit is common at Derali and other places of similar elevation. " I found a nest under a large stone in the middle of a hill foot-path, up and down which people and cattle were constantly passing; the nest contained newly-hatched young. This was the middle of May."

Dr. Scully, writing of the Gilgit district, tells us that this Tit is a denizen of the pine-forests, where it breeds.

Finally Captain Wardlaw Ramsay, writing in the 'Ibis,' states that this Tit was breeding in Afghanistan in May.

Subfamily PARADOXORNITHINÆ.

50. Conostoma æmodium, Hodgs. *The Red-billed Crow-Tit.*

Conostoma æmodium, *Hodgs., Jerd. B. Ind.* ii, p. 10; *Hume, Rough. Draft N. & E.* no. 381.

A nest of the Red-billed Crow-Tit was sent me from Native Sikhim, where it was found at an elevation of about 10,000 feet, in a cluster of the small Ringal bamboo. It contained three eggs, two of which were broken in blowing them.

The nest is a very regular and perfect hemisphere, both externally and internally. It is very compactly made, externally of coarse grass and strips of bamboo-leaves, and internally very thickly lined with stiff but very fine grass-stems, about the thickness of an ordinary pin, very carefully curved to the shape of the nest. The coarser exterior grass appears to have been used when dry; but the fine grass, with which the interior is so densely lined, is still green. It is the most perfectly hemispherical nest I ever saw. Exteriorly it is exactly 6 inches in diameter and 3 in height; internally the cavity measures 4·5 in diameter and 2·25 in depth.

The egg is a regular moderately elongated oval, slightly compressed towards the smaller end. The shell is fine and thin, and has only a faint gloss. The ground-colour is a dull white, and it is sparsely blotched, streaked, and smudged with pale yellowish brown, besides which, about the large end, there are a number of small pale inky purple spots and clouds, looking as if they were beneath the surface of the shell.

The single egg preserved measures 1·11 by 0·8.

A nest sent me by Mr. Mandelli was found, he says, in May, in Native Sikhim, in a cluster of Ringal (hill-bamboo) at an elevation of nearly 10,000 feet. It is a large, rather broad and shallow cup, the great bulk of the nest composed of extremely fine hair-like grass-stems, obviously used when green, and coated thinly exteriorly with coarse blades of grass, giving the outside a ragged and untidy appearance. The greatest external diameter is 5·5, the height 3·2, but the cavity is 4·5 in diameter and 2·2 in depth, so that, though owing to the fine material used throughout except in the outer coating the nest is extremely firm and compact, it is not at all a massive-looking one.

60. Scæorhynchus ruficeps (Bl.). *The Larger Red-headed Crow-Tit.*

Paradoxornis ruficeps, *Bl., Jerd. B. Ind.* ii, p. 5.

Mr. Gammie writes from Sikhim:—"In May, at 2000 feet elevation, I took a nest of this bird, which appears to have been rarely, if ever, taken by any European, and is not described in your Rough Draft of 'Nests and Eggs.' It was seated among, and fastened to, the spray of a bamboo near its top, and is a deep, compactly built cup, measuring externally 3·5 inches wide and the same in depth; internally 2·7 wide by 1·9 deep. The material used is particularly clean and new-looking, and has none of the second-hand appearance of much of the building-stuffs of many birds. The outer layer is of strips torn off large grass-stalks and a very few cobwebs; the lining, of fine fibrous strips, or rather threads, of bamboo-stems. There were three eggs, which were ready for hatching-off. They averaged 0·83 in. by 0·63 in. I send you the nest and two of the eggs.

"Both Jerdon and Tickell say they found this bird feeding on grain and other seeds, but those I examined had all confined their diet to different sorts of insects, such as would be found about the flowers of bamboo, buckwheat, &c. Probably they do eat a few seeds occasionally, but their principal food is certainly insects. Very usually, in winter especially, they feed in company with

Gampsorhynchus rufulus. Rather curious that the two Red-heads should affect each other's society."

The eggs are broad ovals, rather cylindrical, very blunt at both ends. The shell fine, with a slight gloss. The ground is white, and it is rather thinly and irregularly spotted, blotched, and smeared in patches with a dingy yellowish brown, chiefly about the larger end, to which also are nearly confined the secondary markings, which are pale greyish lilac or purplish grey.

61. Scæorhynchus gularis (Horsf.). *The Hoary-headed Crow-Tit.*

Paradoxornis gularis, *Horsf., Jerd. B. Ind.* ii, p. 5.

A nest sent me by Mr. Mandelli as belonging to this species was found, he tells me, at an elevation of 8000 feet in Native Sikhim on the 17th May. It was placed in a fork amongst the branches of a medium-sized tree at a height of about 30 feet from the ground. The nest is a very massive cup, composed of soft grass-blades, none of them much exceeding ·1 inch in width, wound round and round together very closely and compactly, and then tied over exteriorly everywhere, but not thickly, with just enough wool and wild silk to keep the nest perfectly strong and firm. Inside, the nest is lined with extremely fine grass-stems ; the nest is barely 4 inches in diameter exteriorly and 2·5 in height ; the egg-cavity is 2·4 in diameter and 1·2 in depth.

Mr. Mandelli sends me an egg which he considers to belong to this species, found near Darjeeling on the 7th May. It is a broad - oval, very slightly compressed at one end ; the shell dull and glossless ; the ground a dead white, profusely streaked and smudged pretty thickly all over with pale yellowish brown ; the whole bigger end of the egg clouded with dull inky purple and two or three hair-lines of burnt sienna in different parts of the egg. The egg measures 0·8 by 0·61.

Two eggs of this species, procured in Sikhim on the 17th May, are very regular ovals, scarcely at all pointed towards the lesser end. The ground-colour is creamy white, and the markings consist of large indistinct blotches of pale yellow ; round the large end is an almost confluent zone or cap of purplish grey, darker in one egg ; they have no gloss, and both measure 0·82 by 0·61.

Family CRATEROPODIDÆ.

Subfamily CRATEROPODINÆ.

62. Dryonastes ruficollis (J. & S.). *The Rufous-necked Laughing-Thrush.*

Garrulax ruficollis (*J. & S.*), *Jerd. B. Ind.* ii, p. 38; *Hume, Rough Draft N. & E.* no. 410.

Of the Rufous-necked Laughing-Thrush, Mr. Blyth remarks:—
" Mr. Hodgson figures the egg of a fine green colour."

The egg is not figured in my collection of Mr. Hodgson's drawings.

Writing from near Darjeeling, in Sikhim, Mr. Gammie says:—
" I have seen two nests of this bird; both were in bramble-bushes about five feet from the ground, and exactly resembled those of *Dryonastes cærulatus,* only they were a little smaller. One nest had three young ones, the other three very pale blue unspotted eggs, which I left in the nest intending to get them in another day or two, as I wanted to see if more eggs would be laid, but when I went back to the place the nest had been taken away by some one. Both nests were found here in May, one at 3500 feet, the other at 4500 feet.

" I have taken numerous nests of this species from April to June, from the warmest elevations up to about 4000 feet. They are cup-shaped; composed of dry leaves and small climber-stems, and lined with a few fibrous roots. They measure externally about 5 inches in width by 3·5 in depth; internally 3·25 across by 2·25 deep. Usually they are found in scrubby jungle, fixed in bushes, within five or six feet of the ground. The eggs are three or four in number."

Many nests of this species sent me from Sikhim by my friends Messrs. Mandelli and Gammie are all precisely of the same type—deep and rather compact cups, varying from 5 to 6 inches in external diameter; and 3·25 to 3·75 in height; the cavities about 3·25 in diameter and 2·25 in depth. The nest is composed almost entirely of dry bamboo-leaves bound together loosely with stems of creepers or roots, and the cavity is lined with black and brown rootlets, generally not very fine. They seem never to be placed at any very great elevation from the ground.

The eggs of this species, of which I have received a very large number from Mr. Gammie, are distinguishable at once from those of all the other species of this group with which I am acquainted. Just as the egg of *Garrulax albigularis* is distinguished by its very deep tone of coloration, the egg of the present species is distinguished by its extreme paleness. In shape the eggs are moderately broad ovals, often, however, somewhat pyriform, often a good deal pointed towards the small end. The shell is extremely fine and smooth, and has a very fine gloss ; they may be said to be almost white with a delicate bluish-green tinge. In length they vary from 0·95 to 1·1, in breadth from 0·6 to 0·83; but the average of forty-one eggs is 1·02 by 0·75.

65. **Dryonastes cærulatus** (Hodgs.). *The Grey-sided Laughing-Thrush.*

> Garrulax cærulatus (*Hodgs.*), *Jerd. B. Ind.* ii, p. 36; *Hume, Rough Draft N. & E.* no. 408.

A nest of the Grey-sided Laughing-Thrush found by Mr. Gammie on the 17th June near Darjeeling, below Rishap, at an elevation of about 3500 feet, was placed in a shrub, at a height of about six feet from the ground, and contained one fresh egg. It was a large, deep, compact cup, measuring about 5·5 inches in external diameter and about 4 in height, the egg-cavity being 4 inches in diameter and 2¾ inches in depth. Externally it was entirely composed of very broad flag-like grass-leaves firmly twisted together, and internally of coarse black grass and moss-roots very neatly and compactly put together. The nest had no other lining.

This year (1874) Mr. Gammie writes :—" This species breeds in Sikhim in May and June. I have found the nests in our Chinchona reserves, at various elevations from 3500 to 5000 feet, always in forests with a more or less dense undergrowth. The nest is placed in trees, at heights of from 6 to 12 feet from the ground, between and firmly attached to several slender upright shoots. It is cup-shaped, usually rather shallow, composed of dry bamboo-leaves and twigs and lined with root-fibres. One I measured was 5 inches in diameter by 2·5 in height exteriorly ; the cavity was 4 inches across and only 1·3 deep. Of course they vary slightly. As far as my experience goes, they do not lay more than three eggs ; indeed, at times only two."

Dr. Jerdon remarks that " a nest and eggs, said to be of this bird, were brought to me at Darjeeling ; the nest loosely made with roots and grass, and containing two pale blue eggs."

One nest of this species taken in Native Sikhim in July, was placed in the fork of four leafy twigs, and was in shape a slightly truncated inverted cone, nearly 7 inches in height and 5·5 in diameter at the base of the cone, which was uppermost. The leaves attached to the twigs almost completely enveloped it. The nest itself was composed almost entirely of stems of creepers, several of which were wound round the living leaves of the twigs so as to hold them in position on the outside of the nest ; a few bamboo-leaves were intermingled with the creeper's stems in the body of the nest. The cavity, which is almost perfectly hemispherical, only rather deeper, is 3·5 inches in diameter and 2·25 in depth, and is entirely and very neatly lined with very fine black roots. Another nest, which was taken at Rishap on the 21st May, with two fresh eggs, was placed in some small bamboos at a height of about 10 feet from the ground. It is composed externally entirely of dry bamboo-leaves, loosely tied together by a few creepers and a little vegetable fibre, and it is lined pretty thickly with fine black fibrous roots. This nest is about 6 inches in diameter and 3·5 high exteriorly, while the cavity measures 3·5 by 2.

The eggs sent me by Mr. Gammie are a beautiful clear, rather pale, greenish blue, without any spots or markings. They have a slight gloss. In shape they are typically much elongated and somewhat pyriform ovals, very obtuse at both ends : but moderately broad examples are met with. In length they vary from 1·05 to 1·33, and in breadth from 0·76 to 0·86; but the average of thirty-five eggs is 1·18 nearly by 0·82 nearly.

69. **Garrulax leucolophus** (Hardw.). *The Himalayan White-crested Laughing-Thrush.*

Garrulax leucolophus (*Hardw.*), *Jerd. B. Ind.* ii, p. 35; *Hume, Rough Draft N. & E.* no. 407.

According to Mr. Hodgson's notes, the Himalayan White-crested Laughing-Thrush breeds at various elevations in Sikhim and Nepal, from the Terai to an elevation of 5000 or 6000 feet, from April to June. It lays from four to six eggs, which are described and figured as pure white, very broad ovals, measuring 1·2 by 0·9. It breeds, we are told, in small trees, constructing a rude cup-shaped nest amongst a clump of shoots, or between a number of slender twigs, of dry bamboo-leaves, creepers, scales of the turmeric plant, &c., and lined with fine roots.

Dr. Jerdon says :—" I have had the nest and eggs brought me more than once when at Darjeeling, the former being a large mass of roots, moss, and grass, with a few pure white eggs."

One nest taken in July at Darjeeling was placed on the outer branches of a tree, at about the height of 8 feet from the ground. It was a very broad shallow saucer, 8 inches in diameter, about an inch in thickness, and with a depression of about an inch in depth. It was composed of dead bamboo-leaves bound together with creepers, and lined thinly with coarse roots. It contained four fresh eggs. Other similar nests contained four or three eggs each.

From Sikhim, Mr. Gammie writes :—" I have found this Laughing-Thrush breeding in May and June, up to about 3500 feet; I have rarely seen it at higher elevations, and cannot but think that Mr. Hodgson is mistaken in stating that it breeds up to 5000 or 6000 feet. The nests are generally placed in shrubs, within reach of the hand, among low, dense jungle, and are rather loosely built cup-shaped structures, composed of twigs and grass, and lined with fibrous roots. Externally they measure about 6 inches in diameter by 3·5 in depth ; internally 4 by 2·25.

"The eggs are usually four or five in number, but on several occasions I have found as few as two well-set eggs."

Numerous nests of this species have now been sent me, taken

in May, June, and July, at elevations of from 2000 to fully 4000 feet,
and in one case it is said 5000. They are all very similar, large,
very shallow cups, from 6 to nearly 8 inches in external diameter,
and from 2·5 to 3·5 in height; exteriorly all are composed of coarse
grass, of bamboo-spathes, with occasionally a few dead leaves
intermingled, loosely wound round with creepers or pliant twigs,
while interiorly they are composed and lined with black, only
moderately fine roots or pliant flower-stems of some flowering-tree,
or both. Sometimes the exterior coating of grass is not very
coarse; at other times bamboo-spathes exclusively are used, and
the nest seems to be completely packed up in these.

The eggs of this species are broad ovals, pure white and glossy.
They vary from 1·05 to 1·13 in length, and from 0·86 to 0·95 in
width, but the average of eighteen eggs is a little over 1·1 by 0·9.

70. Garrulax belangeri, Less. *The Burmese White-crested Laughing-Thrush.*

Garrulax belangeri, *Less., Hume, Cat.* no. 407 bis.

Mr. Oates, who found the nest of this bird many years ago in
Burma, has the following note:—"Nest in a bush a few feet from the
ground, on the 8th June, near Pegu. In shape hemispherical, the
foundation being of small branches and leaves of the bamboo, and
the interior and sides of small branches of the coarser weeds and
fine twigs. The latter form the egg-chamber lining and are nicely
curved. Exterior and interior diameters respectively 7 and 3½
inches. Total depth 3½ and interior depth 2 inches. Three eggs,
pure white and highly glossy, and they measure 1·14 by ·87, 1·1
by ·88, and 1·03 by ·86."

The nests of this species are large, loosely constructed cups,
much resembling those of its Himalayan congeners. The base and
sides consist chiefly of dry bamboo-leaves with a few dead tree-leaves
scantily held together by a few creepers, while the interior portion
of the nest, which has no separate lining, is composed of fine twigs
and stems of herbaceous plants and the slender flower-stems of
trees which bear their flowers in clusters. The nests vary a good
deal in exterior dimensions as the materials straggle far and wide
in some cases, and the external diameter may be said to vary from
6 to 8 inches, and the height from 3·25 to 4·5; the cavities are more
uniform in size, and are about 3·5 in diameter by 2 in depth.

The eggs are moderately broad ovals, at times somewhat
pointed perhaps towards the small end, pure white and fairly
glossy.

Major C. T. Bingham thus writes of this bird :—" It is very difficult to either watch these birds, unseen yourself, at one of their
dancing parties, or to catch one of them actually sitting on the
nest. Twice had I in the end of March this year come across nests
with one or two of these birds in the vicinity, and yet have had to
leave the eggs in them as uncertain to what bird they belonged.

At last, on the 2nd April, I came in for a piece of luck. I was roaming about in the vicinity of my camp on the Gawbechoung, the main source of the Thoungyeen river, and moving very slowly and silently amid the dense clumps of bamboo, when my ears were saluted by the hearty laughter of a flock of these birds, evidently not far off. Very quietly I crept up, and looking cautiously from behind a thick bamboo-clump, saw ten or twelve of them going through a most intricate dance, flirting their wings and tails, and every now and then bursting into a chorus of shouts, joined in by a few others who were seated looking on from neighbouring bushes. During one of the pauses of the applause, and while the dancers were busy twining in and out, a single rather squeaky 'bravo' came from a bamboo-bush right opposite to me. Looking up I was astonished to see a nest in a fork of the bamboo, and on the nest a *Garrulax* who, probably too busy with her maternal duties to watch the performance going on below her attentively, came in with a solitary shout of approbation at an unseemly time. I watched the performance a few minutes longer, and then frightened the old hen on the nest. The terrific scare I caused by my sudden appearance is beyond description. The dancers scattered with screeches, and the old hen dropped fainting over the side of her nest with a feeble remonstrance, and disappeared in the most mysterious way. After all the nest contained only one egg, very glossy, white, and fresh. The nest was better and stronger built, though very like that of *Garrulax moniliger*, constructed of twigs, and finely lined with black hair-like roots; it measured some 6 inches in diameter, the egg-cavity about 1½ inch deep. Subsequently I took three other nests, on the 4th April and 23rd May. The first contained three, the two latter three and four eggs respectively. A considerable number of eggs measure from 1·22 to 1·06 in length, and from ·92 to ·81 in breadth, and average 1·13 by 0·88."

72. Garrulax pectoralis (Gould). *The Black-gorgeted Laughing-Thrush.*

Garrulax pectoralis (*Gould*), Jerd. B. Ind. ii, p. 39; *Hume, Rough Draft N. & E.* no. 412.

Mr. Oates tells us that he "found the nest of the Black-gorgeted Laughing-Thrush in the Pegu Hills, on the 27th April, containing three fresh eggs; the bird was sitting. The nest was placed in a bamboo-clump about 7 feet from the ground, made outwardly of dead bamboo-leaves and coarse roots, lined with finer roots and a few feathers; inside diameter 6 inches, depth 2 inches. Two eggs measured 1·04 by 0·83 and 0·86. Colour, a beautiful clear blue."

One of these eggs sent by Mr. Oates * seems rather small for

---

* I fear I may have made a mistake in identifying the nest referred to. With this caution, however, I allow my note to stand.—ED.

the bird. It is a very broad, slightly pyriform oval, of a uniform pale greenish-blue tint, and very fairly glossy. It measures 1·05 by 0·87.

This egg appears to me to be an abnormally small one. A nest sent me from Sikhim, where it was found in July, contained much larger eggs, and more in proportion to the size of the bird. The nest I refer to was placed in a clump of bamboos about 5 feet from the ground. It was a tolerably compact, moderately deep, saucer-shaped nest, between 6 and 7 inches in diameter, composed of dead bamboo-sheaths and leaves bound together with creepers and herbaceous stems, and thinly lined with roots. It contained two eggs. These are rather broad ovals, somewhat pointed towards one end, of a uniform pale greenish blue, and are fairly glossy.

These eggs measured 1·33 and 1·30 in length, and 0·98 in breadth.

Mr. Mandelli sent me two nests of this species, both taken in Native Sikhim, the one on the 4th, the other on the 20th July. Each contained two fresh eggs. One was placed in a small tree in heavy jungle, at a height of about 6 feet from the ground, the other in a clump of bamboos a foot lower. Both are large, coarse, saucer-shaped nests, 7 to 8 inches in diameter, and 3·5 to 4 in height externally; the cavities are about 4·5 inches in diameter, and less than 2 in depth; the basal portion of the nests is composed entirely of dry leaves, chiefly those of the bamboo, loosely held together by a few stems of creepers; the sides of the nest are stems of creepers wound round and round and loosely intertwined, and the cavity is lined with rather coarse rootlets, and in one case with fine twigs.

73. Garrulax moniliger (Hodgs.). *The Necklaced Laughing-Thrush.*

Garrulax moniliger (*Hodgs.*), *Jerd. B. Ind.* ii, p. 40; *Hume, Rough Draft N. & E.* no. 413.

Of the Necklaced Laughing-Thrush Dr. Jerdon says:—"I procured both this and the last (the Black-gorgeted Laughing-Thrush) at Darjeeling, and have also seen one or both in Sylhet, Cachar, and Upper Burmah. They both associate in large flocks, and frequent more open forest than most of the previous species. The eggs are greenish blue."

From Sikhim, Mr. Gammie writes :—" In the first week of June I found a nest in low jungle, at 2000 feet, containing four greenish-blue eggs, but, as I did not see the bird, left it until my return a week later. I then saw the female, but in the interval the young had been hatched. The nest closely resembled that of *D. cærulatus* [p. 46], both in shape and composition, and was similarly situated between several upright slender shoots to which it was firmly attached. It was, however, within five feet of the ground, which is lower by 5 feet or so than *D. cærulatus* generally builds.

" I have found this species breeding from April to June, up to elevations not much exceeding 2500 feet. It affects the low, dense scrub growing in moist situations, and usually fixes its nest between several upright sprays, within 5 or 6 feet of the ground. The nest is cup-shaped, made of dry bamboo-leaves, intermixed with a very few pieces of climber-stems, and thickly lined with old leaf-stalks of some pinnate-leaved tree. Externally it measures about 5·5 inches in diameter by 4 in height; internally 3·5 by 2·75.

"The eggs are four or five in number."

Mr. Oates writes:—"On the 27th April I shot a female in the Pegu Hills off her nest. This latter contained one young one, and one deformed egg, which unfortunately got broken; colour a deep blue. The nest was placed in a small seedling bamboo about 6 feet from the ground at a joint where a number of small twigs shot out, inverted umbrella fashion. The nest in every respect closely resembled that of *G. pectoralis.*"

He subsequently remarked:—"Breeds in Lower Pegu chiefly in July. Average of six eggs, 1·16 by ·88; colour, very glossy deep blue. Nest placed in forks of saplings within reach of the hand, massive, cup-shaped, and made of dead leaves and small branches; lined with fine twigs. Outside diameter 7 inches and depth 4; interior 4¼ by 2."

A nest found below Darjeeling in the first week of June on the branch of a good-sized tree, at a height of 12 feet from the ground, was similar to that described by Mr. Gammie, and contained a single fresh egg. This is a moderately broad oval, somewhat pointed towards the small end, and exhibits very little gloss. It is of precisely the same colour as those of the preceding species, but measures only 1·2 in length by 0·9 in breadth.

Writing from Tenasserim, Major C. T. Bingham says:—"Between the 25th March and 28th April I found at least twenty nests of this bird. They were broad, shallow cups of roots and twigs, lined with fine black grass-roots, and placed at heights varying from 4 to 10 feet above the ground, invariably in the forks of low bamboo. The number of eggs varied from 3 to 5; blue in colour, and fairly glossy."

Numerous nests from Sikhim, Pegu, and Tenasserim are all of precisely the same type as described by Mr. Gammie; but some are fully 7 inches in external diameter, and in several the cavity is at least 4 inches in diameter.

The eggs of this species obtained by Mr. Gammie vary very much in size and shape, and somewhat in colour. Some are considerably elongated ovals, with a marked pyriform tendency. Others are particularly broad ovals for this class of egg. The shell is fine and compact, and as a rule they seem to have a fine gloss; but one or two specimens almost want this. In colour they are a pale, clear, slightly greenish blue, unspotted and unmarked. In length they vary from 1·01 to 1·13, and in breadth from 0·81 to 0·9, but the average of thirteen is 1·07 by 0·85.

76. Garrulax albigularis (Gould). *The White-throated Laughing-Thrush.*

Garrulax albogularis (*Gould*), Jerd. B. Ind. ii, p. 38; *Hume, Rough Draft N. & E.* no. 411.

The White-throated Laughing-Thrush breeds throughout the lower southern ranges of the Himalayas from Assam to Afghanistan at elevations of from 4000 to nearly 8000 feet. They lay from the commencement of April to the end of June. The nest varies in shape from a moderately deep cup to a broad shallow saucer, and from 5 to 7 or even 8 inches in external diameter, and from less than 2 to nearly 4 inches in depth internally. Coarse grass, flags, creepers, dead leaves, moss, moss- and grass-roots, all at times enter more or less largely into the composition of the nest, which, though sometimes wholly unlined, is often neatly cushioned with red and black fern and moss-roots. The nests are placed in small bushes, shrubs, or trees, at heights of from 3 to 10 feet, sometimes in forks, but more often, I think, on low horizontal branches, between two or three upright shoots.

Three is, I think, the regular complement of eggs, and this is the number I have always found when the eggs were much incubated. I have not myself observed that this species breeds in company, nor can I ever remember to have taken two nests within 100 yards of each other.

Captain Hutton remarks:—" This is very common in Mussoorie at all seasons, and congregates into large and noisy flocks, turning up the dead leaves, and screaming and chattering together in most discordant concert. It breeds in April and May, placing the nest in the forks of young oaks and other trees, about 7 or 8 feet from the ground, though sometimes higher, and fastening the sides of it firmly to the supporting twigs by tendrils of climbing-plants. It is sometimes composed externally almost entirely of such woody tendrils, intermixed with a few other twigs, and lined with black hair-like fibres of mosses and lichens; at other times it is externally composed of coarse dry grasses and leaves of different kinds of orchids, and lined with fibres, the materials varying with the locality. The eggs are of a deep and beautiful green, shining as if recently varnished, and three in number. In shape they taper somewhat suddenly to the smaller end, which may almost be termed obtusely pointed. The size 1·19 by 0·87 inch. The usual number of eggs is three, though sometimes only one or two are found; but only on one occasion out of more than a dozen nests have I found four eggs. The old bird will remain on the nest until within reach of the hand."

From Murree, Colonel C. H. T. Marshall writes:—" This was the most beautiful egg taken this season, being of a rich, deep, glossy, greenish-blue colour. The nest is composed of fresh ivy-twigs, with the leaves attached, tightly woven together. The birds breed on small trees, not high up, at the end of a branch. While their nests were being examined, they came round in flocks to see what was happening, chattering and making that peculiar laughing note from which this genus takes its name. They are even gregarious in the breeding-season, and all the nests were found pretty near each other about 6000 feet up."

The nest sent me by Colonel Marshall is a broad, shallow cup, or saucer as I should perhaps call it, some 6 inches in diameter, with a central depression of at most 1·5 inch, below which the nest is an inch or 1·5 in thickness. It is very loosely put together, and composed interiorly of moderately fine dry twigs and roots, but exteriorly it is completely wound round with slender green ivy-twigs to which the leaves are attached. It has no lining or pretence for such.

Captain Cock says:—"The White-throated Laughing-Thrush lays one of the most lovely eggs with which I am acquainted. The nest is usually low, never more than 10 feet or so from the ground; and of some fifteen or more nests that I have taken, all were constructed of long stalks of the ground-ivy, twisted round and round into a wreath. The nest is not a deep cup; if anything it is rather shallow, but it is very wide. I always found these nests in thick forest, at high elevations from 6000 to 7000 feet. The birds used to sit close, and when put off their nests would commence their outcries, and from all parts they would assemble and flit about almost within reach of one's hand, making an awful noise, and in the dark shade of the forest their white gorgets had quite a ghostly look. The eggs are always three in number, of a beautiful shining blue-green, sometimes of a very long oval type. I have found the nests at Murree from the 3rd May to quite the end of June."

Colonel G. F. L. Marshall writing of this species says:—"A nest found at Nynee Tal on Ayar Pata, about 7000 feet above the sea, contained two fresh eggs on the 31st May. The eggs were of a rich deep greenish blue, unspotted. The nest was a scanty and loosely-built structure, composed of roots and stems of grass and creepers, cup-shaped, rather shallow, and lined with a curious black creeper, very like coarse hair. The birds were gregarious even though breeding, and were moving about the underwood in parties of three to five. The nest was near the top of an oak-sapling in a dense coppice, placed close against the stem in a bunch of leaves at the top. The only difficulty in finding it lay in the scantiness of the structure rather than in the concealment by the foliage. The bird was on the nest and only moved off about 3 feet, sitting close by and chattering indignantly during my inspection. They are noisy birds, constantly on the move, and their notes, though rather harsh, are very varied and quite *conversational*."

The eggs are long, and pointed at the small end, to which they sometimes taper much. They are very glossy, and vary from a deep dull blue (the blue of a dark oil-paint, very much deeper than that of any other of the Crateropodinæ with which I am acquainted) to a deep intense greenish blue. Possibly other as deeply coloured eggs occur in this family, but I have seen none like them. They are of course entirely unspotted.

In length they vary from 1·16 to 1·25, and in breadth from 0·8 to 0·86; but the average of some twenty eggs measured is 1·22 by 0·83.

78. Ianthocincla ocellata (Vig.). *The White-spotted Laughing-Thrush.*

Garrulax ocellatus (*Vig.*), *Jerd. B. Ind.* ii, p. 41; *Hume, Rough Draft N. & E.* no. 414.

. I know nothing personally of the nidification of the White-spotted Laughing-Thrush, which breeds nowhere, so far as I know, west of Nepal, but I had a nest with a couple of eggs and one of the parent-birds sent me from Darjeeling. The nest was taken in May in one of the low warm valleys leading to the Great Runjeet, and is said to have been placed close to the ground in a thick clump of fern and grass. The nest is chiefly composed of these, intermingled with moss and roots, and is a large loose structure some 7 inches in diameter.

Mr. Blyth remarked in ' The Ibis ' (1867) that this species was " surely a *Trochalopteron* rather than a *Garrulax*," and the eggs seem to confirm this view. These are long, cylindrical ovals, very obtuse even at the smaller end. They are about the same size as those of *Garrulax albigularis*, with a very delicate pale blue ground and little or no gloss. One egg is spotless; the other has a few chocolate-brown specks or spots towards the large end. They measure 1·18 by 0·86 and 1·25 by 0·85.

80. Ianthocincla rufigularis, Gould. *The Rufous-chinned Laughing-Thrush.*

Trochalopteron rufogulare (*Gould*), *Jerd. B. Ind.* ii, p. 47; *Hume, Rough Draft N. & E.* no. 421.

Common as this species is about Simla, I have never yet secured the nest, and know nothing certain about the eggs.

Captain Hutton says :—" This species appears usually in pairs, sometimes in a family of four or five. It breeds in May, in which month I took a nest, at about 6500 feet elevation, in a retired and wooded glen; it was composed of small twigs externally and lined with the fine black fibres of lichens. The nest was placed on a horizontal bough, about 7 feet from the ground, and contained three pure white eggs. Size 1·12 by 0·69; shape ordinary. The stomach of the old bird contained sand, seed, and the remains of wasps."

One egg that I possess of this species I owe to Captain Hutton, and it is of the *Pomatorhinus* type—a long oval, slightly pointed pure white egg, with but little gloss, measuring 1·08 by 0·75.

From Sikhim a nest, said to belong to this species, has been recently sent me. It was found below Darjeeling in July, and was placed in a double fork of the branchlets of a medium-sized tree. It is a moderately deep cup, composed almost entirely of dry, coarser and finer, tendrils of creepers, and is lined with a some black moss-roots and a few scraps of dead leaves. It contained three fresh eggs.

Numerous nests of this species subsequently sent me from Sikhim are all of the same type, all moderately deep cups composed entirely of creeper-tendrils, the cavity only being lined with fine black roots. They appear from the specimens before me to be quite *sui generis* and unlike those of any of its congeners. No grass, no dead leaves, no moss seems to be employed; nothing but the tendrils of some creeper. The nests appear to be always placed at the fork, where three, four, or more shoots diverge, and to be generally more or less like inverted cones, measuring say 4 to 5 inches in height, and about the same in breadth at the top, while the cavities are about 3 inches in diameter and 1·5 to 2 in depth. The nests appear to have been found at very varying heights from the ground from 5 to 15 feet, and at elevations of from 3000 to 5000 feet. They appear to have contained three fresh or more or less incubated eggs.

The eggs were found in Sikhim on different dates between 25th May and 8th September.

Exceptional as the coloration of the eggs of this species may seem, there is no doubt that they are pure white. The shell is thin and fragile, but has generally a decided gloss, and the eggs are typically elongated ovals, obtuse-ended, and more or less pyriform or cylindrical. The eggs vary from 0·92 to 1·13 in length, and from 0·75 to 0·8 in breadth, but the average of eleven eggs is 1·06 by 0·77 nearly.

82. Trochalopterum erythrocephalum (Vig.). *The Red-headed Laughing-Thrush.*

Trochalopteron erythrocephalum (*Vig.*), *Jerd. B. Ind.* ii, p. 43; *Hume, Rough Draft N. & E.* no. 415.

From Kumaon westwards, at any rate as far as the valley of the Beas, the Red-headed Laughing-Thrush is, next to *T. lineatum,* the most common species of the genus. It lays in May and June, at elevations of from 4000 to 7000 feet, building on low branches of trees, at a height of from 3 to 10 feet from the ground.

The nests are composed chiefly of dead leaves bound round into a deep cup with delicate fronds of ferns and coarse and fine grass, the cavities being scantily lined with fine grass and moss-roots. It is difficult by any description to convey an adequate idea of the beauty of some of these nests—the deep red-brown of the withered ferns, the black of the grass- and moss-roots, the pale yellow of the broad flaggy grass, and the straw-yellow of some of the finer grass-stems, all blended together into an artistic wreath, in the centre of which the beautiful sky-blue and maroon-spotted eggs repose. Externally the nests may average about 6 inches in diameter, but the egg-cavity is comparatively large and very regular, measuring about $3\frac{1}{2}$ inches across and fully $2\frac{1}{4}$ inches in depth. Some nests of course are less regular and artistic in their appearance, but, as a rule, those of this species are particularly beautiful.

The eggs vary from two to four in number.

Sir E. C. Buck sent me the following note :—

"I found a nest of this species near Narkunda (about 30 miles north of Simla) on the 26th June. It was placed on the branch of a banj tree, some 8 feet from the ground, and contained two eggs, half set. Nest and eggs forwarded."

Dr. Jerdon says that Shore, as quoted by Gould in his 'Century,' says that "it is by no means uncommon in Kumaon, where it frequents shady ravines, building in hollows and their precipitous sides, and making its nest of small sticks and grasses, the eggs being five in number, of a sky-blue colour." But Shore, as the showman would say, is, so far as eggs and nests are concerned, "a fabulous writer," and the eggs are always more or less spotted, and no nest that I ever saw of this species was composed of "small sticks."

Mr. Blyth says :—"Mr. Hodgson figures a green egg, spotted much like that of *Turdus musicus*, as that of the present species;" but in all Hodgson's drawings this *green* represents a *greenish blue*, as I have tested in dozens of cases.

Colonel G. F. L. Marshall remarks :—"I found a nest of this species on the 15th May at Nynee Tal on the top of Ayar Pata, at an elevation of about 7500 feet above the sea. The nest was a rather deep cup, neatly made and placed about 5 feet from the ground amongst the outer twigs of a thick barberry bush, the leaves of which entirely concealed it. It was composed of a thick layer of dead oak- and rhododendron-leaves, bound round outside with just enough of grass-stems and moss to keep the leaves in place; it had no lining of any description. The egg-cavity was $3\frac{1}{2}$ inches broad by nearly $2\frac{1}{2}$ inches deep. The eggs, two in number, were blue, with a few spots, streaks, and scrawls of brown tending to form a zone at the larger end. They were large for the size of the bird. The ground-colour was like that of the eggs of a Song-Thrush in England.

"Several more nests found subsequently with eggs up to 4th June were similar in structure, but placed in small oak trees from 5 to 15 or 18 feet from the ground.

"I found a nest of this species containing a single hard-set egg on the 17th August; both parent-birds were by the nest; this is unusually late, the chief breeding-month being June."

The eggs are very long ovals, of a delicate pale greenish-blue ground-colour, with a few spots, streaks, and streaky blotches of a very rich though slightly brownish red at the large end. These eggs, though somewhat longer in shape and less freely marked, are exactly of the same type as those of *T. cachinnans* and *T. variegatum*.

The texture of the shell is very fine and compact, and they have a slight gloss. In some eggs the spottings are more numerous, and, besides the primary markings already mentioned, a few purple spots and blotches, mostly very pale, are intermingled with the darker markings. In almost all the eggs that I have seen the markings were absolutely confined to the larger end.

In length the eggs vary from 1·15 to 1·22, and in breadth from 0·8 to 0·86; but the average is about 1·2 by 0·82.

85. **Trochalopterum nigrimentum**, Hodgs. *The Western Yellow-winged Laughing-Thrush.*

Trochalopteron chrysopterum (*Gould*), *apud Jerd. B. Ind.* ii, p. 43; *Hume, Rough Draft N. & E.* no. 416.

The Western Yellow-winged Laughing-Thrush breeds, so far as is yet known, only in Nepal, Sikhim, and Bhootan, from all which localities we have quite young birds, but no eggs.

Dr. Jerdon says :—" The eggs are greenish blue, in a nest neatly made with roots and moss." This, of course, is wrong, as the eggs are now well known to be spotted.

From Sikhim, Mr. Gammie writes:—"The Yellow-winged Laughing-Thrush breeds from April to June at elevations from 5500 feet upwards. It prefers scrubby jungle, and places its nest in bushes about six feet or so from the ground. It is a broad, cup-shaped structure, neatly and strongly made of fine twigs and dry grass-leaves, lined with roots and with a few strings of green moss wound round the outside. Externally, it measures about 6 inches wide, and 4½ deep ; internally 3¼ by 2½.

" The eggs are usually three in number."

Six nests of this species found between the 4th May and 2nd July in Native and British Sikhim were sent me by Mr. Mandelli. They were placed in small trees or dense bushes at heights of from 3 to 8 feet, and contained in some cases two, and in others three fresh or fully incubated eggs, so that sometimes the bird only lays two eggs. Three nests were also sent me by Mr. Gammie, taken in the neighbourhood of the Sikhim Cinchona-Plantations. All are precisely of the same type, all constructed with the same materials, but owing to the different proportions in which these are used some of the nests at first sight seem to differ widely from others. Some also are a good deal bigger than others, but all are massive, deep cups, varying from 5·25 to 6·5 inches in diameter, and from 3 to fully 4 in height externally ; the cavities vary from 3 to 3·5 in diameter, and from 2 to 2·5 in depth. The body of the nests is composed of grass ; the cavity is lined first with dry leaves, and then thickly or thinly with black fibrous roots. Externally the nest is more or less bound together by creepers and stems of herbaceous plants. Sometimes only a few strings of moss and a few sprays of *Selaginella* are to be seen on the outside of the nest; while, on the other hand, in some nests the entire outer surface is completely covered over with green moss, not only on the sides, but on the upper margin, so as to conceal completely the rest of the materials of the nest, and in all the nine nests before me the extent to which the moss is used varies.

The eggs of this species are typically somewhat elongated ovals, some are much pointed towards the small end, others are somewhat pyriform, and others again are subcylindrical. The shell is fine and soft, but has only a moderate amount of gloss. The ground-colour, which varies very little in shade, is a delicate pale, slightly greenish blue, almost precisely the same colour as that of *Trochalopterum erythrocephalum*. The eggs are sparingly (in fact, almost exclusively about the large end) marked with deep chocolate. These markings are in some spots and blotches, but in many assume the form of thicker or thinner hieroglyphic lines. As a rule, three fourths of the egg is spotless, occasionally a single speck or spot occurs towards the small end of the egg. One or two eggs are almost spotless. In length the eggs vary from 1·1 to 1·23, and in breadth from 0·73 to 0·87, but the average of sixteen eggs is 1·17 nearly by 0·82.

87. Trochalopterum phœniceum (Gould). *The Crimson-winged Laughing-Thrush.*

Trochalopteron phœniceum (*Gould*), *Jerd. B. Ind.* ii, p. 48; *Hume, Rough Draft N. & E.* no. 422.

Mr. Gammie says:—"I have found altogether seven nests of the Crimson-winged Laughing-Thrush in and about Rishap, at elevations between 4000 and 5000 feet, and on various dates between the 4th and 23rd May. The locality chosen for the nest is in some moist forest amongst dense undergrowth. It is placed in shrubs, at heights of from 6 to 10 feet from the ground, and is generally suspended between several upright stems, to which it is firmly attached by fibres. It is chiefly composed of dry bamboo-leaves and a few twigs, and lined with black fibres and moss-roots. A few strings of moss are twisted round it externally to aid in concealing it. It is a moderately deep cup, measuring externally about 5 inches in diameter and 4 inches in height, and internally 3½ inches in width and 2 inches in depth.

"The eggs are almost always three in number, but occasionally only two. Of the seven nests taken by me, five contained eggs and two young birds."

The Crimson-winged Laughing-Thrush, according to Mr. Hodgson's notes, breeds in Sikhim, at elevations of from 3000 to 5000 feet, during the months of April, May, and June. The nest is placed in the fork of some thick bush or small tree, where three or four sprays divide, at from 2 to 5 feet above the ground. The nest is a very deep compact cup. One measured *in situ* was 4·5 inches in diameter and the same in height externally, while the cavity was 3 inches in diameter and 2·25 deep. It was very compact and was composed of dry leaves, creepers, grass-flowers, and vegetable fibres, more or less lined with moss-roots and coated externally with dry bamboo-leaves. They lay, we are told, three or four eggs.

Dr. Jerdon says:—" A nest and eggs said to be of this bird were brought to me at Darjeeling; the nest made of roots and grass, and the eggs, three in number, pale blue, with a few narrow and wavy dusky streaks."

The eggs are singularly lovely. In shape they are elongated ovals, generally very obtuse at both ends, and many of them exhibiting cylindrical or pyriform tendencies. The shell is very fine and fairly glossy, and the ground-colour is a most beautiful clear pale sea-green in some, greenish blue in others. The character of the markings is more that of the Buntings than of this family. There are a few strongly marked deep maroon, generally more or less angular, spots or dashes, principally about the large end, and there are a few spots and tiny clouds of pale soft purple, and then there are an infinite variety of hair-line hieroglyphics, twisted and scrawled in brownish or reddish purple, about the egg. The markings are nowhere as a rule crowded, and towards the small end are usually sparse and occasionally wholly wanting. In some eggs a bad pen seems to have been used to scribble the pattern, and every here and there instead of a fine hair-line there is a coarse thick one.

The eggs are pretty constant in size and colour, but here and there an abnormally pale specimen, in which the green has almost entirely disappeared, is met with. In length they vary from 0·98 to 1·15, and in breadth from 0·7 to 0·82, but the average of thirty-one eggs is 1·04 by 0·74.

88. Trochalopterum subunicolor, Hodgs. *The Plain-coloured Laughing-Thrush.*

Trochalopteron subunicolor, *Hodgs., Jerd. B. Ind.* ii, p. 44; *Hume, Rough Draft N. & E.* no. 417.

The Olivaceous or Plain-coloured Laughing-Thrush breeds, according to Mr. Hodgson's notes, in the central region of Nepal from April to June. It nests in open forests and groves, building its nest on some low branch of a tree, 2 or 3 feet from the ground, between a number of twigs. The nest is large and cup-shaped: one measured externally 5·5 inches in diameter and 3·38 in height; internally 2·75 deep and 3·12 in diameter. The nest is composed externally of grass and mosses lined with soft bamboo-leaves. Three or four eggs are laid, unspotted greenish blue. One is figured as 1·07 by 0·7.

90. Trochalopterum variegatum (Vig.). *The Eastern Variegated Laughing-Thrush.*

Trochalopteron variegatum (*Vig.*), *Jerd. B. Ind.* ii, p. 45; *Hume, Rough Draft N. & E.* no. 418 (part).

The Eastern Variegated Laughing-Thrush breeds only at eleva-

tions of from 4000 to 7000 or 8000 feet, from Simla to Nepal,
during the latter half of April, May, and June. The nest is a
pretty compact, rather shallow cup, composed exteriorly of coarse
grass, in which a few dead leaves are intermingled; it has no lining,
but the interior is composed of rather finer and softer grass than
the exterior, and a good number of dry needle-like fir-leaves are
used towards the interior. It is from 5 to 8 inches in diameter
exteriorly, and the cavity from 3 inches to 3·5 in diameter and
about 2 inches deep. The nest is usually placed in some low,
densely-foliaged branch of a tree, at say from 3 to 8 feet from the
ground; but I recently obtained one placed in a thick tuft of
grass, growing at the roots of a young Deodar, not above 6 inches
from the ground. They lay four or five eggs.

The first egg that I obtained of this species, sent me by Sir E.
C. Buck, C.S., and taken by himself near Narkunda, late in June,
out of a nest containing two eggs and two young ones, was a nearly
perfect, rather long oval, and precisely the same type of egg as
those of *T. erythrocephalum* and *T. cachinnans*, but considerably
smaller than the former. The ground-colour is a pale, rather dingy
greenish blue, and it is blotched, spotted, and speckled, almost exclusively at the larger end, and even there not very thickly, with
reddish brown. The egg appeared to have but little gloss. Other
eggs subsequently obtained by myself were very similar, but slightly
larger and rather more thickly and boldly blotched, the majority of
the markings being still at the large end.

The colour of the markings varies a good deal: a liver-red is
perhaps the most common, but yellowish brown, pale purple, purplish red, and brownish red also occur. Here and there an egg is
met with almost entirely devoid of markings, with perhaps only
one moderately large spot and a dozen specks, and these so deep
a red as to be all but black.

The eggs vary from 1·07 to 1·15 in length, and from 0·76 to 0·82
in breadth.

91. Trochalopterum simile, Hume. *The Western Variegated Laughing-Thrush.*

Trochalopterum simile, *Hume; Hume, Cat.* no. 418 bis.

Messrs. Cock and Marshall write from Murree:—" The nidification of this *Trochalopterum* was apparently unknown before. We
found one nest on the 15th June, about twenty feet up a spruce-
fir at the extremity of the bough. Nest deep, cup-shaped, solidly
built of grass, roots, and twigs; the bird sits close. Eggs light
greenish blue, sparingly spotted with pale purple, the same size as
those of *Merula castanea.*"

92. **Trochalopterum squamatum** (Gould).   *The Blue-winged Laughing-Thrush.*

Trochalopteron squamatum (*Gould*), *Jerd. B. Ind.* ii, p. 46; *Hume, Rough Draft N. & E.* no. 420.

From Sikhim my friend Mr. Gammie writes:—" I have never as yet found more than one nest of the Blue-winged Laughing-Thrush, and this one was found on the 18th May at Mongphoo, at an elevation of about 3500 feet. The nest was placed in a bush (one of the *Zingiberaceæ*), growing in a marshy place, in the midst of dense scrub, at a height of about 4 feet from the ground, and was firmly attached to several upright stems. It was composed of dry bamboo-leaves, held together by the stems of delicate creepers, and was lined with a few black fibres. It was cup-shaped, and measured externally 5·7 in diameter by 3·6 in height, and internally 3·7 in width by 2·6 in depth. The nest contained three eggs, which were unfortunately almost ready to hatch off, so that three is probably the normal number of the eggs."

According to Mr. Hodgson's notes the Blue-winged Laughing-Thrush breeds in May and June in the central region of Nepal in forests, at elevations of from 2000 to 6000 feet. The nest is placed in a fork of a branch on some small tree, and is a large mass of dry leaves and coarse dry grass, 7 or 8 inches in diameter externally, mortar-shaped, the cavity about 2·5 deep, and lined with hair-like fibres. The nest, though composed of loose materials, is very firm and compact. They lay four or five eggs, unspotted, verditer-blue, one of which is figured as a broad regular oval, only slightly compressed towards one end, measuring 1·2 by 0·9.

One of the eggs taken by Mr. Gammie (the others were unfortunately broken) is a long, almost cylindrical, oval, very obtuse at both ends and slightly compressed towards the smaller end, so that the egg has a pyriform tendency. It measures 1·25 by 0·82. The colour is an excessively pale greenish blue, precisely the same as that of the eggs of *Sturnia malabarica*; but then this present egg was nearly ready to hatch off when taken, and the fresh eggs are somewhat deeper coloured.

Subsequent to his letter above quoted, Mr. Gammie on the 10th June found a second nest of this species similar to the first, containing three nearly fresh eggs. These are similar in shape to that above described, but in colour are a beautiful clear verditer-blue, altogether a much brighter and richer tint than that of the first. They measure 1·2 and 1·25 by 0·88.

One nest was taken by Mr. Gammie above Mongphoo at an elevation of about 4500 feet on the 30th of April. It was placed in a bush at a height of about 6 feet from the ground, and contained three fresh eggs. It was a loosely put together, massive cup, some 7 inches in diameter and 4 in height externally. It was composed mainly of fine twigs, creeper-stems, and grass, with a few bamboo-leaves intermingled, and the cavity was carefully lined

with bamboo-leaves, and then within that thinly with black fibrous
roots: the cavity measured 3·7 inches in diameter and 2·3 in depth.

The eggs of this species, of which I have now received many,
appear to be typically somewhat elongated ovals, and not unfrequently they are more or less pyriform or even cylindrical. As
a rule, they are fairly glossy, a bright pale, somewhat greenish
blue, quite spotless, and varying a little in tint. In length they
appear to vary from 1·11 to 1·25, and in breadth from 0·82
to 0·91 ; but the average of eleven eggs is 1·2 by 0·87.

93. Trochalopterum cachinnans (Jerd.).   *The Nilghiri Laughing-Thrush.*

Trochalopteron cachinnans (*Jerd.*), *Jerd. B. Ind.* ii, p. 48; *Hume,
Rough Draft N. & E.* no. 423.

The Nilghiri Laughing-Thrush breeds, according to my many
informants, throughout the more elevated portions of the mountains from which it derives its trivial name, from February to the
beginning of June.

A nest of this species sent me by Mr. H. R. P. Carter, who
took it at Coonoor on April 22nd (when it contained two fresh
eggs), is externally a rather coarse clumsy structure, composed of
roots, dead leaves, small twigs, and a little lichen, about 5 inches
in diameter, and standing about 4½ inches high. The egg-cavity
is, however, very regularly shaped, and neatly lined with very fine
grass-stems and a little fine tow-like vegetable fibre. It is a deep
cup, measuring 2½ inches across and fully 3¾ inches in depth.

A nest taken by Miss Cockburn was a much more compact
structure, placed between four or five twigs. It was composed of
coarse grass, dead and skeleton leaves, a very little lichen, and a
quantity of moss. The egg-cavity was lined with very fine grass.
The nest was externally about 5½ inches in diameter and nearly 6
inches in height, but the egg-cavity had a diameter of only about
2½ inches and was only about 2¼ inches deep.

It was Jerdon, I believe, who gave the name of Laughing-
Thrushes to this group, and this name is applicable enough to this
particular bird, the one with which he was most familiar, for it
does *laugh*—albeit, a most maniacal laugh; but the majority of
the group have not the shadow of a giggle even in them, and
should have been designated " Screaming Squabblers."

Mr. J. Darling, Jr., says :—" This bird breeds from February
to May. I have found the nests all over the Nilghiris, at elevations of from 4500 to 7500 feet above the sea. The nest is
placed indiscriminately in any bush or tree that happens to take
the bird's fancy, at heights of from 3 to 12 feet from the ground.

" In shape it is circular, a deep cup, externally some 6 inches in
diameter and 5 or 6 inches in height, and with a cavity 3 to 4 inches
wide and often fully 4 inches in depth. The nest is composed of
moss and small twigs, at times of grass mingled with some spiders'

webs : sometimes there is a foundation of dead leaves. The cavity is lined with fur, cotton-wool, feathers, &c.

" The eggs are two or three in number."

Mr. Wait, writing from Coonoor, says :—" *T. cachinnans* breeds about May, and lays from three to five oval eggs. The ground is bluish, with ash-coloured and brown spots and blotches, and occasionally marks." None of my other correspondents, however, admit that the bird ever lays more than three eggs.

Mr. Davison tells me that "this bird breeds commonly on the Nilghiris, just before the rains set in, in May and the earlier part of June, but it occasionally breeds earlier (in April) or later (in the latter end of June). The nest is cup-shaped, composed of dead leaves, moss, grass, &c., and lined with a few moss-roots or fine grass. It is placed in the fork of a branch about 6 or 8 feet from the ground. The eggs are a bluish green, mottled chiefly towards the larger end, and sometimes also streaked with purplish brown. The normal number of eggs is two ; sometimes, however, three are laid."

From Kotagherry, Miss Cockburn remarks :—" The name ' Laughing-Thrush ' is most applicable to this bird, and its notes are often mistaken for the sound of the human voice. This bird is very shy, except when its nest contains eggs or young, when it becomes extremely bold. I was quite surprised to see a pair whose nest I was taking come so close as to induce me to put out my hand to catch them. The Laughing-Thrush builds a pretty, though large, nest, and generally selects the forked branches of a thick bush, and commences its nest with a large quantity of moss, after which there is a lining of fine grass and roots, and the withered fibrous covering of the Peruvian Cherry (*Physalis peruviana*), the nest being finished with a few feathers, in general belonging to the bird. The inside of the nest is perfectly round, and rarely contains more than two eggs, belonging to the owner. The eggs are of a beautiful greenish-blue colour, with a few large and small brown blotches and streaks, mostly at the large end. I have found the nests of these birds in February, March, and April. Occasionally the Black-and-white Crested Cuckoo, which appears on these hills in the month of March, deposits its eggs (two in number) in the nest of this Thrush. They are easily distinguished, as their colour is quite different from the Thrush's eggs, being entirely dark bluish green."

Mr. Rhodes W. Morgan writing from South India, says, in ' The Ibis':—" It builds a very neat nest of moss, dried leaves, and the outer husk of the fruit of the Brazil Cherry, lined with feathers, bits of fur, and other soft substances. The nest is cup-shaped, and generally contains three eggs, most peculiarly marked with blotches, streaks, and wavy lines of a dark claret-colour on a light blue ground. The markings are almost always at the larger end."

The first specimens that I obtained of the eggs of this species were kindly sent to me by the late Captain Mitchell and Mr. H. R. P. Carter of Madras ; they were taken on the Nilghiris. They are moderately broad ovals, somewhat pointed towards one end, larger than the average eggs of *T. lineatum*, and about the same size as large specimens of the eggs of *Crateropus canorus* and *Argya malcolmi*. The ground-colour is of a delicate pale blue, and towards the large end, and sometimes over the whole surface, they are speckled, spotted, and blotched, but only sparingly, with brownish red and blackish brown, and amongst these markings a few cloudy streaks and spots of dull faint reddish purple are observable. The eggs have not much gloss.

Numerous other specimens subsequently received from Miss Cockburn and others correspond well with the above description. More or less pyriform varieties are common. In some eggs the markings are almost entirely wanting, there being only a very faint brownish-pink freckling at the large end; and in many eggs, even some that are profusely spotted all over, the markings consist only of darker or lighter brownish-pink shades. Occasionally a few, almost black, twisted lines are intermingled with the other markings, and in these cases the lines are frequently surrounded by a reddish-purple nimbus.

The eggs vary in length from 0·92 to 1·08, and in breadth from 0·74 to 0·8, but the average of twenty eggs measured was 1·0 by 0·76.

96. Trochalopterum fairbanki, Blanf. *The Palni Laughing-Thrush.*

Trochalopterum fairbanki, *Blanf., Hume, Cat.* no. 423 bis.

The Rev. S. B. Fairbank, the discoverer of this species, found its nest at Kodai Kanal, in the Palni Hills, in May. The nest was placed in the crotch of a tree, at about 10 feet from the ground, and at an elevation of nearly 6500 feet above the level of the sea. The eggs are moderately elongated ovals, with a fine, fairly glossy shell. The ground is pale greenish blue or bluish green; the markings are spots, small blotches, hair-lines, and hieroglyphic-like scrawls, rather thinly scattered about the surface, and varying in colour through several shades of brownish and reddish purple to bright claret-colour.

The only egg I have measures 1 inch in length by 0·8 inch in breadth.

99. Trochalopterum lineatum (Vig.). *The Himalayan Streaked Laughing-Thrush.*

Trochalopteron lineatum (*Vig.*), *Jerd. B. Ind.* ii, p. 50; *Hume, Rough Draft N. & E.* no. 425 *.

Next to the Common House-Sparrow, the Himalayan Streaked Laughing-Thrush is perhaps the most familiar bird about our

---

* I omit the note on *T. imbricatum* in the 'Rough Draft,' because, as I have shown in the 'Birds of India,' this bird was unknown to Hodgson, and his note refers to *T. lineatum*. Sufficient is now known about the nidification of this latter to render the insertion of Hodgson's note unnecessary.—ED.

houses at all the hill-stations of the Himalayas westward of Nepal and throughout the lower ranges on which these stations are situated; this species breeds at elevations of from 5000 to 8000 feet.

It lays from the end of April to the beginning of September, and very possibly occasionally even earlier and later. I took a nest on the 29th April near Mussoorie; Mr. Brooks obtained eggs in May and June at Almorah; Colonel G. F. L. Marshall at Mussoorie in July and August; and Colonel C. H. T. Marshall at Murree from May to the end of July. I again took them in July and August near Simla, and Captain Beavan found them as late as the 6th of September near the same station.

So far as my own experience goes, the nests are always placed in very thick bushes or in low thick branches of some tree, the Deodar appearing to be a great favourite. Those I found averaged about 4 feet from the ground, but I took a single one in a Deol tree fully 8 feet up. The bird, as a rule, conceals its nest so well that, though a loose and, for the size of the architect, a large structure, it is difficult to find, even when one closely examines the bush in which it is. The nest is nearly circular, with a deep cup-like cavity in the centre, reminding one much of that of *Crateropus canorus*, and is constructed of dry grass and the fine stems of herbaceous plants, often intermingled with the bark of some fibrous plant, with a considerable number of dead leaves interwoven in the fabric, especially towards the base. The cavity is neatly lined with fine grass-roots, or occasionally very fine grass. The cavity varies from 3 inches to 3·5 in diameter, and from 2·25 inches to 2·75 in depth; the walls immediately surrounding the cavity are very compact, but the compact portion rarely exceeds from ·75 to 1 inch in thickness, beyond which the loose ends of the material straggle more or less, so that the external diameter varies from 5·5 inches to nearly 10.

The normal number of eggs appears to me to be three, although Captain Beavan cites an instance of four being found.

Captain Hutton tells us (J. A. S. B. xvii.) that in the neighbourhood of Mussoorie " this bird is met with in pairs, sometimes in a family of four or five, and may be seen under every bush. The nest is placed near the ground, in the midst of some thick low bush, or on the side of a bank amidst overhanging coarse grass, and not unfrequently in exposed and well-frequented places; it is loosely and rather slovenly constructed of coarse dry grasses and stalks externally, lined sometimes with fine grass, sometimes with fine roots. The eggs are three in number, and in shape and size exceedingly variable, being sometimes of an ordinary oval, at others nearly round."

From Almorah and Nynee Tal my friend Mr. Brooks writes to me " that this bird is common everywhere. The nest is generally placed in a low tree or bush where the foliage is thick. It is composed of grass, and lined with finer grass. The eggs are three in number, one inch and one line long by nine lines broad. They are

of a light greenish blue, the tint being much the same as that of
the eggs of *Acridotheres tristis*. They lay from the commencement
of May to the end of June."

Colonel G. F. L. Marshall tells me that "the Streaked Laughing-
Thrush is very common at Mussoorie, where it is called by the
public the Robin of India. It breeds in July and August all about
Landour. The nest is cup-shaped, rather shallow, and loosely put
together, made of grass and fibre with some moss and a few dead
leaves twisted into it; it is placed in a low bush or else on the
ground concealed among the grass-roots on the hill-side. The
eggs, three or four in number, are oval, rather large for the bird,
and of a pure light-blue colour without spots. I took eggs on the
26th and 28th July and on the 16th August."

Sir E. C. Buck writes:—"At Mutianee, three marches north
of Simla, I found on the 28th June a nest in a bush on the side of a
scantily 'jungled' hill. It was 2 feet from the ground, constructed
of grass and stalks externally, and lined with fibrous roots. It contained three fresh eggs. The nest measured—exterior diameter
6 inches, height exteriorly 4 inches; the interior diameter was 3
inches, and the depth of the cavity 2 inches."

The late Captain Beavan tells us that "on the 16th of August,
1866, I found a nest in the garden, in a rose-bush, with four pale
blue eggs in it, like those of *Acridotheres tristis*. The nest is a
large structure, firmly built of dry twigs, bark, sticks, ferns, and
roots. Another nest, with three eggs only, was found in a thick
clump of everlasting peas close to the ground on the 6th of September. The female sat very close, and this may have been the
second nest of the same pair that built the nest mentioned above,
as it was built not far from the first."

Major C. T. Bingham writes:—"Being at Landour for a few
days in May I chanced on a nest of this bird, perhaps the commonest in the hills. It was placed under an overhanging bush on
the side of Lal Tiba hill, and *on the ground*, being constructed
rather loosely of pieces of the withered stem of some creeper,
intertwined with a quantity of oak-leaves, and lined with grass-
roots."

The eggs, of which I must have seen some hundreds, as this is
the commonest Laughing-Thrush about both Mussoorie and Simla,
are typically regular and moderately broad ovals. Abnormally elongated, spherical, and pyriform varieties occur; some are nearly
round like a Kingfisher's, and I have seen one almost as slender
as a Swift's, but, as a rule, the eggs vary but little either in shape
or colour. They are perfectly spotless, moderately glossy, and of
a delicate pale greenish blue, which of course varies a little in
shade and intensity of colour, but which is very much paler on the
average than those of any of the *Crateropi*, and at the same time
less glossy. I am not at all sure whether *T. lineatum* is rightly asso-
ciated with species like *T. cachinnans*, *T. variegatum*, and *T. erythrocephalum*, which all have spotted eggs.

In length the eggs vary from 0·8 to 1·13, and in breadth from

0·63 to 0·8 ; but the average of fifty-eight eggs carefully measured is 1·01 by 0·73.

101. Grammatoptila striata (Vig.). *The Striated Laughing-Thrush.*

Grammatoptila striata (*Vig.*), *Jerd. B. Ind.* ii, p. 11; *Hume, Rough Draft N. & E.* no. 382.

The Striated Laughing-Thrush, remarks Mr. Blyth, " builds a compact Jay-like nest. The eggs are spotless blue, as shown by one of Mr. Hodgson's drawings in the British Museum."

A nest of this species found near Darjeeling in July was placed on the branches of a large tree, at a height of about 12 feet.

It was a huge shallow cup, composed mainly of moss, bound together with stems of creepers and fronds of a *Selaginella*, and lined with coarse roots and broken pieces of dry grass. A few dead leaves were incorporated in the body of the nest. The nest was about 8 or 9 inches in diameter and about 2 in thickness, the broad, shallow, saucer-like cavity being about an inch in depth.

The nest contained two nearly fresh eggs. The eggs appear to be rather peculiarly shaped. They are moderately elongated ovals, a good deal pinched out and pointed towards the small end, in the same manner (though in a less degree) as those of some Plovers, Snipe, &c. I do not know whether this is the typical shape of this egg, or whether it is an abnormal peculiarity of the eggs of this particular nest. The shell is fine, but the eggs have very little gloss. In colour they are a very pale spotless blue, not much darker than those of *Zosterops palpebrosus.*

The eggs measure 1·3 and 1·32 in length, and 0·89 and 0·92 in breadth.

From Sikhim, Mr. Gammie writes :—" In the first week of May I took a nest of the Striated Laughing-Thrush out of a small tree growing in the forest at 5500 feet above the sea. It was fixed among spray about 10 feet up. In shape it is a shallow, broad cup, and is built in three layers: the outer one of twining stems, which besides holding the nest together fastened it to the spray; the middle layer is an intermixture of green moss and fresh fern-fronds, and the inner a thick lining of roots. Externally it measured 7·5 inches broad by 5·25 inches deep; internally 4 inches by 2·75 inches.

" It contained two hard-set eggs."

Several nests of this species that I have now seen have all been of the same type, large nests 9 or 10 inches in diameter, and 4 to 5 in height, the body of the nest composed mainly of green moss interwoven with and bound round about with the stems of creepers and a few pliant twigs, many of which straggle away a good deal outside the limits which I have assigned in stating the dimensions above. The cavities are not quite hemispherical, a little shallower, say 4·5 inches in diameter and 2 inches in depth, closely lined with fine black roots. They have all been placed in the branches of trees at heights of from 8 to 20 feet.

Eggs of this species obtained by Mr. Gammie in May, and Mr. Mandelli in July, are of precisely the same type. They are rather elongated ovals, a good deal pointed towards the small end, near which they are not unfrequently a good deal compressed, so as to render the egg slightly pyriform. The shell is fine and smooth, but has little gloss. The ground-colour is a very pale greenish blue or bluish green, in some almost white; some of them are absolutely spotless, none of them are at all well marked, but some bear from half a dozen to a dozen tiny specks of a dark colour. On one only there is a triangular spot about 0·05 each way, which proves on examination with a microscope to be a deep brownish red. On the other eggs the markings are mere specks.

The eggs vary from 1·25 to 1·35 in length, and from 0·89 to 0·92 in breadth.

104. **Argya earlii** (Blyth). *The Striated Babbler.*

Chatarrhæa earlii (*Blyth*), Jerd. B. Ind. ii, p. 68; *Hume, Rough Draft N. & E. no. 439.*

The Striated Babbler breeds in suitable localities throughout Continental India, from Sindh to Tipperah and Assam, as also in Burmah. Reedy-margined lakes, canals and perennial streams are its favourite haunts, and wherever within the limits above indicated these abound, and the locality is moist and warm, *A. earlii* is pretty sure to be met with.

They lay twice during the year, between the latter end of March and the early part of September, building a neat, compact, and rather massive cup-shaped nest, either between the close-growing reeds, to three or more of which it is firmly bound, or in some little bush or shrub more or less surrounded by high reed-grass. The broad leaves and stringy roots of the reed, common grass, and grass-roots are the materials of which it generally constructs its nest, which varies much in size, according to the situation and fineness of the material used. I have seen them composed almost wholly of reed-leaves, fully 7 inches in diameter and 5 in height, and again built entirely of fine grass-stems not more than 4 inches across and 3 inches in height. When semi-suspended between reeds, they are always smaller and more compact, while when placed in a fork of a low bush they are larger and more straggling. The cavity (always neatly finished off, but very rarely regularly lined, and then only with very fine grass-stems or roots) is usually about 3 inches in diameter by 2 inches in depth.

Colonel G. F. L. Marshall remarks:—"In the Saharunpoor District *A. earlii* commences building about the middle of March, and the young are hatched towards the middle of April. The nest is usually placed in the middle of a tuft of Sarkerry grass, and sometimes in a bush or small tree, generally 3 or 4 feet from the ground. It is a deep cup-shaped structure, rather neatly made of grass without lining, and woven in with the stems if in a clump of grass, or firmly fixed in a fork if in a bush or low tree. The interior diameter is about 3 inches, and the depth nearly 2 inches. The eggs, four in number, are of a clear blue colour without spots of any kind. In shape they are oval, rather thinner at one end; the shell is smooth and thin. The eggs are of the same colour, but considerably larger than those of *Argya caudata*. *Argya earlii* breeds commonly in the Sub-Siwalik District of the Doab; it seems fond of water, as most of the nests I have found were close to the canal bank. It is gregarious even in the breeding-season; small flocks of seven or eight keeping together, fluttering in and out of the low bushes, but seldom alighting on the ground, and occasionally making a noisy chattering cry, especially when disturbed."

From the Pegu District Mr. Oates writes :—" I found two nests on the 24th May, one quite empty though finished, the other containing three eggs.

" The nests were placed a few feet apart in an immensely thick patch of elephant-grass, the undergrowth being fine, once tall, but now dead, grass. It was upon this dead stuff, which in May is much flattened down, that I found the nests. They were not attached to anything, but simply laid in a depressed platform about a foot above the ground, in among the thickest of the stalks of elephant-grass.

" The nest is a bulky structure, some 6 or 8 inches in external diameter, and 4 inches in height, composed chiefly of coarse reeds, becoming finer interiorly till the egg-cup is reached, where the grasses employed are tolerably fine and neatly interwoven. The cavity itself is more than a hemisphere, the diameter being 3 inches and the depth about 2 inches.

" The eggs are of a beautiful blue colour, rather pointed at one end."

Colonel Tickell has the following note on the nidification of this species in the Asiatic Society Journal, 1848, p. 301:—

" *Burra phenga.*—Nest hemispherical, of grasses rather loosely interwoven ; generally on bushes in jungle. Eggs two to four; rather lengthened shape ; clear, full, verditer blue.—June."

Mr. J. R. Cripps writes of this bird in Eastern Bengal :—" Very common, and a permanent resident, keeping to grass-fields in small parties of seven to ten. Very noisy. On the 2nd December, 1877, I found a nest with three slightly-incubated eggs in a small babool bush which stood in a 'sone' grass-field. The nest was a deep cup, whose foundation was a few leaves over which sone-grass was woven rather loosely. Lining of fine grass-roots. The nest was placed in amongst some coarse grass which grew up in the centre of the bush, and was three feet from the ground. External height 4, diameter 4¼, internal diameter 2½, depth 2½ inches. Both Messrs. Marshall and Hume in their works on 'Birds' Nesting' give March and September as the two periods for these birds to lay, but the clutch I found were exceptionally late."

Mr. J. Inglis writes from Cachar:—" The Striated Reed-Babbler is exceedingly common during the whole year. It breeds from March onwards, making its nest in longish grass."

The eggs closely resemble those of *A. caudata* both in colour and shape, but they are conspicuously larger. To judge from Hewitson's figure, for I have never seen the egg, they in shape, size, and colour closely resemble the eggs of *Accentor alpinus*, some I have being very slightly larger, and others exactly the same size as the figure referred to.

In length the eggs vary from 0·78 to 1·01, and in breadth from 0 65 to 0·75, but the average of a large series is 0·88 by 0·7.

105. Argya caudata (Duméril). *The Common Babbler.*

Chatarrhæa caudata (*Dum.*), *Jerd. B. Ind.* ii, p. 67 ; *Hume, Rough Draft N. & E.* no. 438.

The Common Babbler breeds throughout India, not, however, ascending any of our many mountain-ranges to any great elevation.

They lay pretty well all the year round; at any rate from early in March to early in September their eggs are common. Mr. W. Blewitt took a nest at Hansie on the 3rd January, and single nests are recorded by others as found in October, December, and February. They certainly have two broods a year, and perhaps more, the first being hatched from March to May, the second from June to August.

They build in low thorny bushes, and occasionally in clumps of high grass, the nest being rarely more than 3 feet from the ground. The nest itself is cup-shaped, and composed of grass and roots, often unlined, at times lined with very fine grass-stems or horse-hair. As a rule, it is neatly and compactly built, with a deep cavity some 2 to 3 inches in diameter, and 1·75 to 2·25 in depth, but I have seen straggling, ragged, and comparatively shallow nests of this species, having an external diameter of fully 7 inches. Three is the normal number of the eggs, but four are occasionally met with.

Mr. Brooks says :—" This species builds in much the same sort of places as *A. malcolmi*, but it chooses a low thick bush, the nest not being more than 3 feet from the ground. Nest neatly built of grass, roots, hair, &c., and the eggs bright bluish green, very glossy, and much resembling those of *Accentor modularis.*"

Mr. R. M. Adam remarks :—" I took a nest of this bird in Oudh on the 22nd April. It contained a young bird and one unhatched egg. The nest was made of grass not well worked together, and had a lining of finer grass. The ground-work was composed of twigs and stems of creepers interlaced. The exterior diameter of the nest measured 5 inches, and the egg-cavity was 2 inches deep. In one case this bird did not lay till the fifth day after the nest was finished. About Agra this bird breeds during July and August.

" This Bush-Babbler is very common about the Sambhur lake. I have noted it breeding from the beginning of March till the beginning of July. Although this species generally prefers building in the hedges of prickly-pear, I have taken the nests in orange-trees, the karounda, the babool, &c."

Messrs. Davidson and Wenden state that in the Deccan it is "very common and breeds."

Major C. T. Bingham says:—"This bird, uncommon at Allahabad, is plentiful here at Delhi. I found several nests between March and June, all of the Babbler type, deep cups, rather more firmly built than those of the preceding bird, but constructed like them of coarse roots of grass, with finer ones for the inside. They are never placed at any great height from the ground, and generally in some thorny bush. I have found mostly three, rarely four eggs in any one nest."

Mr. Benjamin Aitkin writes:—"I never saw the Common Babbler in Poona, and it certainly does not occur in Bombay. But it is very abundant on the arid plains of Berar, breeding in the low babool-bushes, where large numbers of its eggs are destroyed by lizards. I have found four eggs in a nest oftener than three."

Colonel Butler writes:—"The Common Babbler breeds in the neighbourhood of Deesa principally during the monsoon; but I have found nests occasionally at other seasons of the year, as the following table of dates will show:—

"April 29, 1876. A nest containing 3 fresh eggs.
"May 16, 1876.   „   „   3 fresh eggs.
"May 21, 1876. ·   „   „   2 fresh eggs.
"Nov. 15, 1876.   „   „   4 young birds.

"I found numerous nests from the middle of July to the beginning of September. On the 26th July, 1876, I saw upwards of a dozen nests, some containing fresh eggs, and others incubated. In many instances they contained eggs of *Coccystes jacobinus*. The nest is usually placed 3 or 4 feet from the ground in low thorny bushes (*Zizyphus jujuba* preferred) or in a tussock of sarpat grass. It is built of twigs, roots, grass, &c., loosely put together exteriorly but closely woven interiorly, the lining being composed of fine roots and grass-stems. The eggs vary in number from three to five."

Lieut. H. E. Barnes, writing of Rajputana, says:—"The Striated Bush-Babbler breeds from March to July. The nest is usually placed in a low thorny bush, and is composed of grass-roots and stems; it is deep cup-shaped, neatly and compactly built."

The eggs are typically of a moderately elongated oval shape, slightly compressed towards one end, but more or less spherical and pyriform varieties occur; and I have one specimen, a very long pointed egg, which, so far as size and shape go, might pass for an egg of *Cypselus affinis*; and though this is a peculiarly abnormal shape, I have others which somewhat approach it in form. The eggs are glossy, often brilliantly so, and of a delicate, pure, spotless, somewhat pale blue. The shade of colour in this egg varies very little, and I have never met with either the very pale or very dark varieties common amongst the eggs of *C. canorus* and occasionally found amongst those of *A. malcolmi*. In colour, size, and shape they are not very unlike those of our English

Hedge-Sparrow, whose early eggs formed the prize of our first
boyish nesting-expeditions, but they are slightly larger and typically
somewhat more elongated.

In length they vary from 0·75 to 0·92, and in breadth from 0·6
to 0·7; but the average of one hundred and fifteen eggs measured
was 0·82 by 0·64.

107. Argya malcolmi (Sykes).   *The Large Grey Babbler.*

Malacocercus malcolmi (*Sykes*), *Jerd. B. Ind.* ii, p. 64.
Argya malcolmi (*Sykes*), *Hume, Rough Draft N. & E.* no. 436.

The Large Grey Babbler breeds throughout the central portions
of both the Peninsula and Continent of India from the Nilghiris
to the Dhoon. It does not extend westwards to Sindh or the
North-West Punjab, or eastwards far into Bengal Proper. In
the Central and North-West Provinces it lays from early in March
well into September, having at least two and, as I believe, often
three broods.

It builds on low branches of small trees or in thick shrubs, at
no great elevation from the ground, say at heights of from 4 to 10
feet, a somewhat loosely woven, but yet generally neat, cup-shaped
nest, composed, as a rule, chiefly of grass-roots, but often with an
admixture of thin sticks and grass. Generally there is no lining,
but I have found nests scantily lined with very fine grass and even
horse-hair. Even when, as is the rule, entirely unlined, the inside
is finished off very nicely and smoothly. I have often seen ragged
and untidy nests, but these are the exception. Externally the nest
is some 5 or 6 inches in diameter and 3 or 4 inches in height; the
cavity is from 3 to 4 inches across and from 2 to nearly 3 inches
in depth.

Four is the normal number of the eggs laid, but I have several
notes of finding five.

Mr. Brooks says:—"This species breeds in waste lands overgrown with scanty jungle. The nest is made of sticks, roots, grass,
&c., is rather bulky, and is placed in some moderate-sized bush
about 7 or 8 feet from the ground. The eggs are greenish blue,
bluer and not so brightly coloured as those of *C. terricolor.*"

Mr. R. M. Adam remarks:—"Near Muttra, on the 31st October, I found a pair of birds busy lining the interior of a nest which
they had built in a plum-tree. At the Sambhur lake it is very
common, and commences to breed about the end of March."

Writing from Kotagherry (Nilghiris), Miss Cockburn remarks:—
"Their nests are built of a few twigs and roots, very loosely put
together (on some low branch of a tree), and so few of even these
as hardly to keep the eggs from falling through. These Babblers
lay four oval eggs of a greenish-blue colour, but I once saw a nest
with eight, and as there were several of these birds close to it, I
have no doubt two or three shared it together, perhaps to avoid
the necessity of each pair building for itself. Their nests are
found in the months of March and April.

"It is in the nests of this species and our Common Laughing-Thrush (*T. cachinnans*) that I have chiefly found the eggs of the Pied Crested Cuckoo."

Of this species Colonel G. F. L. Marshall remarks :—"I have taken eggs on the 20th June in Cawnpoor, the 31st July in Bolundshuhur, and the 25th August in Allyghur. The nest is almost always in a keekur tree in a fork about halfway up, and near the end of a branch. It is composed of keekur-twigs and lined with roots. It is thinner in structure than that of *M. terricolor*, but has an outer casing of thorns which the latter wants. They lay four blue eggs, larger and paler than those of *M. canorus*."

Lieut. H. E. Barnes writes that in Rajputana the Large Grey Babbler is "very common. I have found nests in each month from January to December. They have, I believe, several broods in the year; and even when nesting associate in small parties of seven or eight."

Messrs. Davidson and Wenden say :—" Common, and breeds in the Deccan."

Major C. T. Bingham says :—" Breeds both at Allahabad and at Delhi from March to quite the end of August, placing its loosely constructed (rarely firmly built) nest of twigs and fine grass-roots generally at no great height in babool-trees. Twice only I have found them in dense mango-trees at about thirty feet from the ground. The nests are not, I think, as a rule, so deep as those of *Crateropus terricolor*; once or twice I have found the soft down of the Madar (*Calatropes hamiltonii*) incorporated into the lining of grass-roots. The eggs are generally three or four in number."

Mr. Benjamin Aitken writes :—" All the nests which I have seen of the Large Grey Babbler have been on babool-trees. At Akola (Berar) in 1870, a great many had their nests during the month of July. I have recorded two instances of nests placed at a height above the ground of 15 feet and 20 feet. These were at Poona, one on the 21st April, and the other on the 10th May. I could not go up to the nests, but the birds in both cases were sitting closely. I have twice found nests with only three newly-hatched young ones."

Colonel Butler informs us that "the Large Grey Babbler breeds in the neighbourhood of Deesa during the rains. Both the nest and eggs closely resemble those of *C. terricolor*, but the latter differ slightly in being less elongated, not so pointed at the small end, rounder at the large end, and somewhat paler in colour. I have taken nests on the following dates :—

"July 19, 1875.   A nest containing 4 fresh eggs.
"June 30, 1876.   ,,   ,,   4 fresh eggs.
"July 15, 1876.   ,,   ,,   4 fresh eggs.
"July 20, 1876.   ,,   ,,   3 fresh eggs.

"The nest in every instance was similar to that described by Jerdon, viz. :—a loose structure of dead roots, twigs, and grass, the interior being neatly lined with closely-woven roots of 'khus-khus.' The old birds generally select some thorny tree (*Mimosa* &c.) to build on, and the nest is usually from 8 feet to 20 feet from the ground.

"Even in the nesting-season these birds are gregarious, joining a flock generally as soon as they leave the nest."

The eggs of this species do not appear to me to differ perceptibly from those of *Crateropus canorus*. When one first takes a nest or two of each of them, one is apt to draw distinctions and fancy that the eggs of the two species can be discriminated; but after taking forty or fifty nests of each species, it becomes obvious that there is no variety of the one in either colour, shape, or size that cannot be paralleled in the other. All I have said of the eggs of *C. canorus* is applicable to the eggs of this species, and the only difference that, with a huge series of each before me, I can discover is that, as a body, there is less variation in the colour of the eggs of *Argya malcolmi* than in those of *C. canorus*.

In length they vary from 0·88 to 1·1, and in breadth from 0·73 to 0·85; but the average of fifty eggs measured is 0·99 by 0·77.

108. Argya subrufa (Jerd.) *.   *The Large Rufous Babbler.*

Layardia subrufa (*Jerd.*), *Hume, Cat.* no. 437.

The nest is a deep massive cup placed in the fork of twigs, coarsely and roughly but still strongly built. The body of the nest is chiefly composed of leaves, some of which must have been green when used. Outside, the leaves are held in position by blades of grass, creepers, and stems of herbaceous plants, carelessly and roughly wound about the exterior. The cavity is rather more neatly lined with tolerably fine grass-bents. Exteriorly the nest is about 7 inches in height and 5 in diameter. The cavity is about 3½ inches deep by 3 in diameter.

The eggs are precisely like those of the several species of *Argya*, moderately broad ovals rather obtuse at both ends, often with a pyriform tendency. The colour is a uniform spotless clear blue with a faint greenish tinge, and the eggs have usually a fine gloss. The eggs measure 0·98 by 0·75.

110. Crateropus canorus (Linn.)†.   *The Jungle Babbler.*

Malacocercus terricolor (*Hodgs.*), *Jerd. B. Ind.* ii, p. 59; *Hume, Rough Draft N. & E.* no. 432.
Malacocercus malabaricus, *Jerd.*, *Jerd. t. c.* p. 62; *Hume, t. c.* no. 434.

C. terricolor.

The Bengal Babbler breeds throughout the plains of the Bengal

---

* The accompanying incomplete account of the nidification of this bird is all I can find among Mr. Hume's notes. I cannot ascertain who was the discoverer of the nest and eggs described.—ED.

† In the 'Birds of India,' I have united *C. malabaricus* and *C. terricolor.* Mr. Hume probably still considers these two races distinct, and others may agree with him. To avoid confusion, therefore, I have kept the notes appertaining to these two races distinct from each other.—ED.

Presidency (including Bengal, North-Western Provinces, Central Provinces, Oudh, and the Punjab), and I may add in the less desert portions of Sindh, although the race found in that province is not exactly identical with the Bengal bird, and in some respects closely approaches the Malabar race. In Northern Rajpootana it is rare, and further south in the quasi-desert tracts of Central and Western Rajpootana it disappears according to my experience.

Eastward in Cachar and Assam it appears to occur as a mere straggler, but I have no record of its having bred there. It lays from the latter half of March until the close of July, but the great majority lay during the first week after the setting in of the rains, which varies according to locality and season, from the 1st of June to the 15th of July.

They build very commonly in gardens, in thick orange-, citron-, or lime-shrubs, but their nests may be found almost anywhere, in thick shrubs or small trees of any kind, or in thick hedges, at heights of from 4 to 10 feet from the ground, always placed in some fork towards the centre of the shrub or hedge. The nests are rather loosely-put-together cups, composed of grass-stems and roots varying in fineness, and often lined with horse-hair. Some are deep and neatly constructed, others loose, straggling, and shallow, the cavity varying from 3 to more than 4 inches in diameter and from less than 2 to nearly 3 inches in depth.

Three is the normal number of the eggs, but I have repeatedly found four.

Captain Hutton writes to me:—"A nest of this bird was taken in the Dehra Dhoon on the 14th May, and was composed entirely of fine roots, the thinnest being placed within as a lining. Subsequently three others were procured, one of which was externally composed of coarse dry grasses and leaves, with a scanty lining of fine roots; the other two were constructed of the fine woody tendrils of climbing-plants and lined like the others with fine roots. These latter had a strong resemblance to some of the nests of *Garrulax albogularis*, while the difference exhibited in the nature of the materials used arises from the various character of the localities in which the bird may choose to build. Each nest contained four beautiful eggs of a full bright turquoise-green, shining as if varnished. The eggs were nearly all hard-set. This species does not ascend the hills, but appears to be confined to the Dhoon, where it may be seen in small parties in gardens, hedgerows, and low brushwood, turning over the dead leaves in search of seeds and insects. Its flight is low, short, and apparently laboured, from the shortness and rounded form of the wing, but on the ground it hops along with speed. The note is clamorous and chuckling and uttered in concert."

The late Mr. A. Anderson remarked:—"Although one of the most common birds in the North-West Provinces, and in fact verging on a nuisance, its nidification is interesting, inasmuch as its nest (in common with that of *A. malcolmi*) is used as a nursery for the young of *Hierococcyx varius* and *Coccystes melanoleucus*.

"This Babbler builds, as a general rule, during the early part of the rains (June to August), laying usually three or four eggs of a bright greenish-blue colour. The nest itself recalls that of the Blackbird, but it is frequently very clumsily made. On the 21st June last a boy brought me a nest of this species containing *eight* eggs. Two, if not three, of this clutch are easily separable from the others, being more oval and somewhat smaller, and are unquestionably parasitical eggs ; but it is quite impossible to say whether they belong to *H. varius* or *C. melanoleucus.*

"Again, on the 9th July, I took a nest in person, which also contained eight eggs. Seven of these are all alike and are well incubated, while the eighth is quite fresh, and doubtless owes its parentage to one of the above-mentioned Cuckoos.

"Strange to say I have now another nest marked down, which in like manner contains the same number of callow young. It is just possible that the foster-parents may have to perform double duty in this case.

"From the foregoing it may be inferred that *M. canorus* does occasionally lay more than four eggs, or as the birds are gregarious even during the breeding-season, it is possible enough that two birds may occasionally deposit eggs in the same nest.

"I should not think that *H. varius* (the "Brain-fever and Delirium-tremens Bird" as it is frequently called) had much difficulty in depositing her eggs in the nest of the *Malacocerci*, for I have frequently noticed that all the Babblers in the neighbourhood make a clean bolt of it immediately this Cuckoo puts in an appearance, no doubt owing to its great similarity to the Indian Sparrow-Hawk (*M. badius*).

"During the months of September and October I have observed several Babblers in the act of feeding one young *H. varius*, following the bird from tree to tree, and being most assiduous in their attentions to the young interloper."

Mr. R. M. Adam remarks :—" I took a nest of this bird in Agra on the 17th July. It contained five eggs, all of which were nearly hatched. Again on the 21st I took another nest containing only one hard-set egg."

Writing from Calcutta, Mr. J. C. Parker says :—" I found a nest of this bird, near my house in Garden Reach, on the 23rd June. It contained four fresh eggs."

Colonel Butler observes :—" The Bengal Babbler breeds in the neighbourhood of Deesa as a rule, I think, during the rains and in the cold weather, but I have found nests as late as March. The nest is usually placed on the outside branch of some moderate-sized tree (neem &c.). It is a somewhat solidly built structure composed almost entirely of dead twigs, stems of dead leaves, and stalks of coarse dry grass, being lined with a few fine fibrous roots or stems of grass. I found nests on the following dates :—

"July 16, 1875. A nest containing 4 fresh eggs.
"March 20, 1876.   „   „   4 fresh eggs.
"May 29, 1876.   „   „   3 fresh eggs.
"June 17, 1876.   „   „   3 fresh eggs.

"June 17, 1876.   A nest containing 4 young birds.
"Oct. 15, 1876.   „   „   4 fresh eggs.
"Nov. 3, 1876.   „   „   4 slightly incubated.
"In some nests I have noticed a breach upon one side of the nest as if intended for the convenience of the bird's tail. It is not unusual to find an egg of *C. jacobinus* in the nest."

Major C. T. Bingham writes :—"Common both at Allahabad and at Delhi ; I have found this bird breeding from April to the end of July. All nests that I have found have, with the exception of one, been placed in low babool bushes ; once only I found a nest near Delhi in the fork of a low bough of a mango-tree, this was on the 31st July. The nests are more or less loosely constructed cups of slender twigs and grass-roots and inclined."

Mr. J. R. Cripps writing from Eastern Bengal says :—"On the 15th April I found a nest on the very top of a mango-tree about 30 feet off the ground, shooting the male as it flew off the nest."

The eggs of this species are very variable in colour, shape, and size. Typically they are rather broad ovals, somewhat compressed towards one end, and much the shape of, though a good deal smaller than, those of our English Song-Thrush. Some are, however, long and cylindrical ; others more or less spherical. The colour varies from a pale blue, like that of *Trochalopterum lineatum*, to a deep dull blue, recalling, but yet not so dark as, that of *Garrulax albigularis*. The eggs are typically glossy, but it is remarkable that in a large series the deepest coloured are always far the most glossy. Some deep blue eggs of this species are most intensely glossy, more so than almost any other of our Indian eggs, except those of *Metopidius indicus*. I need scarcely say that the eggs are entirely spotless and devoid of all markings, but I may note that each egg is invariably the same colour throughout, and that I have never met with a specimen in which the shade of colour varied in the same egg.

In length the eggs vary from 0·88 to 1·15, and in breadth from 0·75 to 0·82 ; but the average of fifty-one eggs measured is 1·01 by 0·78.

C. malabaricus.

The Jungle Babbler, like the White-headed one, breeds pretty well over the whole of Southern India, but while the latter is chiefly confined to the more open plain country, the former is the bird of the uplands, hills, and forests. Still the Jungle Babbler is found at times in the same localities as the White-headed one, and what is more, specimens occur, as in Cochin, which partake of the distinctive characters of both. A great deal still remains to be done in working out properly this group ; both in Sindh on the west and the Tributary Mehals on the east, and again in some parts of the Nilghiris, races occur quite intermediate between typical *C. terricolor* and typical *C. malabaricus*, while in the south, as already mentioned, forms intermediate between this latter and *C. griseus* seem common. Three distinguishable races again of

C. griseus are met with, but running the one into the other, while intermediate forms between this species and *C. somervillii* (Sykes) are also met with.

Mr. Davison remarks :—" This bird seems to be very irregular in its time of breeding. I have taken the nest in May, June, October, and December. The nest is rather a loose structure of dry grass and leaves, lined with fine dry grass ; it is generally placed in the middle of some thick thorny bush, and cannot generally be got at without paying the penalty of well scratched hands. The eggs, generally five in number, are of a very deep blue with a tinge of green, but of not so decided a tinge as in the eggs of *M. griseus.* It breeds on the slopes of the Nilghiris, not ascending to more than about 6000 feet."

Mr. Wait, writing from Coonoor, says :—" *C. malabaricus* builds a cup-shaped nest in small trees and bushes, and lays from three to five very round oval verditer-blue eggs."

Captain Horace Terry says of this species :—" Rather rare at Pulungi, but very common lower down on the slopes and in the Pittur valley. I got a nest on April 5th at Pulungi with three incubated eggs, and on the 6th one with two incubated eggs, in the Pittur valley. The last was built in a hollow in the top of a stump of a tree that had been broken off some ten feet from the ground."

Mr. I. Macpherson writes from Mysore:—"This bird is occasionally found with *C. griseus* in the bigger scrub forests, but its chief habitat is the larger forests. Its breeding-season is much the same as *C. griseus*, but unlike it, it does not select thorny bushes for building in, its nests being generally found in small trees or bamboo-clumps. Four is the usual number of eggs laid, but five are often found, and the fifth I expect is frequently that of *H. varius.*"

Three eggs sent me by Mr. Carter from Coonoor, in the Nilghiries, are absolutely undistinguishable from those of *Argya malcolmi.* Like these they are a uniform, rather deep greenish blue, devoid of spots or markings, and very glossy. I do not think that, if the eggs of *A. malcolmi, C. malabaricus,* and *C. terricolor* were once mixed, it would be possible to separate them with certainty. Other eggs taken by Mr. Davison are similar but slightly smaller, and, taking them as a whole, I think they average rather darker than those of the two species just mentioned.

The eggs vary in length from 0·93 to 1·02, and in breadth from 0·71 to 0·82; but the average of nine eggs is 0·97 by nearly 0·77.

111. Crateropus griseus (Gm.).   *The White-headed Babbler.*

Malacocercus griseus (*Gm.*), *Jerd. B. Ind.* ii, p. 60 ; *Hume, Rough Draft N. & E.* no. 433.

I should say that the White-headed Babbler breeds all over the plain country of Southern India, not ascending the hills to any great elevation. At the same time, many people would very likely

separate the Madras, Mangalore, and Anjango birds, and insist on
their being different species; but for my part, seeing how the
birds vary in each locality and what a perfect and unbroken chain
of intermediate forms connects the most different-looking examples,
and that all the several races are separable from the other species
of this group by their more or less conspicuously pale heads, I
prefer to keep them all as *C. griseus*.

This species, thus considered, breeds apparently twice a year
from April to June, and again in October and even later.

About Madras the nest is commonly placed in thick thorny
hedges of a shrub locally known as " Kurka-puli," said by Balfour
to be *Garcinia cambogia*, but which does not look like a *Garcinia*
at all. The nest is a loosely-made cup, composed of grass-stems
and roots, and the eggs vary from three to five in number.

Dr. Jerdon says :—" I have often found the nest of this bird,
which is composed of small twigs and roots, carelessly and loosely
put together, in general at no great height from the ground. It
lays three or four blue eggs."

Colonel Butler writes :—" A nest containing four fresh eggs
apparently of this species (it being the common Babbler in this
district) was brought to me by some wood-cutters on the 18th
March, 1880. It was taken in the jungles about six miles from
Belgaum, and measured about 2¾ inches in diameter and about
2 inches deep interiorly, and was of the usual Babbler type,
consisting of dry stems loosely but neatly constructed. The eggs
were highly glossed and deep bluish green, some people might say
greenish blue."

Mr. Iver Macpherson writes of this bird from Mysore :—" I
have found their nests in every month between March and August,
and they possibly breed both earlier and later. The nests are
generally fixed in thorny bushes and at no great height off the
ground. Four is the usual number of eggs laid, but very often
five are found, and I feel much inclined to think that the fifth egg
is often that of *H. varius*."

The eggs of this species that I possess were taken by Mr. Davison in May, in the immediate neighbourhood of Madras. They
are all pretty regular, somewhat cylindrical ovals, excessively
glossy, spotless, and of a deep greenish blue, much deeper than
the eggs of any of the other *Crateropi* are as a rule; in fact,
they approach in colouring to the eggs of *Garrulax albigularis*.

They vary in length from 0·9 to 1·0, and in breadth from 0·62
to 0·74; but I have seen too few eggs to be able to strike any
reliable average.

112. Crateropus striatus (Sw.). *The Southern-Indian Babbler.*

Malacocercus striatus (*Sw.*), *Hume, Cat.* no. 432 bis.

Colonel Legge, writing of this bird's nidification in Ceylon,
says :—" The breeding-season of the 'Seven Brothers' lasts from

March until July. The nest is placed in a cinnamon-bush, shru
or bramble, at about four feet from the ground, and is a compa
cup-shaped structure, usually fixed in a fork and made of stc
grasses and plant-stalks and lined with fine grass, which, in soı
instances I have observed, was plucked green. The interi
measures 2½ inches in depth by about 3 in width. The eggs a
two or three in number, small for the size of the bird, glossy
texture, and of a uniform opaque greenish blue. They measu
from 0·91 to 1·0 in length, by 0·7 to 0·74 in breadth."

113. Crateropus somervillii (Sykes). *The Rufous-tailed Babbler.*

Malacocercus somervillei (*Sykes*), *Jerd. B. Ind.* ii, p. 63; *Hus
Rough Draft N. & E.* no. 435.

Of the nidification of the Rufous-tailed Babbler (which, so :
as I yet know, is confined to the narrow strip of country lyi
beneath the Ghâts for about 60 miles north and south of Bombı
and to the hills or ghâts overlooking this), all I yet know is cc
tained in the following brief note by Mr. E. Aitken; he says:—
"I once found a nest of the Rufous-tailed Babbler at Khandal
I cannot tell the level precisely, but it cannot have been far fro
2000 feet above the sea. It was at the end of May or the ve
beginning of June. The nest was in a small spreading tree
level, open forest country. The situation was just such a one
A. malcolmi generally chooses—the end of a horizontal bran
with no other branches underneath it; but it was not so high
those of *A. malcolmi* usually are, for I could reach it from t
ground. The nest was rather flat and contained three eggs, almı
hatched, of an intense greenish-blue colour.
"In Bombay, where it is far more common, I once, on 1
1st October, saw a pair followed by one young one and a you
Coccystes melanoleucus. This was on a hill, and indeed these bii
seem to confine themselves pretty much to hilly ground."
Mr. Benjamin Aitken writes:—" With reference to your remı
that, as far as you know, the Rufous-tailed Babbler is confir
to the strip of country beneath the Ghâts, I can certainly say tl
they are plentiful on the slopes of Poorundhur hill, eightı
miles south of Poona. It would be interesting to learn on wh
other of the Deccan hills it is found. This species is decidedly fc
of hilly country. It is common on the two ranges of low hills t.
run along the east and west shores of the island of Bombay, 1
never shows a feather in the gardens and groves on the le
ground. I spent the greater part of two days, when I could
spare the time, in searching for the nests, but the birds breed
the date-trees, and it would be hopeless to think of finding a n
without cutting away many of the branches or fronds. Moreov
the bird is extremely wary, and it is by no means easy to gu
on which particular tree it has its nest."

114. Crateropus rufescens (Blyth). *The Ceylonese Babbler.*

Layardia rufescens (*Blyth*), *Hume, Cat.* no. 437 bis.

Colonel Legge writes regarding the nidification of this bird in Ceylon:—" This bird breeds in the Western Province in March, April, and May, and constructs a nest similar to the last [*M. striatus*], of grass and small twigs, mixed perhaps with a few leaves, and placed among creepers surrounding the trunks of trees or in a low fork of a tree. It conceals its habitation, acccording to Layard, with great care ; and I am aware myself that very few nests have been found. It lays two or three eggs, very similar to those of the last species, of a deep greenish blue, and pointed ovals in shape—two which were taken by Mr. MacVicar at Bolgodde measuring 0·95 by 0·75, and 0·92 by 0·74 inch."

115. Crateropus cinereifrons (Blyth). *The Ashy-headed Babbler.*

Garrulax cinereifrons (*Blyth*), *Hume, Cat.* no. 409 bis.

Colonel Legge, in his work on the birds of Ceylon, says :— " The breeding-season of this bird is from April to July. Full-fledged nestlings may be found abroad with the parent birds in August; and from this I base my supposition, for I have never found the nest myself. Intelligent native woodmen, in the western forests, who are well acquainted with the bird, have informed me that it nests in April, building a large, cup-shaped nest in the fork of a bush-branch, and laying three or four dark blue eggs. Whether this account be correct or not, future investigation must decide."

116. Pomatorhinus schisticeps, Hodgs. *The Slaty-headed Scimitar Babbler.*

Pomatorhinus schisticeps, *Hodgs., Jerd. B. I.* ii, p. 29; *Hume, Rough Draft N. & E.* no. 402.

Speaking of the Slaty-headed Scimitar Babbler, Dr. Jerdon says :—" A nest made of moss and some fibres, and with four pure white eggs, was brought to me at Darjeeling as belonging to this bird."

Two nests were sent me by Mr. Mandelli as belonging to this species, the one found near Namtchu on the 3rd April containing four fresh eggs, the other near Yendong on the 15th June, containing three. Another nest which he found on the 22nd April, near the same place as the first, contained four fresh eggs. All were placed on or very near to the ground in brushwood and grass ; all appear to have been large, rather saucer-like nests, from 5·5 to 6·5 inches in diameter externally, and 2·5 to 3 in height. Outside and below they are composed chiefly of coarse grass, dead leaves, especially fern-leaves, while interiorly they are composed of and lined with finer—

in some cases *very* fine—grass. The cavities average, I should guess, 3·75 inches in diameter, and 1·5, or a little more perhaps, in depth.

Mr. J. R. Cripps has the following note on the breeding of this bird in Assam :—" A nest I got was situated at the roots of a clump of bushes, overhanging a small river. A bridge spanning this river was within ten yards, the intervening space being open ; and for such a shy bird to have chosen such an exposed situation to build in astonished me."

From Sikhim Mr. Gammie writes :—" A nest of this Babbler taken on the 20th May much resembled that of *P. ferruginosus*, both in size and structure. The egg-cavity had, however, a lining of at least half an inch in thickness of soft, fibrous material extracted from the bark of some tree, and a little fine grass for the eggs to lie on. It was on the ground, among low jungle, in the Ryeng Valley, at 2000 feet of elevation, and contained four eggs, two of them hatching off and two addled. According to my experience, nests containing so large a proportion of addled eggs are unusual."

Eggs sent by Mr. Mandelli as belonging to this species closely resemble those of *Pomatorhinus ferruginosus*, but are somewhat smaller ; they are oval eggs a good deal pointed towards one end, pure white, and with a high gloss. They were obtained on the 5th and 22nd of April in the neighbourhood of Darjeeling, and measure from 0·95 to 1·04 in length, and 0·72 to 0·73 in breadth. Eggs sent by Mr. Gammie are precisely similar.

Two other eggs of this species subsequently obtained were slightly shorter and broader, and measured 0·95 by 0·77, and 0·98 by 0·78.

118. Pomatorhinus olivaceus, Blyth. *The Tenasserim Scimitar Babbler.*

Pomatorhinus olivaceus, *Blyth, Hume, Cat.* no. 403 bis.

Mr. Davison writes :—" I found a nest of this bird on the morning of the 21st January, 1875, at Pakchan, Tenasserim Province, Burma. It was placed on the ground at the foot of a small screw pine, growing in thick bamboo-jungle ; it was a large globular structure, composed externally of dry bamboo-leaves, and well secreted by the mass of dry bamboo-leaves that surrounded it ; it was in fact buried in these, and if I had not seen the bird leave it, it would most undoubtedly have remained undiscovered. Externally it was about a foot in length by 9 inches in height, but it was impossible to take any accurate measurement, as the nest really had no marked external definition. Internally was a lining about half an inch thick, composed of thin strips of dry bark, fibres, &c. The entrance was to one side, circular, and measuring 2·5 inches in diameter ; the egg-cavity measured 4 inches deep by about 3 in height.

" In the nest were three pure white ovato-pyriform eggs, but so far incubated that they would probably have hatched off before the day was out.

"The measurements of two were 1·1 and 1·09 in length by 0·75 in breadth."

Major C. T. Bingham says:—"This is the *Pomatorhinus* of the Thoungyeen valley, being found from the sources to the mouth of that river. A note recorded two years ago of a nest that I found is given below:—*4th March.*—Having to go over the ground along the southern boundary of the proposed Meplay reserve I had to cut my way through dense bamboo, to go through a long belt of which is hard work. To make it worse in this case several clumps had been burnt by fire and blown down. As I was slowly progressing along, bent almost double, out of a little hollow at my feet a bird flew with a suddenness that nearly knocked me down. I looked into the hollow, and there under the ledge of the sheltering bank was a nest of dry bamboo-leaves lined with strips of the same, shredded fine. It was cup-shaped, loosely made, about 1½ inches in diameter, and the same in depth, containing three pure white eggs, perfectly fresh (measured afterwards two proved respectively, 0·98 × 0·71, 0·99 × 0·73 inch); and gun in hand I watched, hiding myself behind a clump of bamboos about thirty yards off. For an hour I watched, but the bird did not return, so I marked the spot and went on. Returning back the same way just before dusk, I managed to start her again, and to get a hurried shot; she fell and I secured and recognized her as *P. olivaceus.*"

The eggs, which seem small for the size of the bird, are rather broad ovals, some fairly regular, some a good deal compressed just towards the small end, which is, however, always obtuse, never pointed; the shell is fine, compact, and thin, smooth and satiny to the touch, but with scarcely any perceptible gloss. The colour is pure spotless white.

119. Pomatorhinus melanurus, Blyth. *The Ceylonese Scimitar Babbler.*

Pomatorhinus melanurus, *Blyth, Hume, Cat.* no. 404 bis.

Colonel Legge writes of the nidification of this bird in Ceylon:—"This Babbler breeds from December until February. I have observed one collecting materials for a nest in the former month, and at the same period Mr. MacVicar had the eggs brought to him; they were taken from a nest made of leaves and grass, and placed on a bank in jungle. Mr. Bligh has found the nest in crevices in trees, between a projecting piece of bark and the trunk, also in a jungle-path cutting and on a ledge of rock; it is usually composed of moss, grass-roots, fibre, and a few dead leaves, and the structure is rather a slovenly one. The eggs vary from three to five, and are pure white, the shell thin and transparent, and they measure 0·96 to 0·98 in length, by 0·7 in breadth."

120. Pomatorhinus horsfieldii, Sykes.   *The Southern Scimitar Babbler.*

Pomatorhinus horsfieldii, *Sykes, Jerd. B. Ind.* ii, p. 31; *Hume, Rough Draft N. & E.* no. 404.

The Southern Scimitar Babbler breeds throughout the hilly tracts of Southern India, up to an elevation of fully 7000 feet. They are common in Ootacamund, and even on Dodabet as high up as it is wooded.   They seem to breed less plentifully about Kotagherry than they do at Ootacamund itself, Coonoor, Neddivattam, &c.

They lay from February to May, building a largish globular nest of grass, moss, and roots, placed on or very near to the ground in some bush or clump of fern or grass.   They lay five eggs.

A nest of this species which I owe to Mr. Carter, and which was found at Coonoor on the 7th April, 1869, is a huge globular mass of moss and fine moss-roots some 7 inches in diameter, with, on the upper side, an entrance to a small egg-cavity some $3\frac{1}{2}$ inches in diameter, and 2 inches in depth.   It is a most singular nest, a great compact ball of soft feathery moss and very fine moss-roots, which latter predominate in the interior of the cavity, and so form a sort of lining to it.   The great body of the nest is below the cavity, the overhanging dome-like covering of the cavity being comparatively thin.

Mr. Davison remarks :—" The nest of this bird is very peculiar in structure, more like the nest of a field-mouse than of a bird, being in fact merely a ball of grass rather loosely put together, the grass on the exterior being intermingled with dry leaves and other rubbish.   The nest is generally placed either in a clump of fern, or at the roots of some grass-grown bush.   The eggs are pure white, very elongated, and with a remarkably thin and delicate shell.   The normal number appears to be five.   The breeding-season is, I think, the latter end of April and May."

Later, he writes :—" It must, I think, breed twice, as I found a nest on the 10th March with fully-fledged young, and late in April another nest with perfectly fresh eggs."

Writing of this species Dr. Jerdon says :—" I procured its nest near Neddivattam on the Nilghiris, on a bank on the roadside, made with moss and roots, and containing four white eggs of a very elongated form."

Miss Cockburn, of Kotagherry, furnishes me with the following note on the nidification of this species :—" These birds build rather large nests, among the *roots* of bushes, and generally prefer those which grow on the slopes of steep hills.   Their nests are composed of coarse grass, a few roots of the same, and the bark of a bush, which cracks when dry and is very easily pulled off.   These materials are put together into a round nest, and also form a covering above, which makes the inside look very snug indeed. But if any attempts are made to remove the nest, it generally falls to pieces, the materials having no tenacity.   This bird commonly uses no lining to its nest, but lays its eggs (three to five in number) on the coarse grass of which the inside is composed. The eggs are pure white, particularly thin-shelled, and consequently perfectly translucent. They are found during the months of February and March."

Messrs. Davidson and Wenden, writing from the Deccan, remark :—" Very common along tops of ghâts. D. got a nest with two eggs in March."

Mr. T. Fulton Bourdillon writes from Travancore :—" I have been so fortunate as to obtain two nests of this bird lately, though I have never found any before. The first contained three fresh eggs on the 5th December last, and was situated in a bank on the roadside at an elevation of about 3000 feet above sea-level. The nest was very loosely made of grass, with finer kinds of grass for the lining. I endeavoured to preserve it, but it fell to pieces on being taken from its position, and I only succeeded in saving the eggs. As the bird, usually a very shy one, flew off on my approach and remained close by while I was examining the nest, I have no doubt of its identity. Whether she would have laid more eggs I cannot say, but I fancy not; three seems to be the usual number judging from the two clutches taken. The other nest I found on the 8th of this month just completed. It was in much the same position as the last, viz. a bank by the roadside, and as it was near my bungalow I watched to see how the eggs were deposited. The bird laid one egg each day on the 11th, 12th and 13th, and then began to sit, so on the 15th I took the nest. When fresh the eggs are beautifully pink from the thinness of the shell."

Mr. J. Darling, junior, remarks :—

" Mr. Davison makes a very good remark on the nest of this bird, but I found one once under the roots of a tree at Neddivattam, and it was a most beautiful nest, built entirely of the fibrous bark of the Nilghiri nettle, in the shape of an oven, with a hole to go in at one side. It contained four pure white delicate eggs. Another one found near the same place was of the same nature, only resting on some fern-leaves and under a rock, and contained five eggs.

" I found a nest down at Vythery, Wynaad, in a hole in the bank of a road, in December 1874, made entirely of broad grass, very untidy, and containing three eggs."

Mr. Rhodes W. Morgan writing from South India, says :— " Breeds in April, constructing a neat domed nest of leaves on the ground, at the foot of a bush. The nest is lined with fine grasses, and almost always contains three eggs, which, when fresh, are of a beautiful pink colour, owing to the yelk shining through the shell, which is exceedingly fragile. The egg, when blown, is of a very beautiful glossy white. If suddenly approached whilst on its nest, this bird runs out like a rat, and flies when at a distance from the nest. An egg in my collection measures 1·04 by ·7 inch."

The eggs sent me from the Nilghiris by Miss Cockburn and Mr. Carter are nearly perfect ovals, usually much elongated, but sometimes moderately broad, and very slightly compressed towards one end. They are very fragile, and perfectly pure spotless white in colour. Typically, although smooth and satiny in texture, they have but little gloss, but occasionally a fairly glossy egg is to be met with.

In length they vary from 0·98 to 1·12, and in breadth from 0·75 to 0·79 ; but the average seems to be about 1·08 by 0·77.

122. **Pomatorhinus ferruginosus**, Blyth.   *The Coral-billed Scimitar Babbler.*

Pomatorhinus ferruginosus, *Blyth, Jerd. B. Ind.* ii, p. 29; *Hume, Rough Draft N. & E.* no. 401.

The Coral-billed Scimitar Babbler, according to Mr. Hodgson's notes, breeds in Sikhim, at an elevation of 5000 or 6000 feet. Its nest is placed about a foot or 2 feet above the ground, in a bamboo-clump or some thick bush, and is firmly wedged in between the twigs and shoots. It is composed internally of dried bamboo-leaves, grass, and vegetable fibres, outside which bamboo-sheaths are bound on with creepers and fibres of different kinds. The nest is more or less egg-shaped, with the longer diameter horizontal, some 7 inches or so in length and 5 inches in height, and with the entrance at one end, measuring some 3 inches in diameter. Four or five eggs are laid, elongated ovals, somewhat pointed towards the small end, pure white, and measuring about 1·08 by 0·7.

From Sikhim Mr. Gammie writes :—" I took a nest of this bird on the 19th May, at an elevation of about 5000 feet. It was placed on the ground, among low scrub, near the outskirts of a large forest, and was neatly made, for a *Pomatorhinus*, of bamboo-leaves and long grass, with a thin lining of fibry strips torn from old bamboo-stems. In shape it was a cone laid on its side. Externally it measured 9 inches in length by the same in height at front, while the egg-cavity measured 3·5 inches across, and 1·75 in depth. The entrance, which was at the end, measured 3 inches in diameter.

" Next to the lining was a layer of broadish grass-blades, placed lengthways, *i. e.* from base to apex of the cone, then came a cross layer of broad bamboo-leaves succeeded by a second layer of bamboo-leaves placed lengthways. By this arrangement the nest was kept perfectly water-tight. So nicely were these simple materials put together that they held each other in their places without the assistance of a single fibre.

" The nest contained four partially incubated eggs : three of them pointed and exactly alike, but the fourth rounded, and apparently of a different texture, so that it may have been introduced by a Cuckoo."

Two eggs sent by Mr. Gammie are moderately elongated ovals, somewhat obtuse even at the smaller end. The shell is very fine, pure white, and has a fine gloss. They measure 1·1 by 0·83, and 1·06 by 0·78.

125. Pomatorhinus ruficollis, Hodgs. *The Rufous-necked Scimitar Babbler.*

Pomatorhinus ruficollis, *Hodgs., Jerd. B. Ind.* ii, p. 29; *Hume, Rough Draft N. & E.* no. 400.

The Rufous-necked Scimitar Babbler breeds in Nepal, the Himalayas eastward of that State, and in the various ranges running down from Assam to Burmah.

The breeding-season appears to be April and May. They lay five, or sometimes only four, eggs.

From Sikhim Mr. Gammie writes :—" This species breeds, I think, from the middle of April to the middle of May; but I have only as yet taken a single nest, and this I found at Rishap on the 5th May, at an elevation of about 4500 feet. The nest was placed on the ground in open country, but partially concealed by overhanging grass and weeds, and immediately adjoining a deep humid ravine filled with a dense undergrowth. The nest was composed of dry grass, fern, bamboo, and other dry leaves put loosely together and lined with a few fibres. In shape it was domed or hooded, and exteriorly it measured 5·7 inches in height and 5 in diameter. Interiorly the cavity was 2·6 in diameter, and had a total depth of 3·8 measured from the roof, but of only 2 inches below the lower margin of the aperture. This nest contained five eggs, much incubated; indeed, they would have hatched off in one or two days."

The Rufous-necked Scimitar Babbler breeds, according to Mr. Hodgson, in the central portion of Nepal in April and May, building a large, coarse, globular nest of dry grass and bamboo-leaves on the ground in some thick bush or bamboo-clump. The opening of the nest is at the side. They lay four or five white eggs, measuring as figured 0·9 by 0·68.

The eggs sent me by Mr. Gammie are rather elongated ovals, a good deal pointed towards one end, pure white, the shells very fine and fragile, and with a fair amount of gloss.

Ten eggs varied from 0·85 to 1·02 in length, and from 0·62 to 0·74 in breadth, but the average was 0·95 by 0·68.

129. Pomatorhinus erythrogenys, Vigors. *The Rusty-cheeked Scimitar Babbler.*

Pomatorhinus erythrogenys, *Vig., Jerd. B. Ind.* ii, p. 31; *Hume, Rough Draft N. & E.* no. 405.

The Rusty-cheeked Scimitar Babbler breeds from April to June in the Himalayas, at any rate from Darjeeling to the Valley of the Beas, at elevations of from 2000 to 6000 feet. It may be *met* with at double this latter altitude, but I doubt if it *nests* higher.

As a rule, the nest is placed on the ground, in some thick clump of dry fern or coarse grass, amongst dead leaves and moss, but at times I have seen it placed in a thick bush 2 or 3 feet from the ground. It is very common near Kotegurh and below Narkunda, where we found nearly a dozen nests, almost all, however, containing young ones. Typically the nest is domed, and is loosely constructed of the materials at hand—coarse grass, dry fern, dead leaves, moss-roots, and the like, some 6 or 7 inches in diameter and 5 or 6 inches high, with a broad entrance on one side, a good deal above the middle. In some cases, however, where a dense bunch of grass or fern completely curves over the spot selected for the nest, the latter is a mere broad, shallow saucer. There is no regular lining to the nests, but a good many fine roots are at times incorporated in the interior of the cavity. All the nests that I have seen were placed near the edges of clumps of brushwood or scrubby jungle.

I ought here to mention that I am by no means certain that the Nepalese and Sikhim, in fact the eastern race of this species (*P. ferrugilatus*, Hodgs.), will not have to be separated from the more western *P. erythrogenys* of Gould. Long ago Blyth remarked ('Journal Asiatic Society,' 1845, p. 598) that "there seems to be two marked varieties of *P. erythrogenys*, one having white underparts, with merely faint traces of darker spots, the other with the throat and breast densely mottled with greenish olive," or, as I should call it, dingy olive-grey. This is perfectly true, and, as far as I can make out, the latter variety is not one of sex or age, but is local and confined to Kumaon (where the other form also occurs) and the hills eastward of this province. My own remarks above given refer to the true *P. erythrogenys*, and so do Hutton's; but Hodgson's and Mr. Gammie's birds both appear to have been, and the latter's certainly were, grey-throated examples. The eggs are undistinguishable, as, indeed, though they vary somewhat in shape and size, are those of most of the *Pomatorhini*.

Captain Hutton says that this species is "common from 3500 feet up to 10,000 or 12,000 feet, always in pairs, turning up the dead leaves on copsewood covered banks, uttering a loud whistle, answering and calling each other. It breeds in April, constructing its nest on the ground of coarse dry grasses and leaf-stalks of walnut-trees, and is covered with a dome-shaped roof, so nicely blended with the fallen leaves and withered grasses, among which it is placed, as to be almost undistinguishable from them. The eggs are three in number, and pure white; diameter 1·12 by 0·81 inches, of an ordinary oval shape. When disturbed, the bird sprung along the ground with long bounding hops, so quickly that, from its motions and the appearance of the nest, I was led to believe it a species of rat. The nest is placed in a slight hollow, probably formed by the bird itself."

According to Mr. Hodgson's notes, this species would appear to breed at heights of from 2000 to 8000 feet. It lays in May and June. On the 20th May, and again on the 6th June, Mr. Hodgson found nests of this species in thick bushes 3 or 4 feet above the ground. They were broad saucer-shaped nests of coarse vegetable fibres, grass, and grass-roots, 7 inches or so in diameter,

and the cavity, which had no lining, was about 4 inches in diameter by 2 inches in depth. They contained three and four white
eggs respectively. One figured measures 0·98 by 0·73. On
June 8th he found two more nests at Jaba Powah, on the ground,
on edges of brushy slopes close to grassy open plains, the nest a
large mass of grass, oven-shaped, open at one and in one case at
both ends, protected by the root of a tree. There were two and
three white eggs in the nests respectively. The eggs of these nests
are figured as measuring 1·08 by 0·73.

Mr. Gammie remarks :—" I found a nest of this species below
Rungbee, at an elevation of about 2000 feet, on the 17th June. It
was placed on, and partially in a hole in a bank, and contained two
hard-set eggs. It was a large, loose pad of fine grass and dead
fern, with a few broad flag-like grass-leaves incorporated towards
the base, and overhung by a sort of canopy of similar materials.
The basal portion was some 6 inches long and 5 inches broad, and
about 2 inches thick in the thickest part, with a broad shallow
depression for the eggs of about half that depth."

Writing again this year (1874) he says :—" I have only found
two more nests this year, and both in the last week of April ; the
one contained three partially incubated eggs, the other three young
birds. These nests were at Gielle, at an elevation of about 2500
feet. As a rule, these birds nest in open country, immediately
adjoining moist thickly wooded ravines, in which they feed, and
take refuge if disturbed from the nest. The nest is usually placed
on sloping ground, more or less concealed by overhanging herbage,
and is composed, according to my experience, of dry grass sparingly
lined with fibres. It is large ; one I measured in situ was 8 inches
in height and 7 inches in diameter ; the vertical diameter of the
cavity was 4 inches and the horizontal 3½ inches. I have not yet
found more than three eggs or young ones in any nest."

Dr. Scully remarks of this bird in Nipal :—" It lays in May and
June ; two nests, taken on the 30th May and 6th June, were large
loosely-made pads, not domed, and with the egg-cavity saucer-
shaped, each nest contained three pure white eggs."

The eggs of this species are long, and at times narrow, ovals,
pure white and fairly glossy, but occasionally almost glossless,
without any marks or spottings.

In length they vary from 1·0 to 1·2, and in breadth from 0·73
to 0·85, but the average of twenty eggs is about 1·11 by nearly
0·8.

133. Xiphorhamphus superciliaris (Blyth). *The Slender-billed Scimitar Babbler.*

Xiphorhamphus superciliaris (*Blyth*), Jerd. B. Ind. ii, p. 33 ; *Hume, Rough Draft N. & E.* no. 406.

The Slender-billed Scimitar Babbler, according to Mr. Hodgson's
notes, breeds in Sikhim, at elevations of 3000 to 6000 feet, during

the months of May and June. The nest is a large globular one, composed of dry bamboo-leaves and green grass, intermingled and lined with fine roots and fibres. The entrance, which is about 2 to 2·5 inches in diameter, is at one end. A nest containing four eggs, obtained on the 12th June, measured about 7 inches in diameter externally, and it was placed in the crown of a stump from 2 to 3 feet from the ground. Sometimes the nests are placed in tufts of high grass or in thick bushes, but never at any great elevation above the ground. They lay three or four eggs, which are pure white, and one of which is figured as a broad oval, measuring 0·95 by 0·7.

From Sikhim Mr. Gammie writes :—" I took a nest of this Scimitar Babbler on the 29th May, in the middle of the large forest on the top of the Mahalderam ridge, at about 7000 feet elevation. It was built on the ground, on top of a dry bank by the side of a path, and was overhung by a few grassy weeds. In shape it was a blunt cone laid on its side, with the entrance at the wide end. It was loosely made of the dead leaves of a deciduous orchid (*Pleione wallichiana*), small bamboo, chestnut, and grass, intermixed with decaying stems of small climbing-plants. It measured externally 6 inches long, with a diameter of 5·5 at front, and of 1·75 at back. The cavity was quite devoid of lining and measured 3·5 in length by 2·5 wide at entrance, slightly contracting inwards. It contained three partially incubated eggs."

Two eggs of this species obtained by Mr. Gammie are elongated ovals, pure white, and with only a faint gloss. They measure 0·99 and 1·05 in length, by 0·68 and 0·75 in breadth respectively.

Subfamily TIMELIINÆ.

134. Timelia pileata, Horsf.   *The Red-capped Babbler.*

Timalia pileata, *Horsf., Jerd. B. Ind.* ii, p. 24 ; *Hume, Rough Draft N. & E.* no. 396.

Mr. Eugene Oates records that he " found the nest of this bird at Thayetmyo on the 2nd June with young ones a few days old. The nest was placed on the ground in the centre of a low but very thick thorny bush."

Subsequently he wrote from Pegu, further south :—" The nest is placed in the fork of a shrub, very near to, or quite on, the ground, and is surrounded in every case by long grass. A nest found on the 4th July, on which the female was sitting closely, contained three eggs slightly incubated. The breeding-season seems to be in June and July.

" The nest is made entirely of bamboo-leaves and is lined sparingly with fine grass. No other material enters into its composition. It is oval, about 7 inches in height and 4 in diameter, with a large entrance at the side, its lower edge being about the middle of the nest.

" When the bird frequents elephant-grass, where there are no shrubs, it builds on the ground at the edge of a clump of grass, and I have found two nests in such a situation, only a few feet from each other. .

" In looking for the nest a good deal of grass is necessarily trodden down ; the consequance is that if you do not find eggs, there is little chance of their being laid later on. I have found some ten nests, more or less completed, but only three eggs."

And again, later on :—" This bird would appear to have two broods a year, for I procured two sittings of three eggs each this year in April, former nests having been found in June and July. With many eggs before me I find that the density of the markings varies considerably. The size is very constant ; for the length of numerous eggs varies only from ·75 to ·72, and the breadth from ·6 to ·54 inch."

I was, I believe, myself the first to obtain the eggs of this species, but the first of my contributors who sent me eggs, nest, and a note on the nidification of this species was Mr. J. C. Parker. Writing to me in September 1875, he said :—

" On the 14th August I took a nest of *Timelia pileata* on my old ground in the Salt Lakes. I discovered this by a mere accident, for I happened to see a female *Prinia flaviventris* (whose eggs I was in quest of for you) perched on the top of a bush inland about 10 feet from the bank of the canal, and from her movements I thought she must have a nest near at hand.

" Accordingly I landed, although not in trim for wading through a bog. Sure enough I was not mistaken ; the *Prinia* had a nest, but it contained only *one* egg. Close by, however, I saw a nest, from out of which a bird flew, and although I did not shoot it I am quite sure it was *Timelia pileata*. The jungle was particularly thick just about where I stood, indeed impenetrable, and I could not follow the bird, but I soon heard the male bird talking to his mate in that extraordinary way which these birds have, and which once heard cannot be mistaken.

" The nest was placed on the spikes growing from the joints of a species of grass very thick and stiff, and forming a secure foundation for the nest. This latter is 6 inches high and 4 inches broad. Egg-cavity 2 inches, entrance-hole 1½ by 2. The nest itself is very loosely put together with the dead leaves of the tiger-grass twisted round and round, and lined roughly with coarse grass. The nest was quite open to view and about three feet from the ground. I suppose the birds never expected that such a wild swampy spot as they had selected would be invaded by any oologist."

Mr. J. R. Cripps writing from Eastern Bengal says :—" Pretty common. Permanent resident. Oftener found in the patches of cane brushwood jungle found in and around villages than in unfrequented jungle and thickets as Dr. Jerdon says. I have, however, once seen it in a field of jute, which was alongside a village. Its well-known note can be heard a long way off. I have several times found nests in course of construction, but only once secured a clutch of eggs. When the nests are being built, if the bush is at all disturbed the nest is deserted. The earliest date on which I found a nest was the 1st April, 1878; it was half finished, and as I pulled the cane-leaves asunder to see if there were eggs, the birds deserted it. After this I found four nests in cane-clumps on the sides of roads, but they were empty, and as the birds abandoned them in due course I despaired of getting any eggs; but on the 15th June, while going along a road, the edges of which were bounded by the small embankments natives throw up round their holdings, and which are always overgrown with 'sone' grass, I saw one of these birds with a straw in its bill disappear at the root of a small date-tree. The nest could be discerned from the road. On the 20th June I returned and found two fresh eggs; the nest was placed at the junction of the frond and the stem of the date-tree about five inches from the ground, and was an oval deep cup and measured externally 5 inches deep by 3¾ broad. Egg-cavity 2 broad and 1¾ deep, composed exclusively of 'sone' grass with no lining."

The eggs of this species are broad ovals with a tolerably fine gloss. The ground-colour is pure white. The whole of the larger end of the egg is pretty thickly speckled and spotted with brown, varying from an olive to a burnt sienna intermingled with little spots and clouds of pale inky purple, and similar spots and specks chiefly of the former colour, but smaller in size, scattered thinly over the rest of the egg. In size they vary from 0·69 to 0·75 in length, and from 0·55 to 0·6 in breadth.

135. **Dumetia hyperythra** (Frankl.). *The Rufous-bellied Babbler.*

Dumetia hyperythra (*Frankl.*), Jerd. B. Ind. ii, p. 26; *Hume, Rough Draft N. & E. no. 397.*

The Rufous-bellied Babbler breeds throughout the Central Provinces, Chota Nagpoor, Upper Bengal, the eastern portions of the North-West Provinces, parts of Oudh, and even in the low valleys of Kumaon.

It lays from the middle of June to the middle of August, building a globular nest of broad grass-blades or bamboo-leaves some 4 or 5 inches in diameter, sparingly lined with fine grass-roots or a little hair, or sometimes entirely unlined. The nest is placed sometimes on the ground amongst dead leaves, some of which are not unfrequently incorporated in the structure; sometimes in coarse grass or some little shrub a foot or two from the ground, but by preference, according to my experience, in amongst the roots of a bamboo-clump.

Four is the usual number of eggs laid.

Mr. Brooks writes :—" On the 26th June, 1867, in the broken ground above Chunar, I took two nests in the foot of a thick bamboo-bush about 2 feet from the ground. The nests were made of bamboo-leaves rolled into a ball with the entrance at the side, and no lining except a few hairs. There were two eggs in one nest and three in the other. They were all fresh. The eggs in the two nests varied somewhat : the ground of the one was nearly pure white, and it was finely speckled with reddish brown, which at the large end was partly confluent : the other nest had the eggs with a pinkish-white ground, the spots larger and less neatly defined, and with a rather large confluent spot at the large end."

Writing from Hoshungabad, Mr. E. C. Nunn remarks :—" I found two nests of this species, each containing two eggs, on the 20th July and 6th August, 1868. Both nests were ball-shaped, of coarse grass very firmly and compactly twisted together, and with numerous dead leaves incorporated in the body of the nest and towards the base, forming the major portion of the material. They were thinly lined inside with fine grass-roots. One was placed at the root of a small thorny bush : the other on the ground in a thick clump of rank grass." The nest Mr. Nunn sent to me was peculiarly solidly made. The cavity was small, about 2·25 inches in depth and 1·5 in diameter. The bottom of the nest was some 2 inches and the sides 1·25 inch thick.

From Raipoor Mr. F. R. Blewitt tells us that "in July and August four nests of this Babbler were taken ; in two there were four eggs each, in the third, three, and in the fourth, two—thirteen in all. The nests were carefully made on the ground, at the base of clumps of long grass growing very near to bamboo thickets. Three are made exclusively of the dry leaves of the bamboo ; the fourth of coarse grass. They were nearly globular, about 4 inches in diameter, and without any regular lining, although in the interior of the cavity a good deal of fine grass-stems had been incorporated in the nest. They were well hidden in the grass."

Mr. Henry Wenden writes :—" On July 18th, about 15 miles from Bombay, on the line of railway, I found a nest and eggs of the following description : nest, a rough loose ball of soft flat grasses, lined with hard but fine grass-stems, entrance at side near top ; situated in a thorny bush in cactus-hedge, by a narrow lane, not 4 feet wide, through which numerous people passed. The nest, about 3 feet from the ground, was in no way concealed. On the 18th there were two eggs, and on the 20th, when there were four eggs, the bird was snared and nest taken."

The eggs are short, broad ovals, very slightly compressed towards one end. The ground-colour is white or pinkish white, and it is streaked, spotted, and speckled most thickly at the large end (where there is a tendency to form an irregular confluent cap or zone), and thinly towards the small end, with shades of red, brownish red, and reddish purple, varying much in different examples. In some the markings are pretty bold and blotchy, in others they are small and speckly ; in some they are smudgy and ill-defined, in others they are clear and distinct. Some of the eggs are miniatures of some types of *Pyctorhis sinensis*, but many recall the eggs of the Titmouse. They are much about the size of those of *Parus cœruleus* and *P. palustris*, but a trifle less broad than either of these. The eggs have a faint gloss.

In length they vary from 0·63 to 0·7, and in breadth from 0·5 to 0·56; but the average of twenty-four eggs now before me is 0·67 by 0·53.

136. Dumetia albigularis (Blyth). *The Small White-throated Babbler.*

Dumetia albogularis (*Blyth*), Jerd. B. Ind. ii, p. 20; *Hume, Rough Draft N. & E.* no. 398.

Miss M. B. Cockburn, writing from Kotagherry, tells me that "the White-throated Babbler builds its nest in the month of June. One was found by my nest-seekers on the 17th of that month in the year 1873. It was constructed on a coffee-tree, and contained three eggs, which were white, profusely covered with reddish spots of all sizes. The bird was very shy, and would not return to the nest for some hours after it had been discovered; when, however, she did so, she was shot. This year (1874) I found another similar nest on the 9th of June, also containing three eggs."

The nest with which she favoured me was small and nearly globular (say at most 4 inches in external diameter), composed entirely of broad flaggy grass without any lining or any admixture whatsoever of other material. The nest was loosely put together, and had a comparatively narrow circular entrance near the top.

From Mysore Mr. Iver Macpherson writes:—"This is an exceedingly common bird in parts of this district, and their nests are so plentiful that I never now take them.

"I send you all the eggs I have at present, but can procure you any number more next season.

"The birds are to be found in all kinds of wooded country except the heavy forests, and appear to breed from the middle of April to the end of July, and possibly later.

"The nest is a largish globular structure loosely made of either bamboo-leaves or blades of grass, and all that I have ever seen have been lined inside with a few fine fibres.

"Four appears to be the usual number of eggs, but very often there are only three.

"The nests are always built near the ground, sometimes almost touching it, and are fixed in either small bushes, tufts of grass, or young bamboo-clumps."

Mr. J. L. Darling, Jun., states that this bird is very common in Culputty in the Wynaad, at an elevation of about 3000 feet, and that he has found the nests from the end of May to the middle of October. The nest is built in high grass nearly on the ground, or in date-palms, or in arrowroot in the jungle up to heights of 3 feet. The nest is built entirely of grass, lined with finer grass ; a nearly round ball 6 inches in diameter outside and 5 inside, with a hole on the side. The eggs are laid at the rate of one a day, and three are usually found in one nest, occasionally only two. On one occasion after securing the female bird, he found the cock bird sitting on the eggs and he continued to sit there for three days.

Mr. J. Davidson tells us that he found a nest of this bird on the 15th July at Kondabhari with four fresh eggs.

Colonel Legge writes in his ' Birds of Ceylon ' :—" The breeding-season lasts from March until July, the nest being built in a low bush sometimes only a few inches from the ground."

In shape the eggs are moderately elongated ovals. The shell is very fine and smooth, and has in some a rather bright, in some only a very slight gloss. The ground is a China-white. The markings consist of a profusion of specks and spots of a very bright red, which, though spread over the whole surface, are gathered most densely into an imperfect, more or less confluent, cap or zone at the larger end, where also a few purplish-grey spots and specks not usually found on any other part of the egg, are noticeable.

In length the eggs vary from 0·66 to 0·78, and in breadth from 0·5 to 0·55. The average of 28 eggs is 0·72 by 0·53.

139. **Pyctorhis sinensis** (Gm.).   *The Yellow-eyed Babbler.*

Pyctorhis sinensis (*Gm.*), *Jerd. B. Ind.* ii, p. 15 ; *Hume, Rough Draft N. & E.* no. 385.

The Yellow-eyed Babbler breeds throughout the plains of India, as also in the Nilghiris, to an elevation of 5000 feet, and in the Himalayas to perhaps 4000 feet. It lays in the latter part of June, in July, August, and September. Gardens are the favourite localities and in these the little bird makes its compact and solid nest, sometimes in a fork of the fine twigs of a lime-bush, sometimes in a mangoe-, orange-, or apple-tree, occasionally suspended between three stout grass-stems, or even attached to a single stem of the huge grass from which the native pens are made. I have taken a nest, hung between three reeds, exactly resembling in shape and position the Reed-Warbler's nest (*Salicaria arundinacea*), figured in Mr. Yarrell's vignette at page 313, vol. i. 3rd edition.

The nest is typically cone-shaped (the apex downwards), from 5 to 6 inches in depth, and 3 or 4 in diameter at the base ; but it varies of course according to situation, the cone being often broadly truncated. In the base of the cone (which is uppermost) is the egg-cavity, measuring from 2 to 3 inches in diameter, and from 2 to 2·5 inches in depth. The nest is *very* compactly and solidly woven, of rather broad blades of grass, and long strips of fine fibrous bark, exteriorly more or less coated with cobwebs and gossamer-threads. Interiorly, fine grass-stems and roots are neatly and closely interwoven. I once found some horse-hair along with the grass-roots, but this is unusual.

The full number of eggs is, I believe, five. I have repeatedly taken nests containing this number, and have comparatively seldom met with a smaller number of eggs at all incubated.

Colonel G. F. L. Marshall says :—" I found a nest of this species at Roorkee in the early part of July. It contained three eggs and was beautifully made, a deep cup fixed on to an artichoke-stock, and at a little distance much resembled an artichoke."

Mr. E. C. Nunn, writing from near Agra on the 26th September 1867, says :—" I got a *Pyctorhis'* nest yesterday, suspended between two stalks of jowar (*Holcus sorghum*), the nest firmly bound with strips of fibrous bark, at two opposite points of its circumference, to the two stems. This is, I imagine, something out of the usual order of things with these birds. The nests which I have hitherto found have been situated in young mangoe-trees, rose-bushes, or peach- and orange-trees."

From Futtehgurh the late Mr. A. A. Anderson sent me the following note :—

" The nest and eggs of this bird are very beautiful. A pair once built in a pumplenose-tree (*Citrus decumana*) in my garden, laying five long eggs. The nest, still in my collection, was placed in the fork of *four* small upright twigs; it was composed entirely of dry grass-stems (no soft material inside), and laced outwardly, in and out of the twigs, with dry fibre belonging to the plantain-tree.

" The eggs are small for the size of the bird, and scarcely so large as those of the Hedge-Sparrow."

Captain Hutton remarks :—" This likewise is a Dhoon bird; its nest was found there on the 1st July, when it contained four eggs of a dull white colour, thickly speckled and blotched all over with ferruginous spots, forming also an open darker coloured ring at the large end, and intermixed with brown.

" The nest is a deep cup, placed in the trifurcation of the slender upright branch of a low shrub, and is constructed externally of coarse grass-blades held together by cobwebs and seed-down, the lining being fine grass-seed stalks. Diameter of the top $2\frac{1}{2}$ inches; depth within 2 inches; externally $3\frac{1}{2}$ inches."

Mr. F. R. Blewitt tells us that " the Yellow-eyed Babbler breeds from July to September, or, I should say, up to the middle of September. Its selection of a tree for its nest is not confined to any one species, but by preference the bird selects those of small growth, and even frequently high-growing brushwood. The nests are very neatly made, and what is singular is that, as regards build and shape, they are always almost exactly alike. If I have seen one, I must have seen at least fifty this year, all with the same exterior material of closely interlaced vegetable fibre over grass, and the inner lining of fine grass, deep cup-shaped, and in diameter, outer and inner, varying but little. Where it could be effected, the nest was suspended to, or rather fastened between, two forks; or where these were not available, between three twigs. The outer diameters of the nests were from 2·7 to 2·9 inches, inner from 2·3 to 2·5. Four is the regular number of eggs, though occasionally five in one nest have been obtained."

Mr. R. M. Adam remarks :—" This species builds about Agra in May, June, and July. The nest is a beautiful deep cup-shaped structure, almost always fastened to a branch of a low bush. The normal number of eggs appears to be four."

From Kotagherry, near Ootacamund, Miss Cockburn records that " this bird builds a neat cup-shaped nest, generally choosing a branch consisting of three upright sprigs, at the bottom of which the building is placed. The nests (one of which is now before me) are begun with broad grass-leaves, and the inside compactly lined with fine fibres of the same material : to render the whole firm, a few cobwebs are added to the outside, thus fixing the nest securely to the sprigs. These birds build in the months of June and July, and, as far as I have observed, lay only three eggs."

Mr. Philipps, quoted by Dr. Jerdon, says that this bird " *generally* builds on banyan-trees." This is clearly a mistake. I have known of the taking, or have myself taken, altogether upwards of fifty nests in the North-Western Provinces, whence Mr. Philipps was writing, and never yet heard of or saw a nest of this species on a banyan.

Mr. H. Wenden writes :—" At Egatpoora, the top of the Thull Ghât incline, I noticed, on 30th September, a partly-built nest of this species. Watching for some time, I ascertained that both birds shared in the labour of construction. It was situated in the tri-furcated stalk of that plant which bears a clover-like blossom (called Kessara-Hind and Koordoo-Mhar), about 3 feet above the ground, the stalks passing through the side-walls of the nest, which cannot have a better description than that given by Mr. Hume (page 238, ' Rough Draft '). The first egg was laid on 2nd October, and another each succeeding day until there were five. On the 10th the hen-bird was shot and the nest taken.

" On 30th October, in a garden near the same place, another nest was found, on the twigs of a pangra tree, containing three young birds and one egg."

Messrs. Davidson and Wenden say :—" Tolerably common in the Sholapoor District ; more so in the better-wooded parts, and breeds."

Finally, Colonel Butler sends me the following note :—

" Belgaum, 14th September, 1880.—A nest in sugar-cane about 2 feet from the ground, containing five fresh eggs. 17th September : another nest in a sugar-cane field, containing five eggs about to hatch. In both instances the nest was built, not on the blades of sugar-cane, but on a solitary green-leaved weedy-looking plant growing amongst the sugar-cane.

" The Yellow-eyed Babbler breeds during the rains. I have taken nests on the following dates :—

" July 26, 1875. A nest containing 4 fresh eggs.
" July 30, 1875. ,, ,, 3 fresh eggs.
" Aug. 14, 1875. ,, ,, 4 fresh eggs.
" Aug. 21, 1875. ,, ,, 4 fresh eggs.
" July 18, 1876. ,, ,, 4 fresh eggs.

"July 20, 1876.   A nest containing 3 fresh eggs.
"July 28, 1876.   „   „   4 fresh eggs.
"From this date to the end of August I found any number of nests containing eggs of both types.   The nest is usually built in the fork of some low thorny tree from 3 to 7 feet from the ground. The outside of the nest is usually smeared over with cobwebs, reminding one of the nest of a *Rhipidura*."

Mr. Oates writes :—"Breeds abundantly throughout Pegu in June, and probably in the other months of the rains up to September."

The eggs vary a good deal in size and shape, and very much in colouring.   They are mostly of a very broad oval shape, very obtuse at the smaller end.   Some are, however, slightly pyriform, and some a little elongated.   There are two very distinct types of coloration : one has a pinkish-white ground, thickly and finely mottled and streaked over the whole surface with more or less bright and deep brick-dust red, so that the ground-colour only faintly shows through, here and there, as a sort of pale mottling ; in the other type the ground-colour is pinkish white, somewhat *sparingly*, but boldly, blotched with irregular patches and eccentric hieroglyphic-like streaks, often Bunting-like in their character, of bright blood- or brick-dust red.   The eggs of this type, besides these primary markings, generally exhibit towards the large end a number of pale inky-purple blotches or clouds.   There is a third type somewhat intermediate between these, in which the ground-colour, instead of being finely freckled all over as in the former, or sparingly blotched as in the latter, is very coarsely mottled and clouded, as if clumsily daubed over by a child, with a red intermediate in intensity between that usually observable in the two first-described types.   Combinations of these different types of course occur, but fully two thirds can be separated distinctly under the first and second varieties.   Though much smaller, many of the eggs recall those of the English Robin.   The eggs have often a fine gloss.   I have one or two specimens so uniformly coloured that, though perhaps slightly shorter and broader in form, they might almost pass for the eggs of Cetti's Warbler.

In length they vary from 0·65 to 0·8, and in breadth from 0·53 to 0·68 ; but the average of seventy-seven eggs measured is 0·73 by 0·59.

140. Pyctorhis nasalis, Legge.   *The Ceylon Yellow-eyed Babbler*.

Pyctorhis nasalis, *Legge, Hume, Cat.* no. 385 bis.

Colonel Legge writes in his 'Birds of Ceylon' :—"In the Western Province this Babbler commences to breed in February ; but in May I found several nests in the Uva district near Fort Macdonald ; and that month would thus seem to be the nesting-season in the Central Province.   The nest is placed in the fork of a shrub, or in a huge tuft of maana-grass, without any attempt at concealment, about 3 or 4 feet from the ground.   It is a neatly-made compact cup, well finished off about the top and exterior, and constructed of dry grass, adorned with cobwebs or lichens, and lined with fine grass or roots. The exterior is about 2¼ inches in diameter by about 2 in depth. The eggs are usually three in number, fleshy white, boldly spotted, chiefly about the larger end, with brownish sienna; in some these markings are inclined to become confluent, and are at times overlaid with dark spots of brick-red. They are rather broad ovals, measuring, on the average, from 0·76 to 0·79 inch in length, by 0·56 to 0·59 in breadth."

142. **Pellorneum mandellii**, Blauf. *Mandelli's Spotted Babbler.*

Pellorneum nipalensis (*Hodgs.*), *Hume, Rough Draft N. & E.* no. 399 bis.

This species, originally described by Hodgson as *Hemipteron nipalensis*, was confounded by Gray and others with *P. ruficeps*, Swainson, and subsequently rediscriminated and described by Blanford as *P. mandellii*.

Mandelli's Spotted Babbler, according to Mr. Hodgson's notes, begins to lay in April, the young being ready to fly in July. They build a large, more or less oval, globular nest, laid lengthwise on the ground in some bush or clump of rush or reed, composed of moss, dry leaves, and vegetable fibres, and lined with moss-roots. The entrance, which is circular, is at one end. A nest measured by Mr. Hodgson was 6·75 inches in length and 5 in height. The aperture, at one end of the egg-shaped nest, was about 2 inches in diameter, and the cavity was about 2·5 in diameter and nearly 4 inches deep. The eggs are three or four in number, and are figured as broad ovals pointed towards the small end, measuring about 0·86 by 0·65, and having a greyish-white ground, thickly speckled and spotted with more or less bright red or brownish red, and most thickly so at the large end, where the markings are nearly confluent.

A nest said to belong to this species, and found near Darjeeling in July, at an elevation of about 4000 feet, was placed on the ground on the side of a bank—a very dirty untidy nest, more or less cylindrical in shape, composed of dead leaves, including a good many of those of the bamboo, dead twigs, and old roots, and very sparsely lined with black moss-roots. The nest is about 4 inches in diameter externally, and the cavity about 2·5 in diameter.

It contained three fresh eggs, very regular, moderately broad, ovals; the shell fine and compact, with a slight gloss. The ground-colour is white, and the egg everywhere very finely speckled with chocolate- or purplish brown, the markings being by far most dense at the large end, where they form a more or less irregular, and more or less conspicuous, speckly cap.

Two eggs measure 0·86 and 0·9 in length, and 0·65 and 0·66 in breadth.

Another nest, found on the 5th June in Native Sikhim, contained four fresh eggs. It was placed on the ground, and precisely resembled that obtained near Darjeeling in July.

In some eggs the markings are rather bolder and coarser, and in these there are generally some few pale lilac or inky-purple spots intermingled where the markings are densest. Closely looked into, many of the spots in some eggs are rather a pale yellowish brown.

The eggs are clearly all of the same type, and vary very little.

Four eggs varied from 0·84 to 0·9 in length, and from 0·65 to 0·68 in breadth.

144. Pellorneum ruficeps, Swains. *The Spotted Bubbler.*

Pellorneum ruficeps, *Swains., Jerd. B. Ind.* ii, p. 27 ; *Hume, Rough Draft N. & E.* no. 399.

Writing from Kotagherry Miss Cockburn says :—" Spotted Babblers are exceedingly shy. They associate in small flocks except during the breeding-season, when they go about in pairs. I have only known them to frequent small woods and brushwood, a little higher than the elevation of the coffee-plantations.

" Three nests of these birds were found in the months of March and April 1871. The first was placed on the ground, close against a bush. The nest, consisting of dry leaves and grass, appeared to be merely a canopy for the eggs, which were almost on the bare ground, having only a *very few* pieces of straw under them. The eggs were three in number, and covered profusely with innumerable small dark spots, making it difficult to say what the groundcolour really was. The nest was not easily found. The bird left it so quietly as not to be heard, and dropped down the hill like a ball. When the eggs were discovered the bird did not return to them for fully three hours, after which she came very cautiously, but only to meet her doom, poor thing, as she was then shot. The second nest was built in the same way under a bush, and contained three eggs, which were put into my egg-box lined with cotton, but were hatched on the way home. The third nest was constructed under a large stone and with the same materials, and contained two young ones."

An egg of this species, received from Miss Cockburn, is a moderately broad and very regular oval. The ground-colour is a slightly greenish white, and the whole surface of the egg is excessively finely freckled and speckled with lilac or pale purplish grey and a more or less rufous brown. The egg has a slight gloss.

It measures 0·88 by 0·65.

145. Pellorneum subochraceum, Swinh. *The Burmese Spotted Babbler.*

Pellorneum subochraceum, *Swinh., Hume, Cat.* no. 399 sex.

The Burmese Spotted Babbler breeds pretty well over the whole of Pegu and Tenasserim. Mr. Oates writes :—" On the 3rd May

I found a nest on the ground near Pegu. A good many bamboo-leaves had fallen and the nest was imbedded in these. It was formed entirely of these leaves loosely put together, the interior only being sparingly lined with fine grass. The structure *in situ* was tolerably firm, but it would not stand removal. In height it was about 7 inches, and in breadth about 5, the longer axis being vertical. Shape cylindrical with rounded top. Entrance 2½ inches by 1½, placed about the centre. The interior of the nest was a rough sphere of 4 inches diameter.

"There were three eggs, slightly incubated. The ground-colour is pure white, and the whole surface is minutely and thickly speckled with reddish-brown and greyish-purple spots, more closely placed at the thick end, where they coalesce in places and form bold patches.

"On the 29th June, I found another nest of similar construction, placed on the ground in thick forest, at the root of a shrub."

Mr. W. Davison in 1875 gave me the following note :—" On the morning of the 25th March I took at Baukasoon a nest of this species in thick forest; it was placed on the ground and was composed externally of dead leaves, with a scanty lining of fine roots and fibres. It measured externally about 5 inches high by about 4 wide. The egg-cavity was hardly 3 inches in diameter. The nest was only partially domed, and was very loosely and carelessly put together.

"The nest contained three eggs, but these were so far incubated that it was impossible to blow two of them."

The single egg of this species obtained by Mr. Davison is in shape a moderately broad oval, a little pointed towards the small end; the shell is fine, but has little gloss. The ground-colour, so far as this is visible through the thickly-set markings, is white, and it is very finely but densely stippled and freckled (most densely at the large end, where the markings are not unfrequently confluent or nearly so) with dull to bright reddish brown ; here and there, especially about the large end, more or less faint grey or red specks, spots, or tiny clouds may be traced underlying as it were the brown or purplish markings.

The egg sent me from Pegu by Mr. Oates is of precisely the same size and type, but the markings are much less dense and are brighter coloured, The ground-colour is white, and the egg is pretty thickly speckled with a reddish-chocolate brown. Here and there a moderately large irregularly-shaped spot is intermingled with the finer specklings. The markings are rather most dense at the large end, where there is a tendency to form a zone, and here a number of pale purplish-grey streaks and specks are also intermingled.

Major C. T. Bingham says :—" Early on the morning of the 7th April, moving camp from the sources of the Thoungyeen, on the side of a hill at the foot of a bamboo-bush not two feet from the road, I flushed and shot a female of the above species off her nest ; a little loosely-put-together round ball of dry bamboo-leaves, unlined, though domed over, with the entrance at the side, and containing two fresh eggs, white, thickly speckled with brick-red and obscure purple. On the 12th of the same month, I found a second nest behind the zayat or rest-house at Meeawuddy. This was similar to the nest above described, and contained three similar eggs."

The eggs measure from ·78 to ·88 in length, and from ·58 to ·65 in breadth; but the average of twelve eggs is ·82 by ·62.

147. Pellorneum fuscicapillum (Bl.).　*The Brown-capped Babbler*.

Pellorneum fuscocapillum (*Bl.*), *Hume, Cat.* no. 399 quint.

Captain Legge writes, in his 'Birds of Ceylon':—"The nest of this species is exceedingly difficult to find, and scarcely anything is known of its nidification. Mr. Blyth succeeded in finding it in Haputale at an elevation of 5500 feet. It was placed in a bramble about 3 feet from the ground, and was cup-shaped, loosely constructed of moss and leaves; it contained three young."

149. Drymocataphus nigricapitatus (Eyton).　*The Black-capped Babbler*.

Drymocataphus nigricapitatus (*Eyton*), *Hume, Cat.* no. 396 sex.

Mr. W. Davison writes:—"I got one nest of this bird at Klang. I was passing through some very dense jungle, where the ground was very marshy, when one of these birds rose from the ground about a couple of feet in front of me, and alighted on an old stump some few feet away. On examining the place from which the bird rose, I found the nest placed at the base of a small clump of ferns, and concealed by a number of overhanging withered fronds of the fern. The base of the nest, which rested on the ground, was composed of a mass of dried twigs, leaves,·&c.; then came the real body of the nest, composed of coarse fern-roots, the egg-cavity being lined with finer roots and a number of hair-like fibres. It looked compactly and strongly put together, but on trying to remove it, it all came to pieces. When the bird saw me examining the nest it fluttered to within a couple of feet of me, twittering in a most vehement manner, feigning a broken wing to try and draw me away. The nest contained only two eggs, which were slightly set."

These eggs are extremely regular ovals, scarcely smaller, if at all, at one end than at the other. The shell is very fine and fragile, but has only a slight gloss. The ground-colour appears to have been creamy white, but the markings are so thickly set that little of this is anywhere visible. First, pale inky-purple spots and clouds are thickly sprinkled over the surface, and over this the whole egg is freckled with a pale purplish brown. They measured 0·82 in length by 0·62 and 0·63 in breadth.

151. Drymocataphus tickelli. *Tickell's Babbler.*

Trichastoma minus, *Hume* ; *Hume, Cat.* no. 387 bis.

Major C. T. Bingham found the nest of this bird in the valley of the Meplay river, Tenasserim, and he says :—" On the 15th March I found a little domed nest made of dried bamboo-leaves, and lined with fine roots, placed in a cane-bush a foot or so above the ground. It contained three tiny white eggs, with minute pink dottings chiefly at the larger end ; one egg, however, is nearly pure white."

One of these eggs taken by Major Bingham on the 15th March is a very regular, somewhat elongated oval. The shell very fine and delicate, and fairly glossy. The ground is china-white, and it is everywhere speckled and spotted, nowhere very thickly, but most so in a zone near one end, with pale ferruginous. It measured 0·67 by 0·51.

160. Turdinus abbotti (Bl.). *Abbott's Babbler.*

Trichastoma abbotti (*Bl.*), *Jerd. B. Ind.* ii, p. 17.

Abbott's Babbler breeds throughout Burma in suitable localities. Writing from Kyeikpadein, in Southern Pegu, Mr. Oates says :— " On the 22nd May I found a nest with two eggs nearly hatched, and on 23rd of same month another with two eggs, one of which was fresh and the other incubated. This bird builds in thick undergrowth, and the nest is built at a height of about 2 feet from the ground. I have found very many of their nests, but, with the above exceptions, the young had flown. It is generally attached to a stout weed or two, and consists of two portions. First, a platform of dead leaves about 6 inches in diameter and 1 deep, placed loosely, and on this the nest proper is built. This consists of a small cup, the interior diameter of which is 2 inches, and depth 1½. It is formed entirely of fine black fern-roots well woven together. Stout weeds appear favourite sites, but I have found old nests in dwarf palm-trees at the junction of the frond with the trunk, and in one instance I found an old nest on the ground, undoubtedly belonging to this bird. Three eggs measured ·84 by ·66, ·82 by ·67, and ·87 by ·65. They are very glossy and smooth. The ground-colour is a pale pinkish white. At the cap there are a few spots and short lines of inky-purple sunk into the shell, and over the whole egg, very sparingly distributed, there are spots and irregular fine scrawls of reddish brown. A few of the marks are neither spots nor scrawls, but something like knots. The cap is suffused with a darker tinge of pink than are the other parts of the shell.

" A third nest, found on the 10th June, contained three eggs, and differed from those above described in being very massive. It was composed of dead leaves and fern-roots, and measured abou 5 inches in exterior diameter, with the egg-cup about 2½ inches broad and 2 inches deep. It was placed on some entangled small plants about 2 feet from the ground.   Of these eggs I noted that before being blown the shell was of a ruddy salmon colour.   The marks are much as in the others described above."

The eggs are moderately broad ovals, somewhat pointed at times towards the small end, and occasionally slightly pyriform.   The shell is fine and glossy; the ground-colour is pinky white, with a redder shade about the large end.   A few streaks, spots, and hieroglyphics of a deep brownish red, each more or less surrounded by a reddish nimbus, are scattered very thinly about the surface of the egg, while, besides these, a few small greyish-purple subsurface-looking spots may be observed about the larger end.   The average size of the seven eggs I possess is 0·82 by 0·64.

163. Alcippe nepalensis (Hodgs.).   *The Nepal Babbler*.

Alcippe nipalensis (*Hodgs.*), *Jerd. B. Ind.* ii, p. 18; *Hume, Rough Rough Draft N. & E.* no. 388.

The Nepal Babbler, according to Mr. Hodgson's notes, breeds from March to May, building a deep, massive, cup-shaped nest, firmly fastened between two or three upright shoots, and laying three or four eggs, which are figured as measuring 0·7 by 0·55. He has the following note :—

" *Valley, April 1st.*—A pair and nest.   Nest is round, 4 inches deep on the outside and 2 inches within, and the same wide, being of the usual soup-basin shape and open at the top, made of dry leaves bound together with hair-like grass-fibres and moss-roots, which also form the lining, further compacted by spiders' webs, which, being also twisted round three adjacent twigs, form the suspenders of the nest, the bottom of which does not rest upon anything; attached to a low bush 1½ foot from the ground.   The nest contained three eggs of a pinkish-white ground thickly spotted with chestnut, the spots being almost entirely confluent at the large end."

Dr. Jerdon says :—" I had the nest and eggs brought me by the Lepchas.   The nest was loosely made with grass and bamboo-leaves, and the eggs were white with a few reddish-brown spots."

A nest of this species was found near Darjeeling in July, at an elevation of between 3000 and 4000 feet.   It was situated in a small bush, in low brushwood, and placed only about 2 feet from the ground. The nest is a compactly made and moderately deep cup.   The exterior portion of the nest is composed of bamboo-leaves, more or less held in their places by fine horsehair-like black roots, with which also the cavity is very thickly and neatly lined.   Exteriorly the nest is about 3·75 inches in diameter, and nearly 3 in height. The cavity is 2·25 in diameter and 1·6 in depth.

The nest contained three nearly fresh eggs.   The eggs are moderately elongated ovals, very regular and slightly pointed towards the small end.   The shell is fine and exhibits a slight gloss.   The ground-colour is white or pinkish white, and they are *very* minutely speckled all over with purplish red.   The specklings exhibit a decided tendency to form a more or less perfect, and more or less confluent, cap or zone at the large end.

Two of the eggs measure 0·72 and 0·71 in length, and 0·54 and 0·52 in breadth.

From Sikhim, Mr. Gammie writes :—"I have only found this Babbler breeding in May at elevations about 5000 feet, but it doubtless breeds also at much lower elevations, probably down to 2000 feet. The nests are placed within 2 or 3 feet of the ground, between several slender upright shoots, to which they are firmly attached. They are exceedingly neat and compact-built cups, measuring externally about 4 inches across by 2·75 deep, internally 2·15 wide by 1·6 deep. They are composed of dry bamboo-leaves held together by a little grass and very fine, hair-like fern-roots. The egg-cavity is lined with fern-roots.

" The eggs are three or four in number."

Numerous nests of this species kindly sent me by Messrs. Gammie, Mandelli, and others, taken during the months of May and June in British and Native Sikhim, at elevations of from 3000 to 5500 feet, were all of the same type and placed in the same situations, namely amongst low scrub and brushwood, at heights of from 18 inches to 3 feet from the ground. The interior and, in fact, the main body of the nests appear to be in all cases chiefly composed of fine black hair-like roots, with which, in some cases, especially about the upper margin, a little fine grass is intermingled. The cavities are generally much about the same size, say 2 inches in diameter by 1·25 in depth ; but the size of the nests as a whole varies very much. The nest is always coated exteriorly with dry leaves of trees and ferns, broad blades of grass, and the like, fixed together sometimes by mere pressure, but generally here and there held together by fine fibrous roots, and this coating varies so much that one nest before me measures 5·5 in external diameter, and another barely 4, the external covering of fern-leaves, flags, and dry and dead leaves being very abundant in the former, while in the other the covering consists entirely of broad dry blades of grass very neatly laid together. Two, three, and four fresh eggs were found in these several nests, but in no case were more than four eggs found.

Two nests taken by Mr. Gammie contained three and two fresh eggs respectively. The eggs had a delicate pink ground, and were richly blotched, in one egg exclusively, in the others chiefly about the larger end, with chestnut, or almost maroon-red, here and there almost deepening in spots to black, and elsewhere paling off into a rufous haze. The markings are confluent about the large end, and there in places intermingled with a purplish tinge. The other eggs had a china-white ground, with more gloss than the specimens previously described, with numerous small, blackish brownish-red spots and specks, almost exclusively confined to the large end, where they are more or less enveloped in a pinky-red nimbus.

These eggs varied from 0·75 to 0·79 in length, and from 0·56 to 0·6 in breadth.

Other eggs, again, with the same pinky-white ground are thickly but minutely freckled and speckled with rather pale brownish red, most thickly towards and about the large end, where they become confluent in patches, and where tiny purple clouds and spots are dimly traceable.

164. Alcippe phæocephala (Jerd.). *The Nilghiri Babbler.*

Alcippe poiocephala (*Jerd.*), *Jerd. B. Ind.* ii, p. 18 ; *Hume, Rough Draft N. & E.* no. 389.

The Nilghiri Babbler breeds, apparently, throughout the hilly regions of Southern India. It lays from January to June. A nest taken near Neddivattam by Mr. Davison on the 5th April was placed between the fork of three twigs of a bush, at the height of 5 or 6 feet from the ground. It was a deep cup, massive enough but very loosely put together, and composed of green moss, dead leaves, a little grass and moss-roots. It was entirely lined with rather coarse black moss-roots. In shape it was nearly an inverted cone, some 3½ inches in diameter at top, and fully 5 inches in height. The cavity was over 2 inches in diameter and nearly 2 inches in depth. A few cobwebs are here and there intermingled in the external surface, but the grass-roots appear to have been chiefly relied on for holding the nest together.

Another nest found by Miss Cockburn on the 5th June on a small bush, about 7 or 8 feet in height, standing on the banks of a stream, was somewhat different. It was placed in the midst of a clump of leaves, at the tips of three or four little twigs, between which the nest was partly suspended and partly wedged in. It was composed of fine grass-stems, with a few grass- and moss-roots as a lining interiorly, and with several dead leaves and a good deal of wool incorporated in the outer surface, the greater portion of which, however, was concealed by the leaves of the twigs amongst which it was built. It was only about 3½ inches in diameter, and the egg-cavity was less than 2½ inches across, and not above 1½ inch in depth.

Mr. Davison writes :— "This bird breeds on the slopes of the Nilghiris in the latter end of March and April. The nest is uncommonly like that of *Trochalopterum cachinnans*, but is of course smaller ; it is deep and cup-shaped, composed externally of moss and dead leaves, and is lined with moss and fern-roots. It is always (as far as I have observed) fastened to a thin branch about 6 feet from the ground. All the nests I have ever observed were on small trees in the shadiest parts of the jungle, far in, and never near the edge of the jungle or in the open. The eggs are very handsome, and are, I think, the prettiest of the eggs to be found on the Nilghiris and their slopes. The ground-colour is of a beautiful reddish pink (especially when fresh), blotched and streaked with purplish carmine."

Mr. J. Darling, junior, says:—"The Nilghiri Quaker-Thrush breeds on the slopes of the Nilghiri hills, generally in the depths of the forest. I have, however, taken nests in scrub-jungle. I have also found the nest at Neddivattam in April.

"In October I found a nest of this bird at Culputty, S. Wynaad, about 2800 feet above the sea, built at the end of a branch 4 feet from the ground."

Mr. T. F. Bourdillon writes from Travancore:—"This bird breeds commonly with us, and its nest is more often met with than that of any other. The nest is cup-shaped and made of lichen, leaves, and grass. It is usually placed 4 to 8 feet from the ground in the middle of jungle, and is about 2 inches in diameter by 1¾–2 in depth. The full number of eggs is two, and I have obtained on

"April, 1871. 2 fresh eggs.
"Mar. 21, 1873. 2 fresh eggs.
"Feb. 16, 1874. 2 fresh eggs.
"April 11, 1874. 2 young birds, and many nests just vacated."

As in the case of *Pyctorhis sinensis*, the eggs differ much in colour and markings. The two eggs of this species sent me by Miss Cockburn from Kotagherry are moderately broad ovals, very obtuse at the larger end and somewhat compressed towards the smaller. The shell is fine and somewhat glossy. The ground-colour is white or pinkish white, and they are thickly mottled and freckled, most thickly at the larger end, where the markings form a more or less confluent mottled cap, with two shades of pinkish-, and in some spots slightly brownish, red, and towards the large end, where the markings are dense, traces of pale purple clouds underlying the primary markings are observable. In general appearance these eggs not a little resemble those of some of the Bulbuls, and it seems difficult to believe that they are eggs of birds of the same genus as *Alcippe atriceps**, the eggs of which are so much smaller and of such a totally different type. Two eggs of the same species taken by Mr. Davison are moderately broad ovals, somewhat compressed towards one end; have a fine and slightly glossy shell. The ground-colour is a delicate pink. There are a few pretty large and conspicuous spots and hair lines of deep brownish red, almost black, and there are a few large pinkish-brown smears and clouds, generally lying round or about the dark spots; and then towards the large end there are several small clouds and patches of faint inky purple, which appear to underlie the other markings. The character of the markings on some of these eggs reminds one strongly of those of the Chaffinch. Other eggs taken later by Miss Cockburn at Kotagherry on the 21st January are just intermediate between the two types above described.

All the eggs are very nearly the same size, and only vary in length from 0·75 to 0·86, and in breadth from 0·58 to 0·65.

---

* *Alcippe atriceps* and *Alcippe phæocephala*, as they have hitherto been styled by all Indian ornithologists, are not in the least congeneric, as I have pointed out in my 'Birds of India.' I am glad to see my views corroborated by Mr. Hume's remarks on the eggs. There is no reason why these two birds should be considered congeneric, except a general similarity in colour and habits. Their structure differs much.—ED.

165. Alcippe phayrii, Bl. *The Burmese Babbler.*

Alcippe phayrii, *Bl., Hume, Cat.* no. 388 bis.

Major C. T. Bingham writes from Tenasserim :—" In the
half-dry bed of one of the many streams that one has to cross between Kaukarit and Meeawuddy, I found on the 23rd February
a nest of the above species. A firm little cup, borne up some 2
feet above the ground on the fronds of a strong-growing fern, to
three of the leaf-stems of which it was attached. It was made of
vegetable fibres and roots, and lined interiorly with fine black hair-
like roots, on which rested three fresh eggs, in colour pinky white,
blotched and streaked with dull reddish pink, and with faint clouds
and spots of purple. The eggs measure ·79 × ·58, ·78 × ·58, and
·76 × ·59."

Mr. J. Darling, junior, informs us that on the 9th April he
" took three fresh eggs of *Alcippe phayrii*, in heavy jungle, at a very
low elevation, at the foot of Nwalabo in Tenasserim. The nest was
built in a small bush 4 feet from the ground (hanging between
two forked twigs), of bamboo and other leaves, moss, and a few
fine twigs, and lined with moss and fern-roots, 2 inches in diameter,
1½ deep. It was exactly like very many nests of *A. phæocephala*,
taken on the Nilghiri Hills, though some of the latter are much
more compact and pretty."

Mr. W. Davison, also writing of Tenasserim, says :—" On the 1st
March, in a little bush about 2 feet above the ground, I found the
above-mentioned bird seated on a little moss-made nest, and utterly
refusing to move off until I almost touched her, when she hopped
on to a branch a few feet off, and disclosed three little naked fledglings struggling or just struggled out of their shells. I retired a
little way off, and she immediately reseated herself. The eggs, to
judge by the fragments, were of a vinous claret tinge, spotted and
streaked with a darker shade of the same."

These eggs closely resemble those of *A. nepalensis.* They are
neither broad nor elongated ovals, often with a *slight* pyriform
tendency, always apparently very blunt at both ends.

The ground-colour, of which but little is visible, in some eggs
varies from pinky white to pale reddish pink, and the egg is profusely smeared and clouded with pinky or purplish red, varying
much in shade and tint. Here and there, in most eggs, are a few
spots, or occasionally short, crooked or curved lines, where the
colour has been laid on so thick that it is almost black, and such
spots are generally, though not always, more or less surrounded
with a haze of a rather deeper tint than the rest of the smear in
which they occur. The markings are often deepest coloured, or
most conspicuous, about the large end, where occasionally a recog-
nizable cap is formed and there a decided purplish tinge may be
noticed in patches. The general character of the eggs is very uniform ; but the eggs vary to such a degree *inter se*, that it is hopeless
to attempt to describe all the variations. They vary in length

from 0·68 to 0·78 and in breadth from 0·53 to 0·59, but the average of nine eggs is 0·75 by 0·58.

166. Rhopocichla atriceps (Jerd.) *The Black-headed Babbler*.

Alcippe atriceps (*Jerd.*), *Jerd. B. Ind.* ii, p. 19; *Hume, Rough Draft N. & E.* no. 390.

Writing from Coonoor in the Nilghiris, Mr. Wait tells me that the Black-headed Babbler breeds in his neighbourhood in June and July :—" It builds in weeds and grass beside the banks of old roads, at elevations of from 5000 to 5500 feet. The nest is placed at a height of from a foot to 2 feet from the ground, is domed and loosely built, composed almost entirely of dry blades of the lemon-grass, and lined with the same or a few softer grass-blades. In shape it is more or less ovate, the longer axis vertical, and the external diameters 4 and 8 inches. They lay two or three rather broad oval eggs, which have a white ground, speckled and spotted, chiefly at the large end, with reddish brown."

Miss Cockburn sends me a nest of this species which she found on the 17th June amongst reeds on the edge of a stream, about 2 or 3 feet above the water's edge. It appears to have been a glo-bular mass very loosely put together, of broad reed-leaves, between 3 or 4 inches in diameter, and with a central unlined cavity.

Mr. Iver Macpherson, writing from Mysore, says :—" I have only met with this bird in heavy bamboo-forest, and have only found two nests, viz., on the 25th May and 2nd July, 1879. Both nests were fixed low down (2 to 3 feet) in bamboo-clumps, and each contained two eggs, which, for the size of the bird, I con-sidered very large. Nest globular, and very loosely constructed of bamboo-leaves and blades of grass."

An egg sent me from Coonoor by Mr. Wait is a moderately broad, very regular oval, only slightly compressed towards the smaller end. The shell is very fine and satiny, but has only a slight gloss. The ground-colour is white or slightly greyish white, and towards the large end it is profusely speckled with minute dots of brownish and purplish red, a few specks of the same colour being scattered about the rest of the surface of the eggs.

Another egg sent me from Kotagherry by Miss Cockburn exactly corresponds with the above description.

Both are precisely the same in size, and measure 0·75 by 0·55. Other eggs measure from 0·75 to 0·79 in length by 0·53 to 0·58 in breadth*.

---

* Mr. T. Fulton Bourdillon (S. F. ix, p. 300) gives an interesting account of the nest and eggs of a species of *Rhopocichla* which he failed to identify satisfac-torily. It may have been *R. atriceps* or *R. bourdilloni*. Most probably, judging from the locality, it was the latter. As, however, there is a doubt about it, I do not insert the note.—ED.

167. Rhopocichla nigrifrons (Bl.). *The Black-fronted Babbler.*

Alcippe nigrifrons, *Bl., Hume, Cat.* no. 390 ter.

Colonel Legge writes regarding the nidification of the Black-fronted Babbler in Ceylon :—" After finding hundreds of the curious dry-leaf structures, mentioned in ' The Ibis,' 1874, p. 19, entirely void of contents, and having come almost to the conclusion that they were built as roosting-places, I at last came on a newly-constructed one containing two eggs, on the 5th of January last; the bird was in the nest at the time, so that my identification of the eggs was certain. The nest of this Babbler is generally placed in a bramble or straggling piece of undergrowth near a path in the jungle or other open spot ; it is about 3 or 4 feet from the ground, and is entirely made of dead leaves and a few twigs ; the leaves are laid one over another horizontally, forming a smooth bottom or interior. In external form it is a shapeless ball about 8 or 10 inches in diameter, and has an unfinished opening at the side. The birds build with astonishing quickness, picking up the leaves one after another from the ground just beneath the nest. When fresh the eggs are fleshy white, becoming pure white when emptied ; they are large for the size of the bird, rather stumpy ovals, of a smooth texture, and spotted openly and sparingly with brownish red, over bluish-grey specks ; in one specimen the darker markings are redder than in the other, and run mostly in the direction of the axis. Dimensions : 0·74 by 0·56 and 0·74 by 0·55."

169. Stachyrhis nigriceps, Hodgs. *The Black-throated Babbler.*

Stachyris nigriceps, *Hodgs., Jerd. B. Ind.* ii, p. 21 ; *Hume, Rough Draft N. & E.* no. 391.

I have never taken a nest of this species, the Black-throated Babbler, but Mr. Gammie, a careful observer, in whose neighbourhood (Rungbee, near Darjeeling) this bird is very abundant, has taken many nests, two of which he has sent me, with many eggs.

One nest, found at Rishap, on the 14th May, at an elevation of about 4000 feet, contained four nearly fresh eggs. It was a very loose structure, a shallow cup of about 3½ inches in diameter, composed of fine grass-stems without any lining, and coated externally with broad coarse grass-blades.

Another nest taken low down in the valley, at about an elevation of 2000 feet, on the 17th June, contained three fresh eggs. It was placed in a bank at the foot of a shrub. Like the previous one, it was a loose but rather deeper cup, interiorly composed of moderately fine grass, exteriorly of dead leaves. The egg-cavity measured about 2 inches in diameter, and 1½ inch in depth. *In situ,* both probably were more or less domed, the cups more or less overhung by a hood or canopy.

Mr. Gammie remarks :—" I have seen numerous nests of this

species in former years, and have found two this season, but have never seen eggs with 'faint darker spots' as mentioned by Jerdon. Hodgson's description is quite correct. The eggs are a 'pale fawn-colour' *before they are blown*, the shells being so translucent that the yolk shows through partially. The shell is pure white in itself. The cavity of the cup-shaped part of one nest beside me is 2 inches deep by 2 inches wide; outer dimensions $5\frac{3}{4}$ inches deep (from top of hood) by 4 inches wide across the face of entrance. It is loosely though neatly made of bamboo-leaves and fern, lined with dry grass. The bird breeds in May and June, and lays four or five eggs."

Mr. Eugene Oates tells us that he " procured only one specimen of this bird, and that was in the evergreen forests of the Pegu Hills. I shot it off the nest on the 29th April. The nest was on a bank of a nullah well concealed among dead leaves, about 2 feet above the bottom of the bank. The nest is domed, about 7 inches in height and 5 inches in diameter externally, with the entrance at the side near the top. The outside is a mass of bamboo-leaves very loose, being in no way bound together; each leaf is curled to the shape of the nest. The inside, a thin lining only of vegetable fibres. There were three eggs, just on the point of hatching; colour, pure white."

The Black-throated Babbler breeds, according to Mr. Hodgson, in April and May, and builds a large deep cup-shaped nest, either upon the ground in the midst of grass, or at a short distance above the ground between five or six thin twigs; a nest which he measured was externally 4·5 inches in diameter and 3·5 in height, while the cavity was 2·5 in diameter and 2 in depth. The nest is composed of dry bamboo- and other leaves wound together with grass and moss-roots, and lined with these, and is a very firm compact structure, considering the materials. They lay four or five eggs, which are figured as very regular rather broad ovals, of a nearly uniform, very pale *café-au-lait* colour (these were the *unblown* eggs), measuring about 0·75 by 0·58.

Dr. Jerdon remarks :—" A nest and eggs were brought to me at Darjeeling, and said to be of this species. The nest was rather large, very loosely made of bamboo-leaves and fibres, and the eggs were of a pale salmon-colour, with some faint darker spots."

There is no doubt that these must have been the eggs of some other species.

Major C. T. Bingham tells us:—" This little bird, though not at all common, breeds in the Sinzaway Reserve, in Tenasserim. I took five hard-set eggs, placed in a beautiful little domed nest, at the foot of a clump of bamboos, on the bank of a dry choung or nullah. This was on the 20th March. The nest was composed exteriorly of dry bamboo-leaves, and interiorly of fine grass-roots, the entrance being on one side. I shot the female as she crept off the nest."

It does not seem that in the Himalayas this species domes its nest. Numerous other nests that have been sent me from Sikhim,

taken in May, June, and July, were all of the same type—shallow
or deeper cups loosely put together, exteriorly composed of coarse
blades of grass, dead leaves, bamboo-spathes and the like, held
together with a little vegetable fibre or fibrous roots, and interiorly
of fine grass generally more or less mingled with blackish roots,
which in some nests greatly predominate over the grass.

The eggs are broad ovals, somewhat compressed towards one
end, in some cases slightly pyriform. They are pure white, spotless, and fairly glossy.

They vary from 0·68 to 0·84 in length, and from 0·55 to 0·61
in breadth, but the average of thirty-four eggs is 0·76 by somewhat
over 0·58.

170. **Stachyrhis chrysæa**, Hodgs.   *The Golden-headed Babbler.*

Stachyris chrysæa, *Hodgs., Jerd. B. Ind.* ii, p. 22; *Hume, Rough
Draft N. & E.* no. 394.

Mr. Blyth remarks :—" The egg, as figured by Mr. Hodgson, is
pinkish white, and the nest domed and placed on the summit of a
sedge. *S. præcognita* lays a blue egg." (Ibis, 1866, p. 309.)

There is no figure of either the nest or eggs of the Golden-
headed Babbler amongst the drawings of Mr. Hodgson that I
possess.

From Sikhim Mr. Gammie writes :—" I took a nest of this bird
out of a large forest, at 5000 feet elevation, on the 15th May. It
is of an oval shape, neatly made of small bamboo-leaves only,
devoid of lining, and was fixed vertically between a few upright
sprays, within two feet of the ground. It measures externally 5·25
inches in height by 4 in diameter; internally 1·5 in depth, from
lip of egg-cavity, by 1·75 in diameter. The entrance is also 1·75
across.

" The eggs were four in number; three of them well set and the
fourth quite fresh. The set eggs were altogether pure white, but
the fresh egg, unblown, was of a pinky-white colour with a pure
white cap; when blown it exactly resembled the others."

The eggs sent as pertaining to this species by Mr. Gammie are
very regular ovals, pure white, and somewhat glossy, but they are
so small that I can scarcely credit their really belonging to this
species. Their cubit contents are not half those of the average
eggs of *S. nigriceps*. They measure 0·63 by 0·48.

172. **Stachyrhidopsis ruficeps**, Bl.   *The Red-headed Babbler.*

Stachyris ruficeps, *Bl., Jerd. B. Ind.* ii, p. 22; *Hume, Rough Draft
N. & E.* no. 393.

The Red-headed Babbler breeds in Nepal, according to Mr. Hodgson, from April to June, building a large massive cup-shaped nest
amongst bamboos, as a rule, at heights of from 7 to 10 feet from
the ground. The nest is wedged in between half a dozen or more

creepers and shoots, and is composed almost exclusively of dry bamboo-leaves neatly, but rather loosely, interwoven, and lined also with these leaves. One which he measured was rather oval in shape, 5·25 inches in diameter one way, by 4 the other, and 3·6 in height. The leaves used in the rim of the cup were projected a little inwards, so as to make the mouth of the cavity a little smaller than the diameter of this latter within. The diameter of the mouth was 2 inches, that of the cavity 2·5, and the latter is about 1·5 deep. Four eggs are laid, a sort of brownish white, speckled and spotted with brown or reddish brown. The egg figured measures 0·7 by 0·52, and is a moderately broad, regular oval.

Dr. Jerdon says :—"A nest and eggs, said to be of this species, were brought to me at Darjeeling. The nest was a loose structure of grass and fibres, and contained two eggs of a greenish-white colour with some rusty spots."

From Sikhim Mr. Gammie writes :—" I took two nests of this Babbler in April ; one of them at an elevation of 3500 feet, the other at 5000 feet, but it no doubt breeds also both lower and higher. They are of a neat egg-shape, with entrance at side, and were fixed vertically between a few upright sprays, within three feet of the ground, in open situations near large trees. Mr. Hodgson evidently did not take the one he describes with his own hands, for he places it horizontally, which gives a height of 3·6 inches only. The external dimensions are about 5·5 inches in height and 4 in diameter. Internally the diameter is 2 inches, and the depth, from roof, 3·25. The entrance is 2 across. They are composed of dry bamboo-leaves only, put neatly and firmly together, and are lined with a very few grassy fibres. They each contained four well-set eggs."

Mr. Mandelli, however, took a nest of this species at Lebong on the 23rd June, in the middle of a tea-bush which grew at the side of a small ravine, which was neither hooded nor domed. The nest was about 18 inches from the ground and completely sheltered from above by tea-leaves. It was a deep cup composed externally chiefly of bamboo-leaves, but with a good many dead leaves of trees incorporated in the base, and lined with very fine grass-stems. It contained four fresh eggs. It is quite clear that this species, like *S. nigriceps*, only domes its nest in certain situations.

The eggs obtained by Mr. Gammie and Mr. Mandelli are very regular, slightly elongated ovals. The shell is very fine and compact, but has only a faint gloss. The ground is white and round the larger end is a zone or imperfect cap of specks and spots of brownish red, generally intermingled with tiny spots, usually very faint, of pale purple. A few specks and spots brown, yellowish, or reddish brown, and sometimes also pale purple, are scattered about the rest of the egg.

In length the eggs vary from 0·64 to 0·72, and in breadth from 0·50 to 0·53, but the average of eight eggs was 0·68 by 0·52 nearly.

174. **Stachyrhidopsis pyrrhops,** Hodgs. *The Red-billed Babbler.*

Stachyris pyrrhops, *Hodgs.*, *Jerd. B. Ind.* ii, p. 21; *Hume, Rough Draft N. & E.* no. 392.

Accounts differ somewhat as to the eggs of the Red-billed Babbler.

From Murree, Colonel C. H. T. Marshall writes : — " Nest found in low ground, about 100 yards from the River Jheelum, situated in a low bush externally composed of broad dry reed-leaves, and interiorly of fine grass, cup-shaped. Eggs, four in number, long oval, white, with a few reddish specks at the larger end. Length ·7, breadth ·5. Lays in the latter end of June, 4000 feet up."

The nest, which he kindly sent me, is a deep cup, coarsely made interiorly of grass-stems, externally of broad blades of grass, in which a few dead leaves are incorporated; there is no lining. Exteriorly the nest is about 3·5 inches in diameter, and about 3 in depth; the egg-cavity is a little more than 2 inches in diameter, and fully 1·75 in depth.

Mr. Hodgson " found the nest " of this species in Nepal, " at an elevation of about 6000 feet, in shrubby upland." It was " placed in a small shrub about 2 feet from the ground." It was " a very deep cup, about 4 inches in length, and 2·5 in diameter externally, placed obliquely endwise upon cross-stems of the shrub, and opening, as it were obliquely, upwards at one end," the cavity being about 1·5 in diameter. The nest was made of " dry leaves and grass pretty compactly woven." The nest " contained four eggs," which are described as " whitish, with spare and faint fawn-coloured spots," and are figured as measuring 0·65 by 0·47.

Captain Hutton says :—" This is a common species both in the Dhoon and in the hills, and may be found at all seasons, making known its presence among the brushwood by the utterance of a clear and musical note like the ringing of a tiny bell. In the winter time it is often mixed up with flocks composed of *Siva strigula* and *Liothrix luteus*, creeping among the bushes like the *Pari* and *Phylloscopi*. It constructs its nest at the base of bushes, the eggs being three in number, of a faint greenish grey, thickly irrorated with small reddish-brown specks. The nest is composed of dry grass-blades externally, within which is a layer of fine woody stalks and fibres, and lined with black hair. It is cup-shaped, and placed upon a thick bed of dried leaves, which are most probably accumulated beneath the bush by the wind. One nest was taken at Dehra, in a garden, on the 30th July, and others at Mussoorie about the same time."

But the eggs sent by Captain Hutton clearly do not, I think, pertain to this species. Those taken by Colonel Marshall are certainly genuine, and are considerably larger and very differently coloured eggs.

In shape they are moderately broad ovals, some of them slightly compressed towards the small end. The shell is very fine and smooth, but with scarcely any gloss ; the ground is pure white, and they are thinly speckled and spotted, the markings being much more numerous about the large end, where they have a tendency to form an ill-defined cap or zone with brownish red or pinky brown.

In length they vary from 0·62 to 0·69, and in breadth from 0·5 to 0·52.

175. Cyanoderma erythropterum (Blyth). *The Red-winged Babbler.*

Cyanoderma erythropterum, *Bl., Hume, Cat.* no. 396 bis.

Mr. W. Davison found the nest of the Red-winged Babbler at Bankasoon on the 23rd April, just when he was leaving the place. Unfortunately the birds had not yet laid. The nest was a ball composed of dry reed-leaves, about 6 inches in diameter. Externally, with a circular aperture on one side, very like that of *Mixornis rubricapillus* and of *Dumetia*, and again not at all unlike that of *Ochromela nigrorufa*, but placed in a bush about 4 feet high and not on the ground.

176. Mixornis rubricapillus (Tick.). *The Yellow-breasted Babbler.*

Mixornis rubricapilla (*Tick.*), *Jerd. B. Ind.* ii, p. 23; *Hume, Rough Draft N. & E.* no. 395.

This, though said to occur also in Central India, is a purely Indo-Burmese form, found chiefly in the Eastern sub-Himalayan jungles, Assam, Cachar, Burma, and Tenasserim.

It is only from this latter province that I have any information as to the nidification of the Yellow-breasted Babbler.

Mr. Davison writes to me:—"At a small village, called Shymootee or Tsinmokehtee, about 7 miles from the town of Tavoy, and very slightly above the sea-level, say 50 feet, I found on the 6th of May, 1874, a nest of this species. The nest was placed in a dense clump of a very thorny plant (somewhat like a pineapple bush) about a foot from the ground; it was not particularly well concealed. The nest was built of bamboo-leaves, and in general appearance was not at all unlike that of *Ochromela nigrorufa*; but the egg-cavity was very shallow, so that by moving aside an overhanging leaf the eggs were distinctly visible. There were three partially incubated eggs in the nest, a somewhat dull white, spotted with pinkish dots."

The nest is more or less egg-shaped, the longer axis vertical, with a circular aperture on one side near the top.

The exterior diameters are 5 and nearly 4 inches. The aperture about 1·5 in diameter. The cavity is barely 2 inches in diameter, and only 1·25 deep below the lower edge of the entrance.

Both nest and eggs strongly recall those of *Dumetia hyperythra*. The former is composed of the broad, grass-like leaves of the

8*

bamboo, and with only a few stems of grass here and there intermingled as if by accident.   In the sides of the cavity the leaf-
blades are so neatly laid together, side by side, that the interior
seems as if planked, and at the bottom of the cavity there is a
very scanty lining of very fine grass-stems.

Mr. Oates says :—" I found a nest on the 2nd June near Pegu,
with three eggs.   Failing to snare the bird at once, I left the nest
for a short time, and on my return found the eggs gone.   I am
satisfied, however, that the nest belonged to the present species;
for I caught a glimpse of the sitting bird.   The nest was built on
the top of a stump, well concealed by leafy twigs, except the
entrance, which was open to view.   It was a ball of grass with the
opening at the side.

" 28th June.—Nest in a shrub about 10 feet from the ground.
A domed structure with an opening at the side 3 inches high by
2 broad.   Height of nest about 6 and outside width 4.   Made
entirely of bamboo-leaves and lined sparingly with grass.   Eggs 3.

" I have found numerous nests of this species, but always after
the young had flown.   They appear almost always to be placed in
shrubs at heights of 2 to 10 feet from the ground.   One nest,
however, on which I watched the birds at work, was in a pineapple plant between the stalk of the fruit and one of the leaves,
almost on the ground."

The eggs are regular ovals, moderately elongated, only very
slightly compressed towards the smaller end, which is only just
appreciably smaller.

The shell is very fine and delicate, excessively smooth and
fragile, but with only a faint gloss.   The ground is a dead white,
with perhaps the least possible pinkish tinge.   The markings consist of *tiny* specks of brownish or purplish red and pale yellowish
brown, thinly scattered over the rest of the surface, but comparatively densely clustered round the larger end, where they form a
rather conspicuous though irregular and imperfect zone, apparent
enough in all, but much more strongly marked in one egg than in
the others.

In some eggs the markings are all rather bright red and dull
purplish grey; some have a very fair amount of gloss, and a very
pure china-white ground.

The eggs vary in length from 0·65 to 0·71, and in breadth from
0·5 to 0·53.

177. Mixornis gularis (Raffl.).   *The Sumatran Yellow-breasted Babbler.*

Mixornis gularis (*Horsf.*), *Hume, Cat.* no. 395 bis.

The eggs* are very similar to those of *M. rubricapillus*, but are,

---

* I cannot find any note about the nest of this species.   Mr. Davison was
probably the finder of the eggs described.—ED.

perhaps, as a rule, better marked. They are very regular ovals, typically rather slightly elongated, often slightly compressed towards the small end; the shell is very fine and fragile, and has usually a fair amount of gloss. The ground is usually pure white, at times with a pinkish tinge. Round the large end is a more or less conspicuous, more or less continuous zone of specks, spots, and small irregular blotches of two colours, the one varying in different eggs from almost brick-red to brownish orange, the other from reddish purple to purplish grey. In some cases a very few, in others a good many, specks and tiny spots of the same colours are scattered about the other portions of the egg. The eggs measure 0·7 by 0·51.

178. Schœniparus dubius (Hume). *Hume's Tit-Babbler.*

Proparus dubius, *Hume*; *Hume, Cat.* no. 622 bis.

Mr. W. Davison has furnished me with the following note:—

"On the 21st of February I took a nest of this species on Muleyit mountain containing two eggs, and out of the female which I shot off the nest I took another egg ready for expulsion which was in every particular precisely similar to those in the nest.

"The nest was a large globular structure, composed externally of dried reed-leaves, very loosely put together, the egg-cavity deep and lined with fibres. It was placed on the ground close to a rock, and at the foot of a Zingiberaceous plant, and rather exposed to view. The nest was not unlike that of *Pomatorhinus*, but of course considerably smaller, not so much domed, and with the mouth of the egg-cavity pointing upwards.

"A few days later, on the 25th, I took a second nest, quite similar in shape and materials to the first one, but placed several feet above the ground, in a dense mass of creepers growing over a rock. It was quite exposed to view, and from a distance of 3 or 4 feet the eggs were quite visible.

"There were three eggs in the nest, similar to those in the first nest. Both parent birds were obtained. The first nest measured 5 inches long by 4·5 wide, the egg-cavity 3·8 deep by 2·75 wide at the entrance. The other was about half an inch smaller each way.

"The measurements of the six eggs varied from 0·76 to 0·81 in length by 0·56 to 0·6 in width, but the average was 0·78 by 0·59."

The eggs are rather narrow ovals, as a rule, occasionally much pointed towards one end. The shell is very fine and has a faint gloss. The ground-colour is white. The markings, which are difficult to describe, consist first of spots, specks, and hair-line scratches, dark brown, almost black occasionally, and a great amount of irregular clouding, streaking, and smudging of a pale dirty-brown, slightly reddish in some eggs. Besides this, about the large end there is an indistinct irregular zone of faint inky purple spots and small blotches, and a few spots of this same colour may be observed on other parts of the egg.

182. **Sittiparus castaneiceps** (Hodgs.).   *The Chestnut-headed*
Tit-Babbler.

Minla castaneiceps, *Hodgs., Jerd. B. Ind.* ii, p. 255 ; *Hume, Rough Draft N. & E.* no. 619.

Mr. Hodgson's notes inform us that the Chestnut-headed Tit-Babbler breeds in the neighbourhood of Darjeeling in May and June, laying four eggs, which are figured as somewhat elongated ovals, having a very pale greenish-yellow or dingy yellowish-white ground finely speckled, chiefly at the large end, where there is a tendency to form a zone, with red or brownish red, and measuring 0·75 by 0·52. The nest is said to be placed in a thick bush, at a height of about 3 feet from the ground, in a double fork ; to be very broad and shallow, composed of twigs, grass, and moss, and lined with leaves. One, taken on the 18th May, 1846, measured 6 inches in diameter and 2·5 in height externally; the cavity was only 2·1 in diameter and 1 in depth.

From Sikhim Mr. Gammie writes :—" A nest of this bird, with one fresh egg and female, was brought to me in May. The man said he found the nest in the Rungbee forest, at 6000 feet, among the moss growing on the trunk of a large tree, a few feet from the ground. It was a solid cup, made of green moss, with an inner layer of fine dark-coloured roots, and lined with grassy fibres. Externally it measured 4 inches in width by the same in depth ; internally 1·5 wide by 1·25 deep."

Three eggs sent by Mr. Gammie measure 0·7 to 0·75 in length and 0·55 to 0·59 in breadth.

Mr. Davison says :—" On the 20th of February, when encamped just under the summit of Muleyit, on its N.W. slope, I found a nest of this bird containing three eggs, but so hard-set that it was only with the greatest difficulty that I managed to preserve them.

" The nest, a deep cup, was placed about 5 feet from the ground, in a mass of creepers growing up a sapling. It (the nest) was composed externally of green moss and lined with fibres and dry bamboo-leaves.

" On the 29th of the same month I took another nest, also containing three eggs, precisely similar to those in the first nest ; but these were so far incubated and the shell was so fragile that they were all lost. This nest was also composed externally of green moss, beautifully worked into the moss growing on the trunk of a large tree, and it was only with considerable difficulty, and after looking for some time, that I found it. The egg-cavity of this nest was also lined with fibres and dried bamboo-leaves.

" The first nest found was open at the top, and measured 5·5 inches in depth, 3 across the top externally, the egg-cavity 3·5 in depth by 1·8 in diameter at top.

" The second nest was completely domed at the top, and measured externally 7 inches in depth by about 3·5 at top. The egg-cavity was 2·5 inches deep by 1·5 across the mouth.

" Three eggs measured 0·7 to 0·75 in length, and 0·55 to 0·59 in breadth."

The eggs are broad ovals, a little pointed towards the small end, the shell white, almost devoid of gloss. A dense ring or zone of excessively small black spots surrounds the large end, and similar specks are rather sparsely distributed over the whole of the rest of the surface of the egg, having, however, a tendency to become obsolete towards the small end. Sometimes a little brown and sometimes a little lilac is intermingled in the zone.

183. Proparus vinipectus (Hodgs.). *The Plain-brown Tit-Babbler.*

Proparus vinipectus (*Hodgs.*), *Jerd. B. Ind.* ii, p. 257 ; *Hume, Rough Draft N. & E.* no. 622.

The Plain-brown Tit-Babbler is not uncommon in the higher wooded hills between Simla and Kotegurh, and from somewhere near Mutiana Captain Blair sent me a nest and egg, together with one of the old birds which had been caught on the nest.

This latter was a rather compact massive cup, composed of moderately fine blades of grass, measuring externally about 4¼ inches in diameter and standing about 2¼ inches high. The egg-cavity, about 2 inches in diameter and rather more than half an inch deep, was lined with fine blackish-brown grass-roots. Neither nest nor egg is exactly what I should have expected to pertain to this species; but Captain Blair was certain that they belonged to the parent bird which he sent with them, and I therefore describe both with entire confidence in their authenticity.

The egg is a moderately elongated oval, slightly compressed towards one end ; it has a pale-green ground, and near the large end has a strongly marked but very irregular sepia-brown zone, and pale stains of the same colour here and there running down the egg from the zone, as well as a few isolated dark spots of the same tint. Although much smaller, and although the colour of the markings is very different, the ground-colour and the character of the markings much recall those of *Liothrix luteus.* The egg has little or no gloss, and measures 0·73 by 0·55.

Mr. Mandelli obtained two nests of this species—one at Sinchal, near Darjeeling, at an elevation of 9000 feet, on the 2nd June ; the other at Tongloo, at an elevation of 10,000 feet, on the 29th May. The first contained one, the second three fresh eggs, all precisely similar in size and colour to the egg formerly sent me by Capt. Blair, though the nests themselves were rather different in appearance. These nests were both placed amongst the branches of dense brush-wood, at heights of 3 and 4 feet from the ground ; they are very compact, massive little cups, about 3·25 inches in diameter and 2 in height exteriorly ; the cavities are about 2 inches in diameter and 1·25 in depth. The chief materials of the nests are dry blades of grass and bamboo-leaves ; but these are only seen at the bottom of the nests, the sides and upper margins being completely felted over with green moss. Apparently there is a first lining of fine grass and roots ; but very little of this is seen, as the cavity is then thickly covered with black and white hairs.

184. Lioparus chrysæus (Hodgs.).   *The Golden-breasted Tit-Babbler.*

Proparus chrysæus, *Hodgs., Jerd. B. Ind.* ii, p. 256; *Hume, Rough Draft N. & E.* no. 621.

The Golden-breasted Tit-Babbler breeds, according to Mr. Hodgson's notes, near Darjeeling and in the central region of Nepal. It lays from three to four eggs, which are figured as somewhat broad ovals, measuring 0·7 by 0·5, with a pinky-white ground, speckled and spotted thinly, except towards the large end, where there is a tendency to form a cap or zone, with brownish red. The nest is oval or rather egg-shaped, and fixed with its longer diameter perpendicular to the ground in a bamboo-clump between a dozen or so of the small lateral shoots, at an elevation of only a few feet from the ground. One, taken near Darjeeling on the 12th June, measured externally 6 inches in height, 4·5 in breadth, and 3 inches in depth, and on one side it had an oval aperture 2·5 in height and 1·75 in breadth. It appeared to have been entirely composed of dry bamboo-leaves and broad blades of grass loosely interwoven, and with a little grass and moss-roots as lining.

Hodgson originally named this bird *Proparus chrysotis*, but as the bird has *silvery* ears Hodgson himself rejected this name and adopted the one given above. Mr. Gray, however, retains the specific name *chrysotis*. Now, I think a man has a perfect right to change his *own* name; what I object to is other people presuming to do it for him.

Subfamily BRACHYPTERYGINÆ.

187. Myiophoneus temmincki, Vigors.   *The Himalayan Whistling-Thrush.*

Myiophonus temminckii, *Vig., Jerd. B. Ind.* i, p. 500; *Hume, Rough Draft N. & E.* no. 343.

The Himalayan Whistling-Thrush breeds throughout the Himalayas from Assam to Afghanistan, in shady ravines and wooded glens, as a rule, from an elevation of 2000 to 5000 feet, but, at times, especially far into the interior of the hills, up to even 10,000 feet.

It lays during the last week of April, May, and June. The number of eggs varies from three to five.

The nest is almost invariably placed in the closest proximity to some mountain-stream, on the rocks and boulders of which the male so loves to warble; sometimes on a mossy bank; sometimes in some rocky crevice hidden amongst drooping maiden-hair; sometimes on some stream-encircled slab, exposed to view from all sides, and not unfrequently curtained in by the babbling waters of some little waterfall behind which it has been constructed. The

nest is always admirably adapted to surrounding conditions.
Safety is always sought either in inaccessibility or concealment.
Built on a rock in the midst of a roaring torrent, not the smallest
attempt at concealment is made; the nest lies open to the gaze of
every living thing, and the materials are not even so chosen as to
harmonize with the colour of the site. But if an easily accessible
sloping mossy bank, ever bejewelled with the spray of some little
cascade, be the spot selected, the nest is so worked into and coated
with moss as to be absolutely invisible if looked at from below, and
the place is usually so chosen that it cannot well be looked at, at
all closely, from above.

Captain Unwin sent me an unusually beautiful specimen of the
nest of this species, taken early in May in the Agrore Valley—a
massive and perfect cup, with a cavity of 5 inches in diameter and
3 inches deep; the sides fully 2 inches thick; an almost solid mass
of fine roots (the finest towards the interior) externally intermingled with moss, so as to form, to all appearance, an integral
portion of the mossy bank on which it was placed. In the bottom
of the nest were interwoven a number of dead leaves, and the
whole interior was thinly lined with very fine grass-roots and moss.
In this case the nest had been placed on a tiny natural platform
and was a complete cup; but in another nest, also sent by Captain
Unwin, the cup, having been placed on the slope of a bank, wanted
(and this is the more common type) the inner one-third altogether,
the place of which was supplied by the bank-moss *in situ*. In this
case, although the cavity was only of the same size as that above
described, the outer face of the nest was fully 6 inches high, and
the wall of the nest between 3 and 3½ inches thick. The former
contained three much incubated, the latter four nearly fresh eggs.

A nest from Darjeeling which was taken on the 28th July, at an
elevation of about 3500 feet, from under a rock which partly overhung a stream, and contained two fresh eggs, was composed in
almost equal proportions of fine moss-roots and dead leaves with
scarcely a trace of moss. In this case the nest was entirely concealed from view, and no necessity, therefore, existed for coating
it externally with green moss to prevent its attracting attention.

Dr. Jerdon remarks :—" I have had its nest and eggs brought
me (at Darjeeling); the nest is a solid mass of moss, mixed with
earth and roots, of large size, and placed (as I was informed) under
an overhanging rock near a mountain-stream. The eggs were
three in number, and dull green, thickly overlaid with reddish
specks."

" In Kumaon," writes Mr. R. Thompson, " they breed from May
to July, along all the smaller hill-streams, from 1500 up to about
4500 feet. In the cold season it descends quite to the plains—I
mean the Sub-Himalayan plains. The nest is generally more or
less circular, 5 or 6 inches in diameter, composed of moss and mud
clinging to the roots of small aquatic plants or of the moss, and
lined with fine roots and sometimes hair. A deep well-watered
glen is usually chosen, and the nest is placed in some cleft or

between the ledges of some rock, often immediately overhanging some deep gloomy pool."

"On the 16th June," observes Captain Hutton, writing from Mussoorie, "I took two nests of this bird, each containing three eggs, and also another nest, containing three nearly-fledged young ones. The nest bears a strong resemblance to that of the *Geocichlæ*, but is much more solid, being composed of a thick bed of green moss externally, lined first with long black fibrous lichens and then with fine roots. Externally the nest is $3\frac{1}{2}$ inches deep, but within only $2\frac{1}{2}$ inches; the diameter about $4\frac{3}{4}$ inches, and the thickness of the outer or exposed side is 2 inches. The eggs are three in number, of a greenish-ashy colour, freckled with minute roseate specks, which become confluent and form a patch at the larger end. The elevation at which the nests were found was from 4000 to 4500 feet; but the bird is common, except during the breeding-season, at all elevations up to the snows, and in the winter it extends its range down into the Doon. In the breeding-season it is found chiefly in the glens, in the retired depths of which it constructs its nest; it never, like the Thrushes and *Geocichlæ*, builds in trees or bushes, but selects some high, towering, and almost inaccessible rock, forming the side of a deep glen, on the projecting ledges of which, or in the holes from which small boulders have fallen, it constructs its nest, and where, unless when assailed by man, it rears its young in safety, secure alike from the howling blast and the attack of wild animals. It is known to the natives by the name of 'Kaljet,' and to the Europeans as the 'Hill Blackbird.' The situation in which the nest is placed is quite unlike that of any other of our Hill-Thrushes with which I am acquainted. The bird itself is as often found in open rocky spots on the skirts of the forest as among the woods, loving to jump upon some stone or rocky pinnacle, from which it sends forth a sort of choking, chattering song, if such it can be called, or, with an up-jerk of the tail, hops away with a loud musical whistle, very much after the manner of the Blackbird (*M. vulgaris*)."

Sir E. C. Buck says:—"I found a nest at Huttoo, near Narkhunda, date 27th June, 1869, on an almost inaccessible crag overhanging a torrent. It contained three eggs, but two were broken by stones falling in climbing down to the nest. Nest not brought up; one egg secured and forwarded. I saw the bird well, and have no doubt as to its identity."

Writing from Dhurmsalla, Captain Cock informed me that he had obtained several nests in May in and about the neighbouring streams, up to an elevation of some 5000 feet. From Murree, Colonel C. H. T. Marshall remarks:—"Several nests found in June, near running streams, about 4000 feet up."

Dr. Stoliczka tells us that "it breeds at Chini and Sungnum at an elevation of between 9000 and 11,000 feet."

The eggs are typically of a very long oval shape, much pointed at one end, but more or less truncated varieties (if I may use the word) occur. They are the largest of our Indian Thrushes' eggs, and are larger than those of any European Thrush with which I am acquainted. Their coloration, too, is somewhat unique; a French grey, greyish-white, or pale-greenish ground, speckled or freckled with minute pink, pale purplish-pink, or pinkish-brown specks, in most cases thinly, in some instances pretty thickly, in some only towards the large end, in some pretty well all over. In the majority of the specimens there is, besides these minute specks, a cloudy, ill-defined, purplish-pink zone or cap at the large end. In some few there are also a few specks of bright yellowish brown. The eggs have scarcely any gloss.

In length they vary from 1·24 to 1·55 inch, and in breadth from 0·95 to 1·1 inch, but the average of fifty eggs is 1·42 by about 1·0 inch.

188. Myiophoneus eugenii, Hume. *The Burmese Whistling-Thrush.*

Myiophoneus eugenii, *Hume; Hume, Cat.* no. 343 bis.

Major C. T. Bingham contributes the following note to the 'Birds of British Burmah' regarding the nidification of this species in Tenasserim :—" On the 16th April I was crossing the Mehk-haneh stream, a feeder of the Meh-pa-leh, the largest tributary of the Thoungyeen river, near its source, where it is a mere mountain-torrent brawling over a bed of rocks strewed with great boulders. A small tree, drifted down by the last rains, had caught across two of these, and being jammed in by the force of the water, had half broken across, and now formed a sort of temporary V-shaped dam, against which pieces of wood, bark, leaves, and rubbish had collected, rising some six inches or so above the water, which found an exit below the broken tree. On this frail and tottering foundation was placed a round solid nest about 9 inches in diameter, made of green moss, and lined with fine black roots and fibres, in which lay four fresh eggs of a pale stone-colour, sparsely spotted, especially at the larger ends, with minute specks of reddish brown. Determined to find out to what bird they belonged, I sent my followers on and hid myself behind the trunk of a tree on the bank and watched, gun in hand. In about twenty minutes or so a pair of *Myiophoneus eugenii* came flitting up the stream and, alighting near the nest, sat for a time quietly. At last one hopped on the edge of the nest, and after a short inspection sat down over the eggs with a low chuckle. I then showed myself and, as the birds flew off, fired at the bird that had been on the nest, but unfortunately missed. I was satisfied, however, about the identity of the eggs and took them. In shape they are somewhat like those of *Pitta*, and measure 1·45 × 1·02, 1·50 × 1·02, 1·46 × 1·01, and 1·50 × 1·01."

189. **Myiophoneus horsfieldi**, Vigors.   *The Malabar Whistling-Thrush.*

Myiophonus horsfieldii, *Vig., Jerd. B. Ind.* ii, p. 499; *Hume, Rough Draft N. & E.* no. 342.

Mr. W. Davison says:—"The Malabar Whistling-Thrush (rather a misnomer, by the way) breeds on the slopes of the Nilghiris, never ascending higher than 6000 feet. The nest is always placed on some rock in a mountain torrent; it is a coarse and, for the size of the bird, a very large structure, and though I have never measured the nest, I should say that the total height was about 18 inches or more, and the greatest diameter about 18 inches. Exteriorly it is composed of roots, dead leaves, and decaying vegetation of all kinds; the egg-cavity, which is saucer-shaped and comparatively shallow, is coarsely lined with roots. It breeds during March and April."

Miss Cockburn says:—"A nest of this bird was found on the 22nd of March in a hole in a tree situated in a wood at a height of about 40 feet from the ground. Two bamboo ladders had to be tied together to reach it, for the tree had no branches except at the top. The nest consisted of a large quantity of sticks and dried roots of young trees, laid down in the form of a Blackbird's nest. The contents of it were three eggs. They were quite fresh, and the bird might have laid another. The poor birds (particularly the hen) showed great boldness and returned frequently to the nest, while a ladder was put up and a man ascended it."

Such a situation for the nest of *this* bird may seem incredible; but my friend Miss Cockburn is a most careful observer, and she sent me one of the eggs taken from this very nest, and it undoubtedly belonged to this species; moreover, there is no other bird on the Nilghiris that she, who has figured most beautifully all the Nilghiri birds, could possibly have mistaken for this species. At the same time, the situation in which she found the nest was altogether unusual and exceptional.

I now find that such a situation for the nest of this bird is not even very unusual. On the 3rd of July Miss Cockburn took another nest in a hole in a tree, about thirty feet from the ground, containing three fresh eggs, which she kindly sent me; and writing from the Wynaad Mr. J. Darling, jun., remarks that there this species commonly builds in holes in trees. He says:—"*July 22nd.* Nest found near Kythery, S. Wynaad, in a crevice of a log of a felled tree in a new clearing 11 feet from the ground. Nest built entirely of roots. The foundation was of roots from some swampy ground and had a good deal of mud about it. Another nest was in a hole of a dead tree 32 feet from the ground."

Mr. Frank Bourdillon writes from Travancore:—"Very common from the base to near the summit of the hills, frequenting alike jungle and open clearings, though generally found in the neighbourhood of some running stream; I have known this species to build on ledges of rock and in a hollow tree overhanging a stream, in either case constructing a rather loosely put together nest of roots and coarse fibre with a little green moss intermixed. The female lays two to four eggs, and both birds assist in the incubation."

Mr. T. Fulton Bourdillon records the finding of eggs on the following dates :—

"April 29, 1873. Two hard-set eggs.
"May 15, 1873. Three   ,,   ,,
"May 15, 1874. One fresh egg.
"May 30, 1874. Two slightly set eggs."

Col. Butler sent me a splendid nest of this species taken in the cliffs at Purandhur, 15 miles south of Poona. It was placed in the angle between two rocks; it measures in front 7 inches wide, and 1·5 in. high; posteriorly it slopes away into an obtuse angle fitting the crevice in which it was deposited; the cavity is 4 in. in diameter, perfectly circular, and 2·25 in depth. The compactness of the nest is such that it might be thrown about without being damaged. It is composed throughout of fine black roots, only a stray piece or two of light coloured grass being intermixed, and the whole basal portion is cemented together with mud.

He gives the following account of the mode in which he acquired it :—

"I got this nest in rather a singular way which is perhaps worth relating. At a dance last year in Karachi, in a short conversation I had with Colonel Renny, who was then commanding the Artillery in Sind, he mentioned that he had three Blue-winged Thrushes in his house that he had procured at Purandhur the year before. The following day I went over to his bungalow, and after inspecting them and satisfying myself of their identity, ascertained from him where the nest they were taken from was situated and the season at which it was found. Possessed with this information I wrote in May to the Staff Officer at Purandhur, and told him where and when the bird built and asked him if he would kindly assist me in procuring the eggs. In reply I received a very polite letter saying 'that he knew nothing about eggs or birds himself, but that he would be most happy to offer me any assistance in his power in procuring the eggs referred to, and that he would employ a shikarri to keep the hill-side that I had mentioned watched when the breeding-season arrived.' I wrote and thanked him, sending him at the same time a drill and blowpipe by post, with full instructions how to blow the eggs, in case he got any; and to my delight, at the end of July a bhangby parcel arrived one morning with the nest and eggs above described.

"Colonel Renny told me that the birds built on this cliff-side every monsoon."

Mr. E. Aitken has furnished me with the following note :—

"Of this bird I have seen two nests—one containing two hard-set eggs on April 29, 1872, situated in a hole in a tree overhanging a stream about 20 feet from the ground; the other containing three hard-set eggs on May 22nd, 1872, and situated on a ledge of rock in the bed of a stream; both the nests were rather coarsely made of roots. My brother says he has also found three other nests, two placed in holes of trees and the other on a rocky ledge, but

the nests were in every case near to running water. The bird
stays with us all the year, and is one of our commonest species.
Its clear whistle is always to be heard the first thing in the morning before the other birds get up, and during the violent rains of
the S.W. monsoon it seems almost the only bird which does not
lose heart at the incessant downpour. April and May appear to
be the breeding months."

Messrs. Davidson and Wenden remark:—"Scattered all over
the Deccan in suitable localities. W. got two nests, one on the
Bhore Ghât on 5th August, and one on the Thull Ghât on 17th of
same month. That on the Bhore Ghât was built on a ledge of
rock some 15 feet *in* from the face of a railway tunnel where 30 or
40 trains daily passed within a few feet of it. That on the Thull
Ghât was in a cutting at the *entrance* of a tunnel, and about the
same height above and from the rails as the one on the Bhore Ghât.
In both cases the eggs were much discoloured by the smoke from
engines, but on being washed, W. observed that one of the three eggs
in each nest was of a decidedly *greenish blue*, finely speckled and
splashed with pinky brown, while the others were of the *pale
salmon-pink*, as described in Mr. Hume's Rough Draft of ' Nests
and Eggs.' The male bird was sitting on one of the nests and
was shot. W. saw numerous other nests, some high up on cliffs,
beyond the reach of a 15-foot ladder. Two nests in holes in trees
were reported to him, but he could not go to examine them. The
nests were about 4 inches diameter by 2½ inches deep inside and
8 to 10 inches broad outside, and not more than 10 inches high.
The foundation portion contained a great deal of clay and earth,
which seemed to be necessary to secure the nests in positions so
exposed to the heavy gusts of wind which prevail on these
ghâts during the monsoon."

Mr. Rhodes W. Morgan, writing from South India, says:—" I
found the nest of this Thrush on the Seeghoor Ghaut of the Neilgherries. Mr. Davison was with me at the time; and the nest
being built on an open ledge of rock, we both sighted it at the
same moment; and I having managed to make better use of my
legs than my friend, was fortunate enough to secure it, and one
egg, which was of a pale flesh-colour, with a few faint spots and
blotches of claret towards the larger end. The nest was made of
leaves and moss mixed with clay, and lined with fine roots. The
dimensions of the egg are 1·3 inch in length by ·85 in breadth. It
was in May that I found this egg; but the nest had evidently
been deserted for some time; for the egg has a hole in its side,
through which the contents had escaped or been sucked by a snake
or some animal."

Dr. Jerdon says:—" I once procured its nest, placed under a
shelf of a rock on the Burliar stream, on the slope of the Nilghiris.
It was a large structure of roots, mixed with earth, moss, &c., and
contained three eggs of a pale salmon or reddish-fawn colour, with
many smallish brown spots;" and such is unquestionably the usual
situation of the nest.

The eggs of this species, which I have received from Kotagherry and other parts of the Nilghiris, are broad, nearly regular ovals, slightly compressed towards the lesser end; considerably elongated, and more or less spherical, and pyriform varieties occur. The shell is fine, and has a slight gloss; the ground-colour is pale salmon-pink or pinkish-white, occasionally greyish white. The whole egg is, as a rule, finely speckled, spotted, and splashed with pinkish brown or brownish pink. The markings, in most eggs, everywhere very fine, are often considerably more dense at the large end, where they are not unusually more or less underlaid by a pinkish cloud, with which they form an irregular ill-defined and inconspicuous cap.

At times more boldly and richly marked eggs are met with; one now before me is everywhere thickly streaked with dull pink, in places purplish, and over this is thinly but rather conspicuously spotted and irregularly blotched (the blotches being small however) with light burnt sienna-brown.

In length they vary from 1·18 to 1·48 inch, and in breadth from 0·92 to 1 inch.

191. Larvivora brunnea, Hodgs. *The Indian Blue Chat.*

Larvivora cyana, *Gould, Jerd. B. Ind.* ii, p. 145; *Hume, Rough Draft N. & E. no.* 507.

I have never obtained the nest of the Indian Blue Chat. Mr. Davison found it on the Nilghiris. He says:—"I really quite forget the details of that one egg which I brought you along with the skin of the parent, but it was taken in May on the Nilghiris. I remember very well another nest of this species, which I took in the latter end of March or the beginning of April in a shola or detached piece of jungle about 9 miles from Ootacamund.

"The nest was in a hole in the trunk of a small tree, about 5 feet from the ground, and was composed chiefly of moss, but mixed with dry leaves and twigs. It contained three young birds, apparently about four or five days old."

The late Mr. Mandelli sent me a nest of this species which was found at Lebong (elevation 5500 feet) on the 16th May. It contained three eggs, and was placed on the ground amongst grass on a bank made by the cutting of a hill-road. It is a broad shallow nest, composed exteriorly of vegetable fibre, scraps of dead leaves and tiny pieces of moss matted closely together, and is rather thickly lined with black and red hairs, amongst which one or two soft downy feathers are incorporated. The external diameter of the nest is about 4 inches, the height about 1·5, the cavity is about 2·75 inches in diameter, and rather less than 1 in depth.

Two eggs taken by Mr. Darling* are very elongated, somewhat cylindrical ovals, very obtuse at both ends. In both, the shell is fine, and has an appreciable though not brilliant gloss. In one, the

---

* I cannot find any account of the finding of the nest of this bird by Mr. Darling amongst Mr. Hume's notes.—ED.

ground is a pale delicate clay-brown, and the markings consist only of a zone about 0.2 wide round the large end of densely set dull brownish-red specks, and a few similar specks inside the zone only. In the other, the ground has a light greenish tinge, the zone is less marked and merges in a dull brownish-red mottled cap, and a faint marbling, of a paler shade of the cap, is scattered here and there over the whole surface of the egg. They measure 1 by 0·65 and 0·98 by 0·65.

The egg taken by Mr. Davison is an elongated, slightly pyriform oval. The shell is moderately fine, but with only a very slight gloss. The ground-colour is a pale slightly greyish green, and the whole egg is thickly (most thickly so about the large end, where the markings are almost perfectly confluent) mottled and streaked with pale brownish red. It measures 0·98 by 0·67.

193. Brachypteryx albiventris (Fairbank). *The White-bellied Short-wing.*

Callene albiventris, *Fairb.*, *Hume*, *Rough Draft N. & E.* no. 339 bis.

The Rev. S. B. Fairbank, to whom I have owed much useful information and many valuable specimens, kindly sent me the subjoined account of the nidification of the White-bellied Short-wing in the Pulney Hills at an elevation of about 6500 feet :—

"In April, I found a nest in a hole in the side of the trunk of a large tree some 2 feet from the ground. The hole was just large enough for the nest, and was lined with fine roots. I surprised the bird on her nest several times. There were two eggs in the nest when I first found it that were 'hard-set.' A month afterwards she laid two more in the same place, and I took them in good condition. One egg measures 0·9 by 0·68 inch, and another 0·94 by 0·68 inch. The ground-colour is grey, with a tinge of green, and it is thickly covered with small spots of bistre."

Mr. Blanford, who saw the eggs, which I never did, describes them (and by analogy, I should infer more correctly) as "of an olive-brown colour, darker at the larger end, measuring 0·93 by 0·63 inch."

An egg of this species sent me by Dr. Fairbank, measuring 0·93 by 0·66, is a somewhat elongated oval, slightly pointed towards the small end. The shell is fine and fairly glossy; the ground-colour, so far as this is discernible, is greyish green, but it is so thickly clouded and mottled all over with a warm brown, that but little of the ground-colour is anywhere traceable, and the general result when the egg is looked at from a short distance is that of a nearly uniform olive-brown.

Captain Horace Terry also found the nest of this bird on the Pulney Hills. He says :—"I met with it a few times in the big *shola* at Kodikanal, and got two nests, each with two fresh eggs ; the first on the 7th June in a hole in a tree between 4 and 5 feet from the ground, a deep cup of green moss ; the other, in a hole in the bank of a path running through the *shola* was of green moss and a few fine fern-roots. Inside 1·75 inch deep and 2·5 inches across; outside a shapeless mass of moss filling up the hole it was built in. The nest was very conspicuous to any one passing by."

194. Brachypteryx rufiventris (Blyth). *The Rufous-bellied Short-wing.*

Callene rufiventris, *Blyth, Jerd. B. Ind.* i, p. 496; *Hume, Rough Draft N. & E.* no. 339.

I have been favoured with nests of the Rufous-bellied Short-wing by Mr. Carter, who took them from holes or depressions of banks in the Nilghiris in April and May. They closely resemble nests of *Niltava macgrigoriæ* from Darjeeling. They are soft masses of green moss, some 4 or 5 inches in diameter externally, with more or less of a depression towards one side, lined with very fine dark moss-roots. This depression may average about 2½ inches across and ¾ inch in depth; but they vary a good deal. Mr. Carter says:—" I have found the nests of this species about Conoor in May, in holes of banks, on roads running through thick *sholus* (i. e. jungles not amounting to forests). The nests are of moss, shallow, lined with fine root-fibres, the cavity about 3·5 inches in diameter. They lay two eggs, pale olive, shading into a decided brownish red at the larger end. The old birds are very shy in returning to the nest when watched; indeed, they are always shy, hiding in the brushwood of jungles or amongst fallen timber, along which they almost creep."

Mr. Davison informs me that " this species breeds on the Nilghiris from about 5500 feet to about 7000 during April and May, building in holes of trees, crevices of rocks, &c., seldom at any great elevation above the ground. The nest is composed of moss, lined with moss and fern-roots. Two or three eggs are laid."

The few eggs I possess, which I owe to Messrs. Carter and Davison, and which were taken by them in the Nilghiris, have a pale olive-brown ground with, at the large end, an ill-defined mottled reddish-brown cap. In some specimens the mottling extends more or less over the whole egg, though always most dense about the larger end. Though much larger and of a more elongated shape, they not a little resemble some specimens of the eggs of *Pratincola indica* that I possess. In shape they are long ovals, recalling in that respect those of *Myiophoneus temmincki*; they have less gloss than the eggs of most of the Thrushes.

In length they vary from 0·97 to 1·02 inch, and in breadth from 0·65 to 0·69 inch.

197. Drymochares cruralis (Blyth). *The White-browed Short-wing.*

Brachypteryx cruralis (*Bl.*), *Jerd. B. Ind.* i, p. 495; *Hume, Rough Draft N. & E.* no. 338.

According to Mr. Hodgson's notes and drawings, the White-

browed Short-wing breeds in April and May. It constructs its
nest a foot or so above the ground amongst grass and creeping-
plants at the base of trunks of trees; it is composed of moss and
moss-roots, is somewhat globular in shape, and is firmly attached
to the creepers; dried bamboo-leaves and pieces of fern are here
and there fixed to the exterior, and the nest is lined with hair-like
fibres; the entrance is at one side and circular. One nest measured
7 inches in height, 5·5 in width, and 3·38 from front to back.
The aperture was 2 inches in diameter. The eggs (four in number,
or at times three) are pure white, broad ovals, pointed at one end,
measuring 0·9 by 0·65 inch. This species breeds in the central
regions of Nepal and in the neighbourhood of Darjeeling.

Three nests of this species found early in June in Sikhim and
Nepal, at elevations of 5000 to 8000 feet, contained respectively
2, 3, and 4 fresh eggs. They were all placed in brushwood at 2 to
3 feet above the ground, and they are all precisely similar, being
rather massive shallow cups, composed of very fine black roots
firmly felted together, and with a few dead leaves or scraps of moss
in most of them incorporated in one portion or other of the outer
surface. The nests are about 4 inches in diameter and 2 in
height; the cavity is about 2 inches in diameter and 1 in depth;
but, owing to the positions in which they are placed, they are often
more or less irregularly shaped.

Mr. Mandelli obtained three eggs which he considers to belong
to this species, on the 3rd June, near Darjeeling. I rather question the authenticity of these eggs. They are pure white and
devoid of gloss, moderately elongated ovals, only slightly compressed towards the smaller end. They vary from 0·83 to 0·91 in
length and from 0·61 to 0·64 in breadth.

198. **Drymochares nepalensis** (Hodgs.).   *The Nepal Short-wing.*

Brachypteryx nipalensis, *Hodgs., Jerd. B. Ind.* i, p. 494.

From Sikhim Mr. Gammie writes:—"A nest taken by me on the
15th of June at 5000 feet, close to a large forest, contained three
slightly-set eggs. It was placed on the moss-covered trunk of a
fallen tree, and was hooded, with an entrance at the side; rather
neatly made of dry leaves with an outer covering of green moss,
and an inner lining of skeletonized leaves and black fibrous roots.
Externally it measures 5 inches in height by about the same in
width; internally 3 inches high by 2·4 across. The entrance was
2·3 in diameter. The front of the egg-cavity is but slightly depressed below the entrance, gradually sloping backwards to the
depth of nearly an inch."

All the nests of this species that I have seen were of the same
type, more or less globular, more or less hooded or domed, according
to the situation in which they were placed, composed of dry flags
and dead and more or less skeleton leaves, bound together with a
little vegetable fibre and some moss, but chiefly with fine black
fibrous roots, with which the entire cavity is densely lined, inside

which again is a coating of more skeleton leaves; they measure exteriorly 4 or 5 inches in diameter, and the cavities are a little above 2 by 2·5 inches in diameter.

Mr. Mandelli found two of these nests at Lebong (elevation 5500 feet), near Darjeeling, on the 8th July. One contained three fresh eggs, the other three slightly incubated ones. They were about 12 yards apart, in a very shady damp glen, in very dense underwood, to the stems of which they were attached in a standing position about 3 feet from the ground. The entrance was on one side in both cases.

The eggs of this species obtained by Mr. Gammie belong to the same type as those of *Brachypteryx rufiventris* and *B. albiventris*. In shape they are moderately elongated, rather regular ovals, somewhat obtuse at both ends. The shell is fine and compact, and very smooth to the touch, but they have not much gloss. The ground is a pale olive stone-colour, and they are very minutely freckled and mottled, most densely at the large end, with pale, very slightly reddish brown; the freckling is excessively minute and fine.

Two eggs measured 0·8 and 0·82 in length by 0·6 in breadth.

200. Elaphrornis palliseri (Blyth). *The Ceylon Short-wing.*

Brachypteryx palliseri, *Bl.*, *Hume, Cat.* no. 338 bis.

Colonel Legge, writing in his 'Birds of Ceylon,' says:—" Mr. Bligh found a nest at Nuwara Eliva in April 1870; it was placed in a thick cluster of branches on the top of a somewhat densely-foliaged small bush, which stood in a rather open space near the foot of a large tree; it was in shape a deep cup, composed of greenish moss, lined with fibrous roots and the hair-like appendages of the green moss which festoons the trees in such abundance at that elevation. It contained three young ones, plumaged exactly like their parents, who kept churring in the thick bushes close by, but would not show themselves much."

201. Tesia cyaniventris, Hodgs. *The Slaty-bellied Short-wing.*

Tesia cyaniventer, *Hodgs.*, *Jerd. B. Ind.* i, p. 487; *Hume, Rough Draft N. & E.* no. 328.

According to Mr. Hodgson's notes, the Slaty-bellied Short-wing breeds much like the next species. It constructs a huge globular nest of green moss and black moss-roots, which it fixes in any dense dry shrub or clump of shoots, many of which it incorporates in the walls of the nest. The nest measures externally about 7 inches in height and 5 inches in width; it has a circular aperture on one side, a little above the middle, about 2 inches in diameter, and it is placed at a height of one or two feet from the ground. Three or four eggs are laid; these are figured as rather broad ovals, somewhat pointed towards one end, with a whitish ground, profusely speckled and spotted, especially towards the large end,

9*

where the markings are nearly confluent, with bright red, and measuring 0·72 by 0·54 inch.

202. Oligura castaneicoronata (Burt.). *The Chestnut-headed Short-wing.*

Tesia castaneo-coronata (*Burt.*), *Jerd. B. Ind.* i, p. 487 ; *Hume, Rough Draft N. & E.* no. 327.

According to Mr. Hodgson's notes and figures, the Chestnut-headed Short-wing builds a large globular nest, more or less egg-shaped, some 6 inches high and 4 in breadth, composed of moss-roots and fibres, and lined with feathers, and with a circular aperture in the middle of one side about 1·5 inch in diameter. The nest is placed in some clump of shoots or thick bush (the twigs of which are more or less incorporated in the sides of the nest) at a height of 1 or 2 feet from the ground. The birds lay in April and May three or four eggs, which are figured as moderately broad ovals, somewhat pointed at one end, reddish (apparently something like a Prinia's, though this seems incredible), and measuring 0·66 by 0·48 inch.

Dr. Jerdon says :—" A nest made chiefly of moss, with four small white eggs, was brought me as the nest of this bird. It was of the ordinary shape, rather loosely put together, and the walls of great thickness. It was taken from the ground on a steep bank near the stump of a tree."

The three eggs in my museum supposed to belong to this species pertained to this nest, and are excessively tiny, somewhat oval eggs of a pure, dull, glossless unspotted white, very unlike our English Wren's egg and certainly not one half the size. Dr. Jerdon was not quite certain to which species of *Tesia* these eggs belonged, and I therefore only record this " *quantum valeat.*" They measure 0·55 and 0·6 inch in length by 0·4, 0·42, and 0·45 inch in breadth. I am inclined to believe that both nest and eggs belonged to *Pnoepyga pusilla*, Hodgs.

Subfamily SIBIINÆ.

203. Sibia picaoides, Hodgs. *The Long-tailed Sibia.*

Sibia picaoides, *Hodgs., Jerd. B. Ind.* ii, p. 55 ; *Hume, Rough Draft N. & E.* no. 430.

Mr. Gammie obtained a nest of the Long-tailed Sibia from the top of a tall tree, situated at an elevation of about 4000 feet, in the neighbourhood of Rungbee, near Darjeeling. This was on the 17th June, and the nest contained five fresh eggs. The nest is as perplexing as are the eggs ; for the nest is that of a Bulbul, the eggs those of a Shrike or Minivet. The nest is a deep compact cup, about 4½ inches in diameter and 2¾ inches in depth. The egg-cavity is 3 inches across and fully 1¾ inch in depth. Interiorly the nest is composed of excessively fine grass-stems very firmly

interwoven; externally of the stems of some herbaceous plant, a
Chenopod, to which the dry blossoms are still attached, intermingled
with coarse grass, a single dead leaf, and one or two broad grass-
blades more or less broken up into fibres.

The eggs, for the authenticity of which Mr. Gammie positively
vouches, are very unlike what might have been expected. They
are absolutely Shrike's eggs—broad ovals, pointed towards one
end, with a slight gloss, the ground a slightly greyish white, with
a good many small spots and specks of pale yellowish brown and
dingy purple, chiefly confined to a large irregular zone towards the
larger end. They vary in length from 0·86 to 0·93, and in breadth
from 0·7 to 0·73.

204. Lioptila capistrata (Vigors). *The Black-headed Sibia.*

Sibia capistrata (*Vig.*), *Jerd. B. Ind.* ii, p. 54; *Hume, Rough Draft
N. & E.* no. 429.

The Black-headed Sibia lays throughout the Himalayas from
Afghanistan to Bhootan, at elevations of from 5000 to 7000 feet.

It lays during May and June, and perhaps part of July, for I
find that on the 11th of July I found a nest of this species a little
below the lake at Nynee Tal, on the Jewli Road, containing two
young chicks apparently not a day old.

They build on the outskirts of forests, constructing their nests
towards the ends of branches, at heights of from 10 to 50 feet
from the ground. The nest is a neat cup, some 4 or 5 inches in
diameter and perhaps 3 inches in height, composed chiefly of moss
and lined with black moss-roots and fibres. In some of the nests
that I have preserved a good deal of grass-leaves and scraps of
lichen are incorporated in the moss. The cavity is deep, from $2\frac{1}{2}$
to 3 inches in diameter and not much less than 2 inches in depth.

They lay two or three eggs; not more, so far as I yet know.

From Murree, Colonel C. H. T. Marshall tells us that "the egg
of this bird was, we believe, previously unknown, and it was a
mere chance that we found the whereabouts of their nests, as they
breed high up in the spruce firs at the outer end of a bough.
The nest is neatly made of moss, lined with stalks of the maiden-
hair fern. The eggs are pale blue, spotted and blotched with pale
and reddish brown. They are ·95 in length and ·7 in breadth.
This species breeds in June, about 7000 feet up."

Nearly twenty years prior to this, however, Captain Hutton had
remarked:—"At Mussoorie this bird remains at an elevation of
7000 feet throughout the year, but I never saw it under 6500
feet. Its loud ringing note of *titteree-titteree tweëyo*, quickly
repeated, may constantly be heard on wooded banks during summer. It breeds in May, making a neat nest of coarse dry grasses
as a foundation, covered laterally with green moss and wool and
lined with fine roots. The number of eggs I did not ascertain,
as the nest was destroyed when only one egg had been deposited, but

the colour is pale bluish white, freckled with rufous. The nest was placed on a branch of a plum-tree in the Botanical Garden, Mussoorie."

Captain Cock says that he "found this species breeding at Murree, at 6000 feet elevation.

"I took my first nest on the 5th June.

"It builds near the tops of the highest pines, and unless seen building its nest with the glasses, it is impossible to find the nest with the unaided eye.

"The nest is placed on the outer extremity of an upper bough in a pine-tree; is constructed of moss lined with stalks of the maiden-hair fern. Three eggs is the largest number I ever found. The eggs are light greenish white, with rusty spots and blotches principally at the larger end."

From Nynee Tal Colonel G. F. L. Marshall writes :—" This species builds in trees and bushes. The only nest I examined personally was a very compact and thick cup-shaped structure of moss, grass, and roots, lined with grass, and placed amongst the outer twigs of a blackberry bush overhanging a cliff. It was ready for the eggs on the 23rd May. It was found at Nynee Tal on Agar Pata, about 7000 feet above the sea."

From Sikhim Mr. Gammie writes :—" I have only myself taken two nests of this common species. I found both of them the same day (the 21st May), in the Chinchona reserves, at an elevation of about 5000 feet. Both nests were in the forest, built on the outer branches of trees, at heights the one of 15, the other of 40 feet from the ground. The nests were cup-shaped, and very neatly made of moss, leaves and fibres, and lined with black fibres. One measured externally 4·6 in diameter by 2·75 in height, and internally 2·4 in diameter and 1·7 in depth. One nest contained two fresh, the other two hard-set eggs; so perhaps two is the normal number, though the natives say that they lay three. As might be expected from the bird's habit of feeding on the insects on moss-covered trees in moist forests, the nests were in forest by the sides of streams."

The eggs are rather broad, slightly pyriform ovals, often a good deal pulled out as it were at the small end. The shell is fine, but almost entirely devoid of gloss. The ground-colour is a pale greenish white or very pale bluish green. The markings are various and complicated: first there are usually a few large, irregular, moderately dark brownish-red spots and splashes; then there are a very few, very dark, reddish-brown hair-lines, such as one finds on Buntings' eggs; then there is a good deal of clouding and smudging here and there of pale, dingy purplish or brownish red (all these markings are most numerous towards the large end); and then besides these, and almost entirely confined to the large end, are a few pale purple specks and spots. Sometimes the markings are almost wholly confined to the thicker end of the egg. Of course the eggs vary somewhat, and in some specimens the characteristic Bunting-like hair-lines are almost wholly wanting.

The eggs vary in length from 0·95 to 1·0, and in breadth from 0·66 to 0·72.

205. Lioptila gracilis (McClell.). *The Grey Sibia.*

Malacias gracilis (*McClell.*), *Hume, Cat.* no. 429 bis.

Colonel Godwin-Austen is, I believe, the only ornithologist who has as yet secured the nest and eggs of the Grey Sibia. He says :—
" In the pine forest that covers the slopes of the hills descending into the Umian valley in Assam, one of my men marked a nest on June 25th ; I proceeded to the spot soon after I had heard of it, and on coming up to the tree, a pine, saw the female fly off out of the head of it. But the nest was so well hidden by the boughs of the fir, that it was quite invisible from below. The bird after a short time came back, and then I saw it was *Sibia gracilis*; but it was very shy and seeing us went off again, and hung about the trees at a distance of some 50 yards; while thus waiting, some four or five others were also seen. The female, however, would not venture back, and I sent one of my Goorkhas up, to cut off the head of the fir, nest and all, first taking out the eggs. It contained three, of a pale sea-green, with ash-brown streakings and blotchings all over.

" The nest was constructed of dry grass, moss, and rootlets, and the green spinules of the fir were worked into it, fixing it most firmly in its place in the crown of the pine where it was much forked."

206. Lioptila melanoleuca (Bl.). *Tickell's Sibia.*

Malacias melanoleucus (*Bl.*), *Hume, Cat.* no. 429 quart.

Mr. W. Davison was fortunate enough to secure a nest of this Sibia on Muleyit mountain in Tenasserim. He says :—" I secured a nest of this species on the 21st of February, containing two spotless pale blue eggs slightly incubated. The nest, a deep compactly woven cup, was placed about 40 feet from the ground, in the fork of one of the smaller branches of a high tree growing on the edge of a deep ravine.

" The egg-cavity of the nest is lined with fern-roots, fibres and fine grass-stems ; outside this is a thick coating of dried bamboo-leaves and coarse grass, and outside this again is a thick irregular coating of green moss, dried leaves, and coarse fibres and fern-roots.

" Externally the nest measures about 5 inches in height, and nearly the same in external diameter at the top.

" The egg-cavity measures 1·7 deep by 2·7 across.

" The eggs, a pale spotless blue, measure 0·95 and 0·98 in length by 0·66 and 0·68 in breadth."

211. **Actinodura egertoni**, Gould. *The Rufous Bar-wing.*

Actinodura egertoni, *Gould, Jerd. B. Ind.* ii, p. 52; *Hume, Rough Draft N. & E.* no. 427.

There is no figure of the Rufous Bar-wing's nest or eggs amongst the original drawings of Mr. Hodgson now in my custody, but in the British Museum series there appears to be, since Mr. Blyth remarks :—" Mr. Hodgson figures the nest of this bird like that of an English Redbreast, with pinkish-white eggs."

From Sikhim Mr. Gammie writes :—" On the 27th April I took a nest of this Bar-wing in a large forest at an elevation of about 5000 feet. It was placed about 20 feet from the ground, in a leafy tree, between several upright shoots, to which it was firmly attached. It is cup-shaped, mainly composed of dry leaves held together by slender climber-stems, and lined with dark-coloured fibrous roots. A few strings of green moss were twined round the outside to assist in concealment. Externally it measures 4·2 inches wide by 4 deep; internally 2·8 wide and 2·4 deep. It contained but two slightly-set eggs.

" I killed the female off the nest."

Several nests have been obtained and sent me by Messrs. Gammie and Mandelli. One was taken on the 4th May by Mr. Mandelli, at Lebong, at an elevation of 5500 feet, which contained three fresh eggs ; this was placed on the branches of a small tree, in the midst of dense brushwood, at a height of about 4 feet from the ground.

Another, taken in a similar situation at the same place on the 22nd May, contained two fresh eggs, and was at a height of about 12 feet from the ground.

These nests vary just in the same way as do those of *Trochalopterum nigrimentum* ; some show only a sprig or two of moss about them, while others have a complete coating of green moss. They are cup-shaped, some deeper, some shallower ; the chief material of the nest seems to be usually dry leaves. One before me is composed entirely of some *Polypodium*, on which the seed-spores are all fully developed; in another, bamboo-leaves have been chiefly used ; these are all held together in their places by black fibrous roots ; occasionally towards the upper margin a few creeper-tendrils are intermingled. The whole cavity is lined more or less thickly, and the lip of the cup all round is usually finished off with these same black fibrous roots ; and then outside all moss and selaginella are applied according to the taste of the bird and, probably, the situation—a few sprigs or a complete coating, as the case may be.

Two eggs of this species sent me by Mr. Gammie are regular, slightly elongated ovals, with very thin and fragile shells, and fairly but not highly glossy. The ground is a delicate pale sea-green, and they are profusely blotched, spotted, and marked with curious hieroglyphic-like figures of a sort of umber-brown ; while about the larger end numerous spots and streaks of pale lilac occur.

These eggs measure 0·98 in length by 0·65 and 0·68 in breadth.

Other eggs obtained by Mr. Mandelli early in June are quite of the same type, but somewhat shorter, measuring 0·85 and 0·93 in length by 0·68 and 0·7 in breadth. But the markings are rather more smudgy and rather paler, and there are fewer of the hair-like streaks and hieroglyphics.

213. Ixops nepalensis (Hodgs.). *The Hoary Bar-wing.*

Actinodura nipalensis (*Hodgs.*), *Jerd. B. Ind.* ii, p. 53; *Hume, Rough Draft N. & E.* no. 428.

The Hoary Bar-wing is said in Mr. Hodgson's notes to breed from April to June in Sikhim and the central region of Nepal up to an elevation of 4000 or 6000 feet. The nest is placed in holes, in crevices between rocks and stones; is circular and saucer-shaped. One measured externally 3·62 in diameter by 2 inches in height; the cavity measured 2·5 in diameter and 1·37 in depth. The nest is composed of fine twigs, grass, and fibres, and externally adorned with little pieces of lichen, and internally lined with fine moss-roots. The birds are said to lay from three to four eggs, which are not described, but they are figured as pinky white, about 0·85 in length and 0·55 in width. Mr. Blyth, however, remarks :— " One of Mr. Hodgson's drawings represents a white egg with ferruginous spots, disposed much as in that of *Merula vulgaris.*"

Clearly there is some mistake here. Most of the drawings I have are the originals, taken from the fresh specimens when they were obtained, with Mr. Hodgson's own notes, on the reverse, of the dates on and places at which he took or obtained the eggs, nests, and birds figured, with often a description and dimensions of the two former, and invariably full dimensions of the latter. On the other hand, the drawings in the British Museum are mostly more finished and artistic *copies* of these originals; so how the spots got on to the eggs of the British-Museum drawing I cannot say; there is no trace of such in mine.

219. Siva strigula, Hodgs. *The Stripe-throated Siva.*

Siva strigula, *Hodgs., Jerd. B. Ind.* ii, p. 252; *Hume, Rough Draft N. & E.* no. 616.

The nest of the Stripe-throated Siva is placed, according to Mr. Hodgson, in the slender fork of a tree at no great elevation from the ground. It is composed of moss and moss-roots, inter-mingled with dry bamboo-leaves, and woven into a broad compact cup-shaped nest. One such nest, taken on the 27th May, with three eggs in it, measured exteriorly 4·25 in diameter and 3 inches in height, with a cavity (thickly lined with cow's hair) about 2·5 in diameter and 2·25 in depth. The birds lay in May and June. The eggs are three or sometimes four in number; they are pale greenish blue or bluish green, and vary in length from 0·8 to 0·9, and in breadth from 0·6 to 0·65, and are, some thickly, some thinly, speckled and freckled, usually most densely towards the large end, with red or brownish red. His nests were taken both in Sikhim and Nepal.

221. Siva cyanuroptera, Hodgs. *The Blue-winged Siva.*

Siva cyanouroptera, *Hodgs., Jerd. B. Ind.* ii, p. 253 ; *Hume, Rough Draft N. & E.* no. 617.

The Blue-winged Siva breeds, according to Mr. Hodgson's notes, in the central regions of Nepal, and in the neighbourhood of Darjeeling, in May and June. The nest is placed in trees, at no great elevation above the ground, and is wedged in where three or four slender twigs make a convenient fork. A nest taken on the 2nd June was a large compact cup, measuring exteriorly 4·75 in diameter and 3·75 in height, and having a cavity 2·6 in diameter and 1·87 in depth. It was composed of fine stems of grass, dry leaves, moss, and moss-roots, bound together with pieces of creepers, roots, and vegetable fibres, and closely lined with fine grass-roots. They lay from three to four eggs, which are figured as moderately broad ovals, considerably pointed towards the small end, 0·85 in length by 0·6 in width, having a pale greenish ground pretty thickly speckled and spotted, especially on the broader half of the egg, with a kind of brownish brick-red.

Mr. Mandelli found a nest of this species at Lebong (elevation 5500 feet) on the 28th April. It contained four fresh eggs ; it was placed in a fork of a horizontal branch of a small tree at a height of only 3 feet from the ground. The nest is, for the size of the bird, a large cup, externally entirely composed of green moss firmly felted together. This outer shell of moss is thickly lined with the dead leaves of a *Polypodium*, and this again is thinly lined with fine grass. The nest was about 4 inches in diameter, and 2·5 in height externally ; the cavity was about 2·5 broad and 1·5 deep.

The nests of this species are very beautiful cups, very compact and firm, sometimes wedged into a fork, but more commonly suspended between two or three twigs, or sometimes attached by one side only to a single twig. They are placed at heights of from 4 to 10 feet from the ground in the branches of slender trees, and are usually carefully concealed, places completely encircled by creepers being very frequently chosen. The chief materials of the nest are dead leaves, sometimes those of the bamboo, but more generally those of trees ; but little of this is seen, as the exterior is generally coated with moss, and the interior is lined first with excessively fine grass, and then more or less thinly with black buffalo- or horse-hairs. The cups are about 3 inches in diameter and 2 in height externally, the cavities barely 2 in diameter and perhaps 1·5 in depth ; but they vary somewhat in size and shape according to the situation in which they are placed and the manner in which they are attached, some being considerably broader and shallower, and some rather deeper.

Eggs of this species sent me from Mr. Mandelli, which were obtained by him in the neighbourhood of Darjeeling, are decidedly elongated ovals, fairly glossy, and with a pale slightly greenish-blue ground. A number of minute red or brownish-red or yellowish-brown specks and spots occur about the large end, sometimes irregularly scattered, sometimes more or less gathered into an imperfect zone. The rest of the egg is either spotless or exhibits only a few tiny specks and spots. The eggs measure 0·75 and 0·76 by 0·51 and 0·52.

223. **Yuhina gularis**, Hodgs. *The Stripe-throated Yuhina.*

Yuhina gularis, *Hodgs., Jerd. B. Ind.* ii, p. 261; *Hume, Rough Draft N. & E.* no. 626.

The Stripe-throated Yuhina breeds, according to Mr. Hodgson's notes, from April to July, building a large massive nest of moss, lined with moss-roots, and wedged into a fork of a branch or between ledges of rocks, more or less globular in shape, and with a circular aperture near the top towards one side. A nest taken on the 19th June, near Darjeeling, was quite egg-shaped, the long diameter being perpendicular to the ground, and measured 6 inches in height and 4 inches in breadth, the aperture, 2 inches in diameter, being well above the middle of the nest; the cavity was lined with fine moss-roots. The eggs are figured as rather elongated ovals, 0·8 by 0·56, with a pale buffy or *café au lait* ground-colour, thickly spotted with red or brownish red, the markings forming a confluent zone about the large end.

225. **Yuhina nigrimentum** (Hodgs.). *The Black-chinned Yuhina.*

Yuhina nigrimentum (*Hodgs.), Jerd. B. Ind.* ii, p. 262; *Hume, Rough Draft N. & E.* no. 628.

A nest of the Black-chinned Yuhina, taken by Mr. Gammie on the 17th June below Rungbee, at an elevation of about 3500 feet, was placed in a large tree, at a height of about 10 feet from the ground, and contained four hard-set eggs. It is a mere pad, below of moss, mingled with a little wool and moss-roots, and above, that is to say the surface where the eggs repose, of excessively fine grass-roots.

Dr. Jerdon says:—"A nest was once brought me which was declared to belong to this species; it was a very small neat fabric, of ordinary shape, made with moss and grass, and contained three small pure white eggs. The rarity of the bird makes me doubt if the nest really belonged to it."

The eggs are tiny little elongated ovals, pure white, and absolutely glossless.

Two sent me by Mr. Gammie measure 0·58 by 0·42 and 0·57 by 0·43.

226. Zosterops palpebrosa (Temm.). *The Indian White-eye.*

Zosterops palpebrosus (*Temm.*), *Jerd. B. Ind.* ii, p. 265; *Hume, Rough Draft N. & E.* no. 631.

The Indian White-eye, or White-eyed Tit as Jerdon terms it, breeds almost throughout the Indian Empire, sparingly in the hotter and more arid plains, abundantly in the Nilghiris and other ranges of the Peninsula to their very summits, and in the Himalayas to an elevation of 5000 or 6000 feet.

The breeding-season extends in different localities from January to September, but I think that everywhere April is the month in which most eggs are to be met with.

Sometimes they have two broods; whether this is always the case I do not know.

The nest is placed almost indifferently at any elevation. I have taken one from amongst the topmost twigs of a huge mohwa tree (*Bassia latifolia*) fully 60 feet high, and I have found them in a tiny bush not a foot off the soil. Still I think that perhaps the majority build at low elevations, say between 2 and 6 feet from the ground.

The nest is always a soft, delicate little cup, sometimes very shallow, sometimes very deep, as a rule suspended between two twigs like a miniature Oriole's nest, but on rare occasions propped in a fork. The nest varies much in size and in the materials with which it is composed.

Fine grass and roots, tow, and a variety of vegetable fibres, thread, floss silk, and cobwebs are all made use of to bind the little nest together and attach it to the twigs whence it depends. Grass again, moss, vegetable fibre, seed-down, silk, cotton, lichen, roots and the like are used in the body of the nest, which is lined with silky down, hair, moss, and fern-roots, or even silk, while at times tiny silvery cocoons or scraps of rich-coloured lichen are affixed as ornaments to the exterior.

One nest before me is a very perfect and deep cup, hung between two twigs of a mohwa tree and almost entirely hidden by the surrounding leaves. The exterior diameter of the nest is 2½ inches, and the depth 2 inches. The egg-cavity measures scarcely more than 1½ inch across and very nearly as much in depth. It is composed of very fine grass-stems and is thinly coated exteriorly with cobwebs, by which also it is firmly secured to the suspending twigs, and externally numerous small cocoons and sundry pieces of vegetable down are plastered on to the nest. Another nest, hung between two slender twigs of a mango tree, is a shallow cup some 2½ inches in diameter, and not above an inch in depth externally. The egg-cavity measures at most 1½ inch across by three-fourths of an inch in depth. The nest is composed of fine tow-like vegetable fibres and thread, by which it is attached to the twigs, a little grass-down being blended in the mass, and the cavity being very sparsely lined with very fine grass-stems. In another nest, somewhat larger than the last described, the nest is made of moss slightly tacked together with cobwebs and lined with fine grass-fibres. Another nest, a very regular shallow cup, with an egg-cavity 2 inches in diameter and an inch in depth, is composed almost entirely of the soft silky down of the *Calatropis gigantea,* rather thickly lined with very fine hair-like grass, and very thinly coated exteriorly with a little of this same grass, moss, and thread. Another, with a similar-sized cavity, but nearly three-fourths of an inch thick everywhere, is externally a mass of moss, moss-roots, and very fine lichen, and is lined entirely with very soft and brilliantly white satin-like vegetable down. Another, with about the same-sized cavity, but the walls of which are scarcely one-fourth of an inch in thickness, is composed *entirely* of this satiny down, thinly coated exteriorly and interiorly with excessively fine moss-roots (roots so fine that most of them are much thinner than human hair); a few black horsehairs, which look coarse and thick beside the other materials of the nest, are twisted round and round in the interior of the egg-cavity. Other nests might be made entirely of tow, so far as their appearance goes; and in fact with a very large series before me, no two seem to be constructed of the same materials.

I have nests before me now, taken in September, March, June, and August, all of which when found contained eggs.

Two is certainly the normal number of the eggs; about one fifth of the nests I have seen contained three, and once only I found four.

From Murree Colonel C. H. T. Marshall informs us that he took the eggs in June at an elevation of about 6000 feet.

Colonel G. F. L. Marshall says:—"I have taken eggs of this species at Cawnpore in the middle of June. I found six nests, five of which were in neem-trees. I also found the nest in Naini Tal at 7000 feet above the sea, with young in the middle of June; one only of all the nests I have seen was lined, and that was lined with feathers; they were, as a rule, about eight feet from the ground, but one was nearly forty feet up."

Capt. Hutton gives a very full account of the nidification of this species. He says:—"These beautiful little birds are exceedingly common at Mussoorie, at an elevation of about 5000 feet, during summer, but I never saw them much higher. They arrive from the plains about the middle of April, on the 17th of which month I saw a pair commence building in a thick bush of *Hibiscus,* and on the 27th of the same month the nest contained three small eggs hard-set. I subsequently took a second from a similar bush, and several from the drooping branches of oak-trees, to the twigs of which they were fastened. It is not placed on a branch, but is suspended between two thin twigs, to which it is fastened by floss silk torn from the cocoons of *Bombyx huttoni,* Westw., and by a few slender fibres of the bark of trees or hair according to circumstances.

"So slight and so fragile is the little oval cup that it is aston-

ishing the mere weight of the parent bird does not bring it to the
ground, and yet within it three young ones will often safely out-
ride a gale that will bring the weightier nests of Jays and Thrushes
to the ground.

" Of seven nests now before me four are composed externally of
little bits of green moss, cotton, and seed-down, and the silk of the
wild mulberry-moth torn from the cocoons, with which last
material, however, the others appear to be bound together within.
The lining of two is of the long hairs of the yak's tail, two of which
died on the estate where these nests were found, and a third is
lined with black human hair. The other three are formed of
somewhat different materials, two being externally composed of
fine grass-stalks, seed-down, and shreds of bark so fine as to
resemble tow; one is lined with seed-down and black fibrous
lichens resembling hair, a second is lined with fine grass, and a
third with a thick coating of pure white silky seed-down. In all
the seven, the materials of the two sides are wound round the twigs,
between which they are suspended like a cradle, and the shape is
an ovate cup, about the size of half a hen's egg split longitudinally.
The diameter and depth are respectively 2 inches and 1½ inch by
three-fourths of an inch. The eggs are usually three in number."

Mr. Brooks, writing from Almorah, says :—" This morning, 28th
April, I found a nest of *Zosterops palpebrosa* containing two fresh
eggs. Yesterday I found one of the same bird containing three
half-fledged young ones. Near the Tonse River, in the Allahabad
District, I found these birds in July nesting high in a mango-tree,
the nest suspended like an Oriole's to several leaves; now I find it in
low bushes, at heights of from 3 to 5 feet from the ground. The
eggs, as before, skim-milk blue, without markings of any kind."

From Gurhwal Mr. R. Thompson says :—" A small cup-shaped
elegant nest is built by this bird suspended by fastenings from the
fork of a low branch. The nest is about 2½ inches in diameter and
three-fourths of an inch in depth, composed of cobwebs, fine roots,
hairs, &c., neatly interwoven and lined internally with vegetable
down. The eggs, two, three, or four in number, are of a pale
whitish-blue, oval, and somewhat larger than those of *Arachnechthra
asiatica*. The birds select all kinds of trees, but the nest is always
suspended. The breeding-season is about March and April, and
the brood is quickly hatched and fledged.

" A nest found by me on the 22nd April, and containing four
eggs, was built most ingeniously in a creeper that hung from a
small tree. The birds had arranged it so that the long down-
bearing tendril of the creeper blended with the nest, which in the
main was composed of the material surrounding it.

" Another nest found on the 26th contained three young ones.
It was built in a low branch of a large mango-tree, and might
have been 12 feet from the ground. It was a neat compact
structure, deeply hollow, and made up of cobwebs, fine straw, and
hair, and lined with vegetable down, closely and neatly interwoven.

" The parent birds were evidently feeding the young on the ripe

fruit of the *Khoda* or *Chumroor* (*Ehretia lævis*). I got one fruit
from the old birds, being anxious to know what the young ones
were getting for their dinner.

"The pairing-season commences about the end of March, when
the males may be heard uttering a feeble kind of rambling
song, which in reality is merely modified repetitions of a single
note."

Mr. A. Anderson remarked that "the White-eye breeds throughout the North-Western Provinces and Oudh during the months
of June, July, and August. The nest is a beautiful little model
of the Oriole's; and according to my experience it is invariably
suspended, and *not fixed in the fork of small branches* as stated
by Jerdon. I have on several occasions watched a pair in the
act of building their nest. They set to work with cobwebs,
and having first tied together two or three leafy twigs to which
they intend to attach their nest, they then use fine fibre of the *sun*
(*Crotalaria juncea*), with which material they complete the outer
fabric of their very beautiful and compact nest. As the work
progresses more cobwebs and fibre of a silky kind are applied
externally, and at times the nest, when tossed about by the wind
(sometimes at a considerable elevation), would be mistaken by a
casual observer for an accidental collection of cobwebs. The inside
of the nest is well felted with the down of the madar plant, and then
it is finally lined with fine hair and grass-stems of the softest kind.
Sometimes the nest is suspended from only two twigs, exactly after
the fashion of the Mango-birds (*Oriolus kundoo*); and in this case
it is attached by means of silk-like fibres and fine fibre of *sun* for
about 1½ inch on each side; at others it is suspended from several
twigs; and occasionally I have seen the leaves fixed on to the sides
of the nest, thus making it extremely difficult of detection.

"In shape the nest is a perfect hollow hemisphere; one now
before me measures (inside) 1·5 in diameter. The wall is about
0·3 in thickness.

"Almost all my nests have been built on the neem tree, the
long slender *petioles* of which are admirably adapted for its suspension.

"As a rule the nest is built at a considerable height, and owing
to its situation there is not a more difficult nest to take. Great
numbers get washed down in a half-finished state in a heavy fall
of rain.

"The eggs are, exactly as Jerdon describes them, of a pale blue,
'almost like skimmed milk,' and the usual number is three, though
four are frequently laid."

"On the 7th September," writes Mr. R. M. Adam, "in my
garden in Lucknow, I discovered a nest of this bird in course of
construction, but when it was nearly finished the birds left it.
The nest was a beautiful little cup made of fine grass and cobwebs.
It was situated in a slender fork of a mango-tree about 15 feet
from the ground."

Major C. T. Bingham says:—"Common both at Allahabad and

at Delhi; breeds in both places in May, June, and July. All nests I have seen have been finely made little cups of fibres, bits of thread and cobwebs, lined interiorly with horsehair, generally suspended between two slender twigs at no great height from the ground."

Mr. E. Aitken writes :—" I have only actually taken one nest of the White-eye. That was in Poona (2000 feet above the sea) on the 21st July. The bird, however, builds abundantly in Poona about gardens, trees on the roadside, &c.

" This particular nest was fixed to a thin branch of a tamarind-tree on the side of a lane among gardens. It was within reach of my hand, and was attached both to the thin branch itself and to two twigs. It was well sheltered among leaves.

" The nest was a cup rather narrower at the mouth than in the middle. Its external diameter at the top was $2\frac{1}{4}$ inches; internal diameter $1\frac{1}{2}$ inch; depth $1\frac{1}{4}$ inch internally. It was composed of a variety of fibres closely interwoven with some kind of vegetable silk, and was lined principally with horsehair and very fine fibres. It contained three eggs."

Mr. Davison tells us that " the White-eye breeds on the Nilghiris in February, March, April, and the earlier part of May.

" The nest is a small neat cup-shaped structure suspended between a fork in some small low bush, generally only 2 or 3 feet from the ground, but sometimes high up, about 20 or 30 feet from the ground. It is composed externally of moss and small roots and the down from the thistle; the egg-cavity is invariably sparingly lined with hair. The eggs, two in number, are of a pale blue, like skimmed milk."

From Kotagherry Miss Cockburn remarks :—" Their nests are, 1 think, more elegantly finished than those of any of the small birds I have seen up here. They generally select a thick bush, where, when they have chosen a horizontal forked branch, they construct a neat round nest which is left quite open at the top. The materials they commence with are green moss, lichen, and fine grass intertwined. I have even found occasionally a coarse thread, which they had picked up near some Badagar's village and used in order to fasten the little building to the branches. The inside is carefully lined with the down of seed-pods. White-eyes' nests are very numerous here in the months of January, February, and March. They are extremely partial to the wild gooseberry bush as a site to build on. One year I found ten out of eleven nests on these bushes, the fruit of which is largely used by the aborigines of the hills. A pair once built on a thick orange-tree in our garden. We often stood quite close to one of them while sitting on the eggs, and it never showed the slightest degree of fear. They lay two eggs of a light blue colour."

Mr. Wait, writing from Conoor, says that " *Z. palpebrosa* breeds in April and May, building in bushes and shrubs, and making a deep round cup-shaped nest very neatly woven in the style of the Chaffinch, composed of moss, grass, and silk cotton, and sparsely lined with very fine grass and hair. The eggs are two in number, of a roundish oval shape, and a pale greenish-blue colour."

Finally Colonel Legge informs us that this species breeds in Ceylon in June, July, and August.

The eggs are somewhat lengthened ovals (occasionally rather broader), and a good deal pointed towards the small end. The shell is very fine but almost glossless; here and there a somewhat more glossy egg is met with. They are normally of a uniform very pale blue or greenish blue, without any markings whatsoever, but once in a way an egg is seen characterized by a cap or zone of a somewhat purer and deeper blue. Abnormally large and small specimens are common. They vary in length from 0·53 to 0·7, and in breadth from 0·42 to 0·58; but the average of thirty-eight eggs is 0·62 by 0·47, and the great majority of the eggs are really about this size.

229. Zosterops ceylonensis, Holdsworth. *The Ceylon White-eye.*

Zosterops ceylonensis, *Holdsw., Hume, Cat.* no. 631 bis.

Colonel Legge, referring to the nidification of the Ceylon White-eye, says:—"This species breeds from March until May, judging from the young birds which are seen abroad about the latter month. Mr. Bligh found the nest in March on Catton Estate. It was built in a coffee-bush a few feet from the ground, and was a rather frail structure, suspended from the arms of a small fork formed by one bare twig crossing another. In shape it was a shallow cup, well made of small roots and bents, lined with hair-like tendrils of moss, and was adorned about the exterior with a few cobwebs and a little moss. The eggs were three in number, pointed ovals, and of a pale bluish-green ground-colour. They measured, on the average, ·64 by ·45 inch."

231. Ixulus occipitalis (Bl.). *The Chestnut-headed Ixulus.*

Ixulus occipitalis (*Bl.*), *Jerd. B. Ind.* ii, p. 259; *Hume, Rough Draft N. & E.* no. 624.

A nest of this species, taken by Mr. Gammie out of a small tree below Rungbee, at an elevation of about 3000 feet, was a small, somewhat shallow cup, composed almost entirely of very fine moss-roots, but with a little moss incorporated in the outer surface. Externally the nest was about 3½ inches in diameter and 2 inches in height. The egg-cavity was about 2¼ inches by barely 1¼ inch. This nest was found on the 17th June and contained three hard-set eggs, *which* were thrown away!

232. Ixulus flavicollis (Hodgs.). *The Yellow-naped Ixulus.*

Ixulus flavicollis (*Hodgs.*), *Jerd. B. Ind.* ii, p. 259; *Hume, Rough Draft N. & E.* no. 623.

I have never taken a nest of the Yellow-naped Ixulus.

Mr. Gammie says :—" I have only as yet found a single nest of this species, and this was one of the most artfully concealed that I have ever seen. I found it in forest in the Chinchona reserves, at an elevation of about 5000 feet, on the 14th May. It was a rather deep cup, composed of moss and fine root-fibres and thickly lined with the latter, and was suspended at a height of about six feet amongst the natural moss, hanging from a horizontal branch of a small tree, in which it was entirely enveloped. A more beautiful or more completely invisible nest it is impossible to conceive. It contained three fresh eggs. The cup itself was exteriorly 3·7 inches in diameter and 1·9 in depth, while the cavity was 2·5 in diameter and 1·5 in depth."

The Yellow-naped Ixulus breeds, according to Mr. Hodgson's notes, in the central region of Nepal and the neighbourhood of Darjeeling, laying during the months of May and June. It builds on the ground in tufts of grass, constructing its nest of moss and moss-roots, sometimes open and cup-like and sometimes globular, and lining it with sheep's wool. Mr. Hodgson figures one nest suspended from a branch, and although neither the English nor the vernacular notes confirm this, it is supported to a certain extent by Mr Gammie's experience. At the same time, though the situation and surroundings of both seem to have been similar, Mr. Hodgson figures his nest, not cup-shaped, but egg-shaped, and with the longer diameter horizontal. Seven nests are recorded as having been taken, and all on the ground. One, cup-shaped, taken on the 7th June, 1846, which is also figured, in amongst grass and leaves on the ground, measured externally 3·5 inches in diameter, 2·5 in height, and internally 2 inches both in diameter and depth.

The full complement of eggs is said to be four. Two types of eggs are figured, both rather broad ovals, measuring about 0·75 by 0·6. The one has a buffy-white ground and is thinly speckled and streaked, except quite at the broad end, where the markings are nearly confluent, with pale dingy yellowish brown ; the other has a pale earthy-brown ground, and is spotted similarly to the one just described, but with red and purple. This latter egg appears on the same plate with the suspended nest, and is, I think, doubtful.

Several nests of this species, which I owe to Captain Masson of Darjeeling, are very beautiful structures, moderately shallow and rather massive cups, externally composed of moss, and lined thickly with fine black moss-roots. The cavity of the nests may have been about 1¾ inch in diameter by less than 1½ inch in depth, but the sides of the nests are from one inch to 2 inches in thickness, constructed of firmly compacted moss.

Other nests of this species that have since been sent me show that the bird very commonly suspends its nest to one or two twigs, not unfrequently making it a complete cylinder or egg in shape, with the entrance at one side, but always using moss, in some cases fine, in some coarse, according to the nature of the moss growing where the nest is placed, as the sole material, and lining the cavity thickly with fine black moss and fern-roots.

Dr. Jerdon tells us that at Darjeeling he has repeatedly had the nest brought to him. "It is large, made of leaves of bamboos carelessly and loosely put together, and generally placed in a clump of bamboos. The eggs are three to five in number, of a somewhat fleshy-white, with a few rusty spots."

I cannot but think that in this case wrong nests had been brought to Dr. Jerdon. The eggs that I possess are all of one type—rather elongated ovals with scarcely any gloss, and strongly recalling in shape, size, and appearance densely marked varieties of the eggs of *Hirundo rustica*, but with the markings rather browner and slightly more smudgy.

The eggs are typically rather elongated ovals, often slightly compressed towards the small end, sometimes rather broader and slightly pyriform. The shell is extremely fine and compact, but has scarcely any gloss; the ground-colour is sometimes pure white, sometimes has a faint brownish-reddish or creamy tinge. The markings are invariably most dense about the large end, where they form a zone or cap, regular, well defined and confluent in some specimens, irregular, ill-defined and blotchy in others. As a rule these markings, which consist of specks, spots, and tiny blotches, are comparatively thinly scattered over the rest of the egg, but occasionally they are pretty thickly scattered everywhere, though nowhere anything like so densely as at the large end. The colour of the markings is rather variable. It is a brown of varying shades, varying not only in different eggs, but there being often two shades on the same egg. Normally it is I think an umber-brown, yellower in some spots, but varying slightly in tinge, leaning to burnt umber, sienna, and raw sienna.

Other eggs subsequently obtained by Mr. Gammie are of much the same character as those already described, but one is a good deal shorter and broader, and the markings are more decided red than are some of the yellowish-brown spots observable in the eggs first obtained.

In length the eggs seem to vary from 0·76 to 0·8, and in breadth from 0·54 to 0·58.

Subfamily LIOTRICHINÆ.

235. Liothrix lutea (Scop.). *The Red-billed Liothrix.*

Leiothrix luteus (*Scop.*), *Jerd. B. Ind.* ii, p. 250.
Leiothrix callipyga (*Hodgs.*), *Hume, Rough Draft N. & E.* no. 614.

The Red-billed Liothrix breeds from April to August, at elevations of from 3000 to 6000 feet, throughout the Himalayas south, as a rule, of the first snowy range and eastward of the Sutlej; west

10*

of the Sutlej I have not heard of its occurrence. It also doubtless breeds throughout the hill-ranges running down from Assam to Burmah.

Mostly the birds lay in May, affecting well-watered and jungle-clad valleys and ravines. They place their nests in thick bushes, at heights of from 2 to 8 feet from the ground, and either wedge them into some fork, tack them into three or four upright shoots between which they hang, or else suspend them like an Oriole's or White-eye's nest.

The nest varies from a rather shallow to a very deep cup, and is composed of dry leaves, moss, and lichen in varying proportions, bamboo-leaves being great favourites, bound together with slender creepers, grass-roots, fibres, &c., and lined with black horse- or buffalo-hair, or hair-like moss-roots. The nests differ much in appearance: I have seen one composed almost entirely of moss, and another of nothing but dry bamboo-sheaths, with a scrap or two of moss. They are always pretty substantial, but sometimes they are very massive for the size of the bird.

Three is certainly the usual complement of eggs.

According to Mr. Hodgson's notes, this species breeds in the central mountainous region of Nepal, and lays from April to August. The nest, which is somewhat purse-shaped, is placed in some upright fork between three or four slender branches, to all of which it is more or less attached. It is composed of moss, dry leaves, often of the bamboo, and the bark of trees, and is compactly bound together with moss-roots and fibres of different kinds; it is lined with horse-hair and moss-roots, and contains generally three or four eggs.

The following note I quote *verbatim* :—

" *Central Hills, August 12th.*—Male, female, and nest. Nest in a low leafy tree 5 cubits from the ground in the Shewpoori forest ; partly suspended and partly rested on the fork of the branch ; suspension effected by twisting part of the material round the prongs of the fork ; made of moss and lichens and dry leaves, well compacted into a deep saucer-shaped cavity; 3·62 high, 4·5 wide outside, and inside 2·25 deep and 3 inches wide ; eggs pale verditer, spotted brown, and ready for hatching. The bird found in small flocks of ten to twelve, except at breeding-season."

A nest sent to me last year by Mr. Gammie was found by him on the 24th April, at an elevation of about 5000 feet, in the neighbourhood of Rungbee. It was built by the side of a stream in a small bush, at a height of about 3 feet from the ground, and contained three eggs. The nest is a deep and, for the size of the bird, very massive cup, exteriorly composed entirely of broad flag-like grass-leaves, with which, however, a few slender stems of creepers are intermingled, internally of grass-roots; the egg-cavity being thinly lined with coarse, black buffalo-hair. Externally the nest is more than 5 inches in diameter and nearly 4 inches high ; but the egg-cavity, which is very regularly shaped, is $2\frac{1}{2}$ inches in diameter and 2 inches in depth.

This year Mr. Gammie writes to me :—" I have taken many nests of the Red-billed Liothrix here in our Chinchona reserves, at all elevations from 3500 to 5000 feet. They breed in May and June, amongst dense scrub, placing their nests in shrubs, at heights of from 3 to 5 feet from the ground, and either suspending them from horizontal branches, or hanging them between several upright stems, to which they firmly attach them. The nest itself is cup-shaped and composed principally of dry bamboo-leaves held together by a few fibres, and a few strings of green moss wound round the outside. The lining consists of a few black hairs, and the usual number of eggs is three. A nest I recently measured was extern-ally 4 inches in diameter and 2·7 in height, while the cavity was 2·6 across by 1·9 in depth."

Mr. Gammie subsequently found a nest on the very late date of 17th October at Rishap, Darjeeling. It contained three eggs, two of which were addled.

Dr. Jerdon says that at Darjeeling he " got the nest and eggs repeatedly; the nest made chiefly of grass, with roots and fibres, and fragments of moss, and usually containing three or four eggs, bluish white, with a few purple and red blotches. It is generally placed in a leafy bush at no great height from the ground. Gould, quoting from Mr. Shore's notes, says that the eggs are black spotted with yellow: this is of course erroneous. I have taken the nest myself on several occasions, and killed the bird, and in every case the eggs were coloured as above."

I wish to add here, as I have abused him occasionally, that Mr. Shore was, I understand, a most excellent man, and that I have now come to the conclusion that the extraordinary fictions that he recorded about the eggs of birds can only have been due to colour-blindness of a peculiarly aggravated nature. It is not that he mistook eggs, but that he describes *impossible* eggs—Kingfishers' eggs variegated black and white, and here in this case black eggs spotted with yellow! Why, there *are* no such eggs in the whole world, I believe. On the other hand, his whole life proves that he could not have deliberately set to work to invent falsehoods. To return.

The eggs vary a good deal in shade and size, but are more or less long ovals, slightly pointed towards the lesser end. The ground-colour is a delicate very pale green or greenish blue, in one, not very common type, almost pure white, and they are pretty boldly blotched or spotted and speckled as the case may be, and clouded, most thickly towards the large end, and very often almost ex-clusively in a zone or cap round this latter, with various shades of red or purple and brown. Some blotches in some eggs are almost carmine-red, but the majority are brownish red or reddish brown, varying much in depth and intensity of colour. There is some-thing Shrike-like in the markings of many eggs; and where the markings are most numerous, namely at the large end, they are commonly intermingled with streaks and clouds of pale lilac. The smaller end of the egg is often entirely free from markings. I should mention that all the eggs have a faint gloss, and that some are decidedly glossy.

They vary in length from 0·76 to 0·95, and in breadth from 0·59 to 0·66; but the average of thirty-four eggs is 0·85 by 0·62.

237. Pteruthius erythropterus (Vig.). *The Red-winged Shrike-Tit.*

Pteruthius erythropterus (*Vig.*), *Jerd. B. Ind.* ii, p. 245; *Hume, Rough Draft N. & E.* no. 609.

Writing from Murree, Colonel C. H. T. Marshall says :—" There is no record about the nidification of this species. Its nest is exceedingly difficult to find, and it was only by long and careful watching through field-glasses that Captain Cock discovered that there was a nest at the top of a very high chestnut-tree, to and from which the birds kept flying with building-materials in their beaks. The nest is most skilfully concealed, being at the top of the tree, with bunches of leaves both above and below. The nest, like that of the Oriole, is built pendent in a fork. It is somewhat roughly made of moss and hair. The eggs are pinky white, blotched with red, forming in some a ring round the larger end. They average ·9 in length and ·65 in breadth. We were fortunate enough to secure two nests; both were more than 60 feet from the ground. Breeds in the end of May, at an elevation of 7000 feet."

Captain Cock says :—" I first found this bird building its nest on the top of a high chestnut-tree at Murree in the month of May. When the nest was ready I took my friend Captain C. H. T. Marshall to be present at the taking of it, as it had never, I think, been taken before. We took the nest on the 30th May.

" It was an open flattish cup, like the nest of *O. kundoo* in structure, only shallower. It contained three eggs, pinky white, covered with a shower of claret spots that at the larger end formed a cap of dark claret colour. Another nest, which I took in June from the top of an oak, contained two eggs."

To Colonel Marshall and Captain Cock I am indebted for a nest and egg of this species.

The nest is a moderately deep cup, suspended between two prongs of a horizontal fork. Externally it is about 4 inches in diameter and about 3 inches in depth. The egg-cavity is nearly hemispherical, 3 inches in diameter and 1·5 in depth. It is a very loosely made structure, composed internally of not very fine roots and externally coated with green moss. Along the lines of suspension a good deal of wool is incorporated in the structure, and it is chiefly by this wool that the nest is suspended. The fork is a slender one, the prongs being from 0·3 to 0·4 in diameter.

The egg is a broad oval, a good deal pointed towards the small end. The shell is very fine and compact, and has a fine gloss. The ground-colour is white or pinky white, and is pretty thickly speckled and finely spotted all over with brownish red and a little pale inky purple. Just towards the large end the markings are very dense, and form more or less of a confluent cap of mingled brownish red and pale lilac, the latter everywhere appearing to underlie the former.

The egg was taken on the 10th June, and measures 0·9 by 0·68.

239. Pteruthius melanotis, Hodgs.   *The Chestnut-throated Shrike-Tit.*

Allotrius œnobarbus, *Temm. apud Jerd. B. Ind.* ii, p. 246.
Allotrius melanotis, *Hodgs., Hume, Rough Draft N. & E.* no. 611.

According to Mr. Hodgson's notes and figures, the Chestnut-throated Shrike-Tit breeds in Sikhim and Nepal up to an elevation of 6000 or 7000 feet. The nest is placed at a height of 6 to 10 feet from the ground, between some slender, leafy, horizontal fork, between which it is suspended like that of an Oriole or White-eye. It is composed of moss and moss-roots and vegetable fibres, beautifully and compactly woven into a shallow cup some 4 inches in diameter, and with a cavity some 2·5 in diameter and less than 1 in depth. Interiorly the nest is lined with hair-like fibres and moss-roots; exteriorly it is adorned with pieces of lichen. The eggs are two or three in number, very regular ovals, about 0·77 in length by 0·49 in width. The ground-colour is a delicate pinky lilac, and they are speckled and spotted with violet or violet-purple, the markings being most numerous towards the large end, where they have a tendency to form a mottled zone.

243. Ægithina tiphia (Linn.).   *The Common Iora.*

Iora zeylonica (*Gm.*) *et* I. typhia (*Linn.*), *Jerd. B. Ind.* ii, pp. 101, 103.
Ægithina tiphia (*Linn.*), *Hume, Rough Draft N. & E.* nos. 467, 468.

I have already on several occasions (see especially ' Stray Feathers,' 1877, vol. v, p. 428) recorded my inability to distinguish as distinct species *Æ. tiphia* and *Æ. zeylonica.* I am quite open to conviction; but believing them, so far as my present investigations go, to be inseparable, I propose to treat them as a single species in the present notice.

The Common Iora (the genus, though possibly nearly allied, is too distinct from *Chloropsis* to allow me to adopt, as Jerdon does, one common trivial name for both) breeds in different localities from May to September. I have taken nests and eggs of typical examples of both supposed species, and have had them sent me with the parent birds by many correspondents; and though both vary a good deal, I am convinced that all the variations which occur in the nests and eggs of one race occur also in those of the other. If one gets only two or three clutches of the eggs of each, great differences, naturally attributed to difference of species (see Captain Cock's remarks, *infrà*), may be detected;

but I have seen more than 'fifty, and, so far as I am concerned, I have no hesitation in asserting that, as in the case of the birds so in that of their nests and eggs, no constant differences can be detected if only sufficiently large series are compared.

The birds build usually on the upper surface of a horizontal bough, at a height of from 10 to 25 feet from the ground. Sometimes, when the bough is more or less slanting, the nest assumes somewhat more of a pocket-shape. Occasionally it is built between three or four slender twigs, forming an upright fork; but this is quite exceptional.

As a rule nests of the Iora very closely resemble those of *Leucocerca*, so much so that when I sent a beautiful photograph of a nest, which I had myself watched building, of the latter species to Mr. Blyth, he unhesitatingly pronounced it to be a nest of the former. There is, however, a certain amount of difference; the Iora's nests are looser and somewhat less compact and firm. My experience does not confirm Mr. Brooks's remarks (*vide infrà*) that they are usually shallower; on the contrary all those now before me are, as indeed all the many I can remember to have seen were, deep, thin-walled cups, which had been placed on more or less horizontal branches, not uncommonly where some upright-growing twig afforded the nest additional security. The egg-cavity averages about 2 inches in diameter, and varies from an inch to $1\frac{1}{4}$ inch in depth; the walls, composed of vegetable fibres, and varying in different specimens from only one eighth to three eighths of an inch in thickness, are everywhere thickly coated externally with cobwebs, by which also the nest is firmly attached to the branch on which it is seated, as well as, where such adjoin the nest, to any little twig springing from that branch. Interiorly they are more or less neatly lined with very fine grass-stems. The bottom of the nest in its thinnest part is rarely above one eighth of an inch in thickness, but running, as it so often does, down the curving sides of the branch, it becomes a good deal thicker, and where placed on a small branch, say not exceeding an inch in diameter, the lateral portions of the bottom of the nest are sometimes more than half an inch in thickness.

One nest which I obtained recently in the Botanical Gardens at Calcutta was built in an upright fork of four slender twigs; and in this case the bottom of the nest was obtusely conical, and at its deepest point may have been nearly an inch in depth. I have never seen a similar nest.

The eggs are normally three in number, but I have at times found only two, and these more or less incubated.

Mr. Brooks, writing of a nest he took in the Mirzapoor District, says:—" Did you ever get particulars of the nest of *Iora zeylonica* on the forked branch of a mango-tree 12 or 14 feet from the ground? Nest composed of the same materials as that of *Leucocerca albifrontata*, but not quite so neat and much more shallow; eggs salmon-coloured and spotted with pale reddish brown, intermixed with a few larger dashes of purple-grey. The bird lays in July;

three eggs. This is the only nest I have not taken since I came to India the second time."

From Raipoor, Mr. F. R. Blewitt remarks :—" The Iora breeds from July to September, and certainly *not*, as Dr. Jerdon supposes, twice a year. Both birds assist in the building of the nests, and there evidently appears to be no choice of any particular kind of tree on which to build. I have found them indiscriminately on the mango, mowah, neem, and other trees. The nest is invariably made either just above or between the. fork of two outshooting slender horizontal branches. It is very neatly made, deeply cup-shaped, of grass and fibres, with spider's web on the exterior. The maximum number of eggs is three ; they are of a pale whitish colour, marked generally, chiefly at the broad end, with brownish spots. The brown spots vary in size on different eggs. I secured the first eggs on the 12th July, and the last on the 2nd September. A pair of birds were on this last date just completing their nest, which unfortunately was destroyed by the heavy rains."

Captain Cock says :—" *Iora tiphia* is tolerably common at Seeta-poor (Oudh), and I have several times taken their nests and eggs. I may here mention that I have taken eggs of *Iora zeylonica* at Etawah, and that knowing the birds well, I can say that it is quite a distinct bird; although in the marking of its eggs there is a slight resemblance, yet the nests of the two species are quite different. On the 13th May I observed a nest of *I. tiphia* on a young mango-tree, at the edge of a croquet-ground in our garden. I shot both male and female and took the eggs ; the nest was placed on the upperside of a sloping bough, was covered outside with cob-web, and lined with thin dry grass. It contained two fresh eggs of a delicate pink colour, with broad irregularly-shaped dashes of light brown down the sides of the shell, not tending to coalesce in any way at either apex. Another pair also built their nest on the edge of the same ground in another tree ; but unfortunately in a weak moment I pointed out the nest to a lady friend, and as there-after no one ever played croquet on the ground without staring at the nest, the birds got disgusted and soon deserted it."

To this I need merely add that *of course* typical *Æ. tiphia* and typical *Æ. zeylonica* are very distinct, but that as every intermediate form occurs, they are not, according to my views of what consti-tutes a species, entitled to specific separation, and that as regards nest and eggs, according to my experience, every variety in the one is to be found in the other.

Dr. Jerdon, speaking of Southern India, remarks :—" I have seen the nest and eggs on several occasions. The nest is deep, cup-shaped, very neatly made with grass, various fibres, hairs, and spiders' webs ; and the eggs, two or three in number, are reddish white, with numerous darker red spots, chiefly at the thicker end. . It breeds in the south of India in August and September ; perhaps, however, twice a year."

Writing from South Wynaad, Mr. J. Darling (Junior) says :—" I found the nest, which with the eggs and both parents I have now sent you, in the Teriat Hills on the 24th May, at an elevation of about 2300 feet. It was placed on, and near the extremity of, a bough, at a height of about 10 feet from the ground. It is round, about 2 inches in height and the same in diameter, and the cavity was about an inch or a trifle more in depth. It is built of grass and reed-bamboo-fibres, and is coated with spider's web. It only contained two eggs."

Both parents (sexes ascertained by dissection) are in the typical *tiphia* plumage, without one particle of black on either head, nape, or back.

Mr. Davidson writes:—" In the Satara and Sholapur districts the cock puts on his summer plumage in May and the whole back of head, neck, and back (not rump) is glossy and black.

" This bird lays from the end of June to beginning of August. It is very shy when building and is easily caused to forsake its nest; if a single egg is taken from the nest it does not forsake it, however, but lays on (three instances this year)."

Mr. W. E. Brooks has favoured me with the following very interesting note on the habits of this Iora:—

" Ioras are very numerous and have such a variety of notes that I thought at first there were several sorts; but as far as I can see there is but one species. Iora spreads its tail in a wonderful manner, and comes spinning round and round towards the ground looking more like a round ball than a bird. All the time it descends it utters a strange note, something like that of a frog or cricket, a protracted sibilant sound. This bird is close to *Liothrix* and *Stachyrhis*, although it belongs to the plains."

Colonel Butler writes:—" A nest on the 17th August, 1880, on the outside branch of a silk-cotton tree in Belgaum about 12 feet from the ground, containing three fresh eggs.

" I found many other nests building all through the hot weather and rains; but in every single instance except the present one they were deserted before they were completed."

Major Bingham writes from Tenasserim:—" This species is common throughout the country. As a rule its nest is well hid, but one I saw in the compound of a house in Maulmain was placed in the exposed leafless fork of a tree, not above six feet from the ground. It contained no eggs when I examined it, and was deserted a day or two after. This was in the beginning of May."

Mr. Oates remarks on the breeding of this bird in Pegu:—" Nests are found chiefly in June and July, but the birds probably lay also in May."

In shape the eggs are moderately broad ovals, slightly pointed towards one end. They vary, however, a good deal, some being much more elongated than others. They are almost entirely devoid of gloss. The ground-colour is generally greyish white, but some have creamy and some a salmon tinge; typically they have numerous long streaky pale brown or reddish-brown blotches, chiefly confined to the large end, where they often seem to spring from an irregular imperfect zone of the same colour. The colour of the blotches varies a good deal. In some it is a pale greyish or purplish brown; in others decidedly reddish, or even well-marked and somewhat yellowish brown. Some pale, purplish streaks and clouds generally underlie the brown blotches where they are thickest, and there form a kind of nimbus. In some eggs the markings are confined to a narrow imperfect zone of pale purplish specks or very tiny blotches round the large end, and some of the eggs remind one of those of *Leucocerca albifrontata.* The peculiar streaky longitudinal character of the markings, almost wholly confined to the large end, best distinguishes the eggs of the Ioras from those of any other Indian bird with which they are likely to be confounded.

In length they vary from 0·63 to 0·76, and in breadth from 0·51 to 0·57: but the average of forty-seven eggs measured is 0·69, nearly, by a trifle more than 0·54.

246. Myzornis pyrrhura, Hodgs. *The Fire-tailed Myzornis.*

Myzornis pyrrhoura, *Hodgs., Jerd. B. Ind.* ii, p. 263; *Hume, Rough Draft N. & E.* no. 629.

I have received a single egg said to belong to the Fire-tailed Myzornis from Native Sikhim, where it was found in May in a small nest (unfortunately mislaid) which was placed on a branch of a large tree at no great height from the ground. The place where it was found had an elevation of about 10,000 feet. Although the parent bird was sent with the egg, I cannot say that I have any great confidence in its authenticity, and only record the matter *quantum valeat.*

The egg is a very regular, rather elongated oval. The egg was never properly blown and has been consequently somewhat discoloured. It may have been pure white, and it may have been fairly glossy when fresh, but it is now a dull ivory-white with scarcely any gloss. It measured 0·68 in length by 0·5 in breadth.

252. Chloropsis jerdoni (Bl.). *Jerdon's Chloropsis.*

Phyllornis jerdoni, *Bl., Jerd. B. Ind.* ii, p. 97; *Hume, Rough Draft N. & E.* no. 463.

I have never myself found the nest of Jerdon's Chloropsis, but my friend Mr. F. R. Blewitt has sent me numerous specimens of both nests and eggs from Raipoor and its neighbourhood.

In that part of the country July and August appear to be the months in which it lays; but elsewhere its eggs have been taken in April, May, and June, so that its breeding-season is much the same as that of many of the Bulbuls. The nest is a small, rather shallow cup, at most 3½ inches in diameter and 1½ in depth; is composed externally entirely of soft tow-like vegetable fibre, which appears to be worked over a light framework of fine roots and slender tamarisk-stems, amongst which some little pieces of lichen are intermingled. There is no attempt at a lining, the eggs being laid on the fine grass and slender twigs (about the thickness of an ordinary-sized pin) which compose the framework of the nest.

The eggs as a rule appear to be two in number.

Mr. Blewitt remarks :—" The Green Bulbul breeds in July and August. The bird does not preferentially select any one description of tree for its nest, though the greater number secured were taken from mowah trees (*Bassia latifolia*). The nest is generally firmly affixed at the fork of the end twigs of an upper branch from 15 to 25 feet from the ground. Sometimes, however, eschewing twigs, the bird constructs its nest on the *top* of the main branch itself, cunningly securing it with the material to the rough exterior surface of the branch. Three is certainly the maximum number of eggs. During the period of nidification the parent birds are very watchful and noisy, and their alarm and over-anxiety on the near approach of a stranger often betray the nest."

The late Captain Beavan recorded the following interesting note in regard to this species :—

" This handsome bird is very abundant in Manbhoom, where it is called ' Hurrooa ' by the natives. Its note is so much like that of *Dicrurus ater* that I have frequently been deceived by the resemblance. It breeds in the district. A nest with two eggs was brought to me at Beerachalee on April 4th, 1865. It is built at the fork of a bough and neatly suspended from it, like a hammock, by silky fibres, which are firmly fixed to the two sprigs of the fork, and also form part of the bottom and outside of the nest. The inside is lined with dry bents and hairs. The eggs (creamy white with a few light pinky-brown spots) are rather elongated, measuring 0·85 by 0·62. Interior diameter of nest 2·25, depth 1·5. The cry of alarm of this species is like that of *Parus major*."

Dr. Jerdon remarked (' Illustrations of Indian Ornithology '), writing at the time from Southern India :—

" I have seen a nest of this species in the possession of S. N. Ward, Esq. It is a neat but slightly cup-shaped nest, composed chiefly of fine grass, and was placed near the extremity of a branch, some of the nearest leaves being, it was said, brought down and loosely surrounding it. It contained two eggs, white, with a few claret-coloured blotches. Its nest and eggs, I may remark, show an analogy to that of the Orioles."

Mr. Layard tells us that this species is " extremely common in the south of Ceylon, but rare towards the north. It feeds in small flocks on seeds and insects, and builds an open cup-shaped nest. The eggs, four in number, are white, thickly mottled at the obtuse end with purplish spots."

And Sir W. Jardine says :—" For the interesting nest and eggs of *Phyllornis jerdoni*, Blyth, we are indebted to E. S. Layard, Esq., Magistrate of the district of Point Pedro (the northernmost extremity of Ceylon), in which district we understand it to have been procured. A large groove along the underside of the nest indicates it to have been placed upon a branch; the general form is somewhat flat, and it is composed of very soft materials, chiefly dry grass and silky vegetable fibres, rather compactly interwoven with some pieces of dead leaf and bark on the outside, over which a good deal of spider's web has been worked. It contains four eggs, white, abruptly speckled over with dark bistre mingled with some ashy spots." Layard is not generally reliable where eggs are concerned, for he did not usually take them with his own hands and natives *will* lie; and I doubt the *four* eggs here, but I think, so far as the nest goes, that he was right in this case.

The eggs are rather elongated ovals; some of them a good deal pointed towards one end, others again slightly pyriform. The shell is very delicate; the ground-colour white to creamy white; as a rule almost glossless, in some specimens slightly glossy. They are sparingly marked, usually chiefly at the large end, with spots, specks, small blotches, hair-lines, or hieroglyphic-like figures, which are typically almost black, but which in some eggs are blackish, or even reddish, or purplish brown. In no specimens that I have seen were the markings at all numerous, except just at the large end; and in some they consist solely of a few tiny specks, scattered about the crown of the egg.

The eggs vary from 0·8 to 0·92 in length, and from 0·56 to 0·63 in breadth; but the average of a dozen was 0·86 by 0·6.

254. Irena puella (Lath.). *The Fairy Blue-bird.*

Irena puella (*Lath.*), *Jerd. B. Ind.* ii, p. 105; *Hume, Rough Draft N. & E.* no 469.

Mr. Frank Bourdillon favoured me with an egg of the Fairy Blue-bird, which with other rare eggs he obtained on the Assamboo Hills. So little is known of this range that I quote his remarks upon this locality.

"I must premise that the specimens were obtained along the Assamboo Range of hills, between the elevations of 1500 and 3000 feet above sea-level. This range of hills, running in a north-westerly and south-easterly direction from Cape Comorin to 8° 33′ north latitude, forms the boundary line between Travancore and the British Territory of Tinnevelly, the average height of the range being about 4000 feet, while some of the peaks are as high as 5500 feet. The general character of the hills is dense forest, broken here and there by grass ridges and crowned by precipitous rocks, above which lies an almost unexplored table-land, varying in width from a mile to 12 or 15 miles, at an elevation of almost 4000 feet."

"The egg of the Fairy Blue-bird," he adds, "was taken slightly set on the 28th February, 1873, from a loose sparsely-built nest situated in a sapling about 12 feet from the ground. The nest was composed of dead twigs lined with leaves, and was about 4 inches broad and very slightly indented."

As will be remembered, Dr. Jerdon states that "Mr. Ward obtained, what he was informed were, the nest and eggs; the nest was large, made of roots and fibres and lined with moss; and the eggs, two in number, were pale greenish, much spotted with dusky:" and I have no doubt that Mr. Ward's eggs were genuine.

The egg is an elongated oval, compressed almost throughout its entire length, very blunt at both points; a long cone, the apex broadly truncated and rounded off obtusely, seated on half a very oblate spheroid. In no one single point—shape, texture of shell, colour or character of markings—does this egg approach to those of either the Oriole or the Chloropsis. This shell is very close-grained and fine, but only moderately glossy. The ground is pale green, and it is streaked and blotched with pale dull brown. The markings are almost entirely confluent over the large end (where they appear to be underlaid by dingy, dimly discernible greyish blotches), and from the cap thus formed they descend in streaky mottlings towards the small end, growing fewer and further apart as they approach this latter, which is almost devoid of markings.

It is impossible to generalize from a single specimen as to the position this bird *should* hold, but this one egg renders it quite certain to my mind that the nearest allies of *Irena* are neither *Oriolus* nor *Chloropsis*, and that it is quite impossible to place it with the *Dicruridæ*. The eggs of *Psaroglossa spiloptera* are not very dissimilar, and I expect that it is somewhere between the *Paradiseidæ*, *Sturnidæ*, and *Icteridæ* that *Irena* will ultimately have to be located.

The egg measures 1·1 by 0·73.

Mr. Fulton Bourdillon writes:—"The last note I have to send you at present is that of a Blue-bird's nest (*Irena puella*). Of this there can be no possible doubt, as my brother and I shot both the male and female birds, and I took the nest with my own hands. It was in a pollard tree beside a stream among some thick branches about 20 feet from the ground. The nest was neatly but very loosely constructed of fresh green moss, which formed the bulk of the nest, and lined with the flower-stalks of a jungle shrub. It was very well concealed, and was about 4 inches broad with a cavity not more than 1½ inch deep. It contained two eggs slightly set, measuring respectively 1·11 × ·84 and 1·16 × ·81. These eggs tally very fairly in colour, shape, and size with those sent last year; of the identity of which I was doubtful at the time, though now I think there can be no mistake.

"Since writing last I have had another nest of *Irena puella* brought me with two fresh eggs. The nest was very loosely put together and similar in all respects to the one last sent. The eggs measure ·95 × ·81 and ·92 × ·79, with the same well-defined ring round the larger end. The nest was in a small tree about 10 feet from the ground and was well concealed. It was composed of twigs, without any lining."

The nest sent me by Mr. Bourdillon is a very flimsy affair,

reminding one much of the nest of *Graucalus macii* and not in the
smallest degree of that of an Oriole. A mere pad, some 4 inches
in diameter, composed of very thin twigs or dry flower-stalks with
a couple of dead leaves intermingled, and an external coating of
green moss.

Major C. T. Bingham has favoured me with the following notes
from Tenasserim :—" At the sources of the Winsaw stream, a
feeder of the Thoungyeen river, on the 30th April I found a nest
of this bird, a mere irregularly roundish pad of moss with very little
depression in the centre, containing two fresh eggs, and placed
12 feet or so above the ground in the fork of an evergreen sapling.
The eggs measure 1·18 × 0·86 and 1·19 × 0·86 respectively, and are
so thickly spotted and blotched with brown as to show very little
of the ground-colour, which latter, however, appears to be of a
greenish white.

" On the 11th April I was slowly clambering along a very steep
hill-side overlooking the Queebaw choung, a small tributary of the
Meplay stream, when from a tree whose crown was below my feet
I startled a female *Irena puella* off her nest. I could see the nest
and that it contained two eggs, so I shot the female, who had
taken to a tree a little above me. On getting the nest down, I
found it a poor affair of little twigs, with a superstructure of moss,
shaped into a shallow saucer, on which reposed two eggs, large for
the size of the bird, of a dull greenish white, much dashed, speckled,
and spotted with brown. They were so hard-set that I only
managed to save one, which measured 1·09 by 0·77 inch."

Mr. Davison writes :— " At Kussoom, in some moderately thin
tree-jungle I found the nest of *Irena puella*. The nest was placed
in the fork of a sapling some 12 feet from the ground. The nest
externally was composed of dry twigs, carelessly and irregularly
put together. The egg-cavity was shallow, not more than 1·5 inch
at its deepest part, and it was lined with finer twigs, fern-roots,
and some yellowish fibre. The nest contained two fresh eggs."

Two eggs, taken by Mr. Davison at Kussoom in the north of
the Malay Peninsula, to which the Malayan form does not extend,
are rather elongated ovals, with a slightly pyriform tendency.
The shell is fine, smooth, and compact, and has a perceptible gloss.
The ground-colour is greenish white; round the large end is a
huge, smudgy, irregular zone of reddish brown and inky grey, the
one colour predominating in the one egg, the other in the other.
Inside the zone are specks and spots of the same colours, and below
the zone streaks and spots of these same colours, thinly set, stretched
downwards towards the small end of the egg.

Other eggs subsequently received are very similar to that first
sent by Mr. Bourdillon, except that in shape they are more
regular ovals, and that the brown markings in some have a reddish
and in some a purplish tinge, and that in some eggs the mottlings
and markings are pretty thick even at the small end.

In length they seem to vary from 1·08 to 1·2 inch and in breadth
from 0·73 to 0·88 inch.

In some eggs the ground appears to have no green tinge, but is simply a greyish white. In one egg the markings are all of one colour, a sort of chocolate-brown, a dense almost confluent mass of mottlings in a broad irregular zone round the large end and elsewhere pretty thickly set over the entire surface of the egg. They have always a certain amount of gloss, but are never very glossy.

257. Mesia argentauris, Hodgs. *The Silver-eared Mesia.*

Leiothrix argentauris (*Hodgs.*), *Jerd. B. Ind.* ii, p. 251.
Mesia argentauris, *Hodgs., Hume, Rough Draft N. & E.* no. 615.

According to Mr. Hodgson's notes, the Silver-eared Mesia breeds in the low-lands of Nepal, laying in May and June. The nest is placed in a bushy tree, between two or three thin twigs, to which it is attached. It is composed of dry bamboo and other leaves, thin grass-roots and moss, and is lined inside with fine roots. Three or four eggs are laid : one of these is figured as a broad oval, much pointed towards one end, measuring 0·8 by 0·6, having a pale green ground with a few brownish-red specks, and a close circle of spots of the same colour round the large end.

Dr. Jerdon brought me two eggs from Darjeeling, which he believed to belong to this species. They much resemble those of *Liothrix lutea.* They are oval, scarcely pointed at all towards the lesser end, and are faintly glossed. The ground-colour of one is greenish, the other creamy, white, and both are spotted and streaked, chiefly in an irregular zone near the large end, with different shades of red and purple. The markings are smaller than those of the preceding species. Further observations are necessary to confirm the authenticity of the eggs.

They measure 0·85 and 0·87 by 0·65.

From Sikhim Mr. Gammie writes :—" I have taken about half a dozen nests of this bird. They closely resemble those of *Liothrix lutea* in size and structure and are similarly situated, but instead of having the egg-cavity lined with dark-coloured material, as that species has, all I found had light-coloured linings ; such was even the case with one nest I found within three or four yards of a nest of the other species.

" The eggs are usually four in number."

Other eggs obtained by Mr. Gammie correspond with those given me by Dr. Jerdon. They are as like the eggs of *L. lutea* as they can possibly be, and if there is any difference, it consists in the markings of the present species being as a body smaller and more speckled than those of *L. lutea.*

The six eggs that I have vary in length from 0·82 to 0·9, and in breadth from 0·6 to 0·65.*

---

* There is in the Tweeddale collection a skin of a young nestling of this species procured by Limborg on Muleyit mountain in Tenasserim in the second week of April. On the label attached to the specimen is a note to the effect that the nest from which the nestling was taken was made of moss.—ED.

258. **Minla igneitincta**, Hodgs. *The Red-tailed Minla.*

Minla ignotincta, *Hodgs.*, *Jerd. B. Ind.* ii, p. 254; *Hume, Rough Draft N. & E.* no. 618.

The Red-tailed Minla, according to Mr. Hodgson's notes and figures, breeds in the central region of Nepal and near Darjeeling, during May and June. It builds a beautiful rather deep cup-shaped nest of mosses, moss-roots, and some cow's hair, lined with these two latter. The nest is placed in the fork of three or four slender branches of some bushy tree, at no great elevation from the ground, and is attached to one or more of the stems in which it is placed by bands of moss and fibres. A nest taken on the 24th May measured externally 3·28 inches in diameter and 2·25 in height; internally the cavity was 2 inches in diameter and 1·62 in depth. They lay from two to four eggs, of a pale verditer-blue ground, speckled and spotted pretty boldly with brownish red. An egg is figured as a regular rather broad oval, measuring 0·78 by 0·55.

On the other hand, Dr. Jerdon says :—" Its nest has been brought to me, of ordinary shape, made of moss and grass, and with four white eggs, with a few rusty red spots."

260. **Cephalopyrus flammiceps** (Burton). *The Fire-cap.*

Cephalopyrus flammiceps (*Burt.*), *Jerd. B. Ind.* ii, p. 267; *Hume, Rough Draft N. & E.* no. 633.

Writing from Murree, Colonel C. H. T. Marshall tells us :—" On the 25th May we found the nest of this species (the Fire-cap) in a hole in a rotten sycamore-tree about 15 feet from the ground. The nest was a neatly made cup-shaped one, formed principally of fine grass. We were unfortunately too late for the eggs, as we found four nearly fledged young ones, showing that these birds lay about the 15th April. Elevation, 7000 feet."

Captain Cock says :—" I found a nest in the stump of an old chestnut-tree at Murree. The nest was about 13 feet from the ground near the top of the stump, placed in a natural cavity; it was constructed of fine grass and roots carefully woven and was of a deep cup shape. It contained five fully fledged young ones. The end of May was the time when I found this, and I have never yet succeeded in finding another."

261. **Psaroglossa spiloptera** (Vigors). *The Spotted-wing.*

Saroglossa spiloptera (*Vig.*), *Jerd. B. Ind.* ii, p. 336; *Hume, Rough Draft N. & E.* no. 691.

Personally I know nothing of the nidification of the Spotted-wing.

Captain Hutton tells us that "this species arrives in the hills about the middle of April in small parties of five or six, but it does not appear to ascend above 5500 to 6000 feet, and is therefore more properly an inhabitant of the warm valleys. I do not remember seeing it at Mussoorie, which is 6500 to 7000 feet, although at 5200 feet on the same range it is abundant during summer. Its notes and flight are very much those of the Starling (*Sturnus vulgaris*), and it delights to take a short and rapid flight and return twittering to perch on the very summit of the forest trees. I have never seen it on the ground, and its food appears to consist of berries.

"Like the two species of *Acridotheres*, it nidificates by itself in the holes of trees, lining the cavity with bits of leaves. The eggs are usually three, or sometimes four or five, of a delicate pale sea-green speckled with blood-like stains, which sometimes tend to form a ring near the larger end; shape oval, slightly tapering."

The eggs are so different in character from those of all the Starlings that doubts might reasonably arise as to whether this species is placed exactly where it ought to be by Jerdon and others. I possess at present only three eggs of this bird, which I owe to Captain Hutton. They are decidedly long ovals, much pointed towards the small end, and in shape and coloration not a little recall those of *Myiophoneus temmincki*. The eggs are glossless, of a greenish or greyish-white ground, more or less profusely speckled and spotted with red, reddish brown, and dingy purple. In two of the eggs the majority of the markings are gathered into a broad irregular speckled zone round the large end. In the third egg there is just a trace of such a zone and no markings at all elsewhere. In length they vary from 1·03 to 1·08, and in breadth from 0·68 to 0·74.*

---

* **HYPOCOLIUS AMPELINUS**, Bonap.   *The Grey Hypocolius.*

Hypocolius ampelinus, *Bp.*, *Hume*, *Cat.* no. 269 quat.

Although this bird has not yet been found breeding within Indian limits, the following account of its nidification at Fao, in the Persian Gulf, by Mr. W. D. Cumming (Ibis, 1886, p. 478) will prove interesting:—

"It is not till the middle of June that they breed.

"In 1883, first eggs were brought by an Arab about the 13th of June, and on the 15th of the same month I found a nest containing two fresh eggs. In 1884, on the 14th of June a nest was brought me containing four fresh eggs, and on the 15th I found a nest containing also four fresh eggs.

"2nd July, I came across four young birds able to fly. On the 3rd, three nests were brought, one containing two fresh eggs, another three young just fledged, and the other four eggs slightly incubated. On the 9th, another nest, containing four young just fledged was brought. On the 15th I saw a flock of small birds well able to fly; on the 18th I found a nest containing four young about a couple of days old, and on the 20th a nest containing three eggs well incubated was brought from a place called 'Goosba' on the opposite bank (Persian side) of the river.

"The nests are generally placed on the leaves of the date-palm, at no very great height. The highest I have seen was built about ten feet from the ground, but from three to five feet is the average height.

"They are substantial and cup-shaped, having a diameter of about 3¼ inches by 2¼ inches in depth, lined inside with fine grass, the soft fluff from the willow when in seed, wool, and sometimes hair.

"The eggs are of a glossy leaden white, with leaden-coloured blotches and spots towards the larger end, sometimes forming a ring round the larger end, and at times spreading over the entire egg.

Subfamily BRACHYPODINÆ.

263. Criniger flaveolus (Gould). *The White-throated Bulbul.*

Criniger flaveolus (*Gould*), *Jerd. B. Ind.* ii, p. 83; *Hume, Rough Draft N. & E.* no. 451.

A nest of this species sent me from Darjeeling was found in July, at an elevation of about 3000 feet.

It was placed on the branches of a medium-sized tree, at a height of only about 5 feet from the ground.

The nest was a compact, rather shallow saucer, 5·5 inches in diameter and about 2 inches in height externally. The cavity was about 3·5 in diameter and an inch in depth. The greater portion of the nest was composed of dead leaves bound together firmly by fine brown roots; inside the leaves was just a lining of rather coarser brown roots, and again an inner lining of black horsehair-like roots and fine stems of the maiden-hair fern.

The nest contained three fresh eggs. These eggs vary from broad to somewhat elongated ovals, are more or less pointed towards the small end, and exhibit a fine gloss.

The ground is a beautiful salmon-pink, and it is thinly spotted, blotched, and marked with irregular lines of deep maroon-red. Most of the markings in one egg are gathered into a very irregular straggling zone round the large end, and the other egg exhibits a tendency to form a similar zone. Besides these primary markings a few spots and clouds of dull purple, looking as if beneath the surface of the shell, are thinly scattered about the egg, chiefly in the neighbourhood of the zone.

These eggs vary from 0·9 to 1·0 in length, and from 0·7 to 0·72 in breadth.

Several nests of this species sent me by the late Mr. Mandelli and obtained by him in British and Native Sikhim during July and the early part of August are all precisely of the same type. They each contained two fresh eggs; they were all placed in the branches of small trees in the midst of dense brushwood or heavy jungle, at heights of from 4 to 10 feet from the ground. The nests are broad and saucer-like, nearly 5 inches in diameter, but not much above 2 in height externally; the cavities average about 3·25 in diameter and about 1 in depth. The body of the nest is composed of dead leaves, the sides are more or less felted round with rich brown fibrous, almost wool-like roots; inside the leaves fine twigs and stems of herbaceous plants, all of a uniform brown tint, are wound round and round, apparently to keep the leaves in their places interiorly, and then the cavity is lined with jet-black horsehair-

---

"On rare occasions I have noticed a greenish tinge in very fresh eggs. This, I think, is due to the colour of the inner membrane, which is generally a very light green, in some very faint and in others more decided; this tinge seems to disappear after the egg is blown.

"Very rough measurements are as follows:—0·9×0·63; 0·83×0·63; 0·83× 0·6; 0·83×0·66; 0·86×0·66."

like vegetable fibres. What these are I do not know, but they are precisely like horsehair to look at, only they are comparatively brittle. The contrast of colour between the jet-black lining and the rich brown of the lip of the saucer, which is constant in all the nests, is very striking.

The eggs of this species sent me by Mr. Mandelli, obtained by him in Sikhim at elevations of from 2000 to 4000 feet in July and the early part of August, possess a very distinctive character. They are broad ovals, much pointed towards the small end, and they are more glossy than the eggs of any other of this family with which I am acquainted. The ground-colour is pink. The markings consist of curious hair-line scratches, clouded blotches, and irregular spots—in some eggs all very hazy and ill-defined, in others more scratchy and sharp. The great majority of the markings seem to be gathered together into an irregular and imperfect zone round the large end. In colour the markings vary from a deep brownish maroon to a dull brickdust-red, sometimes they are slightly more purplish. In some eggs a few faint clouds or small spots of subsurface-looking dusky purple may be noticed mingled with the rest of the markings.

These eggs are totally unlike the eggs of *Criniger ictericus*. I have never had an opportunity of verifying the eggs myself, but as three different nests have now been taken, all containing precisely similar eggs, I believe there can be no doubt of their authenticity.

269. **Hypsipetes psaroides**, Vigors. *The Himalayan Black Bulbul.*

Hypsipetes psaroides (*Vig.*), *Jerd. B. Ind.* ii, p. 77; *Hume, Rough Draft N. & E.* no. 444.

The Himalayan Black Bulbul breeds throughout the outer and lower ranges of the Himalayas, at any rate from Bhootan to Afghanistan, at elevations varying from 2000 to 6000 feet.

They lay mostly in May and June, but eggs may occasionally be met with during the latter half of April.

The nest of *Hypsipetes psaroides* is usually made of rather coarse-bladed grass, with exteriorly a number of dry leaves, and more or less moss incorporated, and lined with very fine grass-stems and roots of moss. A good deal of spider's web is often used exteriorly to bind the nest together, or attach it more firmly to the fork in which it rests. Its general shape is a moderately deep cup, the cavity measuring some 2½ inches in diameter by 1½ inch in depth. The sides, into which leaves and moss are freely interwoven, vary from an inch to a couple of inches in thickness. The bottom, loosely put together, is rarely more than from a quarter to half an inch in depth. It appears to be generally placed on the fork of a branch, at a moderate height from the ground.

Four is the normal number of eggs, but I have more than once found three partially incubated eggs in a nest.

From Darjeeling Mr. Gammie remarks:—"A nest of this bird, which I took on the 17th June, at a height of nearly 50 feet from the ground, on one of the topmost branches of a tree, contained three hard-set eggs. This was below Rungbee, at an elevation of about 3000 feet. The nest was a compact, moderately deep cup, composed of very fine twigs and stems, and with a quantity of dead leaves incorporated in the structure, especially towards its lower surface; it had no lining, but the stems used towards the interior of the nest were somewhat finer than the rest. Exteriorly the nest had a diameter of about 4·5 inches, and a height of about 2·5; interiorly a diameter of about 2·5, and a depth of nearly 1·5."

Mr. Hodgson, writing from Nepal, says :—

"*May 20th, Jaha Powah.*—Two nests on the skirts of the forest in medium-sized trees, placed on the fork of a branch. They are made of moss and dry fern and dry elastic twig-tops, and lined with long elastic needles of *Pinus longifolia*. They are compact and rather deep, half pensile, that is to say, partly slung between the branches of the fork to which they are attached by bands of vegetable fibres. Each contained four eggs, pinkish-white, thickly spotted with dark sanguine." Another year he wrote :—

"*May 9th, in the Valley.*—A mature female with nest and eggs. Nest saucer-shaped, the cavity 3·5 wide by 2·5 deep, made of slender twigs and grass-fibres, with no lining. Eggs three, pale pink, blotched all over with sanguine brown."

Writing from Almorah, Mr. Brooks tells us that "the nest and eggs were found by Mr. Horne on the 27th May near Bheem Tal."

Colonel G. F. L. Marshall also found a nest in the same place. He says :—"I have only myself found the nest once at Bheem Tal (4000 feet); it was situated in a thicket. The nest of this species is similar in shape but much more substantial than those of the Common Bulbul. The eggs are much larger and more elongated in shape, but the colouring is similar to those of the Bulbul, and in many cases the blotches have a tendency to form a zone near the thick end. The nest I found was taken on the 10th June and contained fresh eggs.

"On the 30th May, 1875, I found a nest of this species at Naini Tal on Ayarpata, over 7000 feet above the sea. I record the circumstance, as their breeding at so great an elevation is exceptional. The nest contained three fresh eggs; it was made of leaves and moss, lined with bents of grass, between two branches but partially resting on a third, in a bush at the outskirts of a forest on a steep bank and about eight feet from the ground."

From Mussoorie, Captain Hutton recorded the following very full and interesting note :—

"They breed during April, May, and June, making a rather neat cup-shaped nest, which is usually placed in the bifurcation of a horizontal branch of some tall tree; the bottom of it is composed of thin dead leaves and dried grasses, and the sides of fine woody stalks of plants, such as those used by the White-cheeked Bulbul, and they are well plastered over externally with spiders' webs; the lining is sometimes of very fine tendrils, at other times of dry grasses, fibrous lichen, and thin shavings of the bark of trees left by the wood-cutters. I have one nest, however, which is externally formed of green moss with a few dry stalks, and the spiders' webs, instead of being plastered all over the outside, are merely used to bind the nest to the small branches among which it is placed. The lining is of bark-shavings, dry grasses, black fibrous lichens, and a few fine seed-stalks of grasses. The internal diameter of the nest is 2¾ inches, and it is 1½ inches deep. The eggs are usually three in number, of a rosy or purplish white, sprinkled over rather numerously with deep claret or rufescent purple specks and spots. In colours and distribution of spots there is great variation, sometimes the rufous and sometimes the purple spots prevailing; sometimes the spots are mere specks and freckles, sometimes large and forming blotches; in some the spots are wide apart, in others they are nearly, and sometimes in places quite, confluent; while from one nest the eggs were white, with widely dispersed dark purple spots and dull indistinct ones appearing under the shell. In all the spots were more crowded at the larger end."

Colonel C. H. T. Marshall remarks :—" Numerous nests of this species were found at Murree, agreeing well with Hutton's description. They breed in May and June, never above 6000 feet."

The eggs are rather long ovals. Typically a good deal pointed towards the small end, and more or less pyriform, but at times nearly perfect ovals. They have little or no gloss. The ground-colour varies from white, very faintly tinged with pink, to a delicate pink, and they are profusely speckled, spotted, blotched, or clouded with various shades of red, brownish red, and purple. The markings vary much in character, extent, and intensity of colour. There seem to be two leading types, with, however, almost every possible intermediate variety of markings. The one is thickly speckled over its whole surface with minute dots of reddish purple, no dot much bigger than the point of a pin, and no portion of the ground-colour exceeding 0·1 in diameter free from spots. In these eggs the specklings are most dense, as a rule, throughout a broad irregular zone surrounding the large end, and this zone is thickly underlaid with irregular ill-defined streaky clouds of dull inky purple. In some eggs of this type, the smaller end is comparatively free from specks. In the other type, the surface of the egg is somewhat sparingly, but boldly, blotched and splashed, first with deep umber, chocolate, or purple-brown, and, secondly, with spots and clouds of faint inky purple, recalling not a little the style of markings of the eggs of *Rhynchops albicollis*. Then there are eggs partly speckly and partly blotched, some in which the markings are all rich red and where no secondary pale purple clouds are observable, and others again in which all the markings are dull purplish brown. Generally it may be said that the markings have a tendency to form a cap or zone at the large end.

A nest of three eggs recently obtained from Mussoorie were more richly coloured than any I have yet seen, and were decidedly glossy. The ground-colour is a rich rosy pink, boldly, but sparingly, blotched and spotted with deep maroon, underlaid by clouds and spots of pale purple, which appear as if beneath the surface of the shell. In all the eggs the markings are far more numerous at the large end, where in one they form a huge confluent maroon-coloured patch, mottled lighter and darker.

An egg recently obtained in Cashmere on the 20th June was a somewhat elongated oval, more or less compressed towards one end; a delicate glossy white ground with a faint pink tinge; a rich zone of reddish-purple spots and specks round the large end; a few similar markings scattered sparingly over the rest of the surface of the egg, and a multitude of very faint streaks and clouds of very *pale* inky purple underlying the primary markings.

In length the eggs vary from 0·9 to 1·15, and in breadth from 0·7 to 0·78; but the average of twenty-five eggs measured is 1·03 by 0·75.

271. **Hypsipetes ganeesa**, Sykes. *The Southern-Indian Black Bulbul.*

Hypsipetes neilgherriensis, *Jerd.*; *Jerd. B. Ind.* ii, p. 78; *Hume, Rough Draft N. & E.* no. 445.
Hypsipetes ganeesa, *Sykes, Jerd. t. c.* p. 78.

Mr. Davison tells me that " this species breeds from April to about the middle of June. The nest is generally placed from 12 to 20 feet from the ground, in some dense clump of leaves; favourite sites are the bunches of parasitic plants with which nearly every acacia, and in fact nearly every other tree about Ootacamund, is covered. The nest is composed exteriorly of moss, dry leaves, and roots, lined with roots and fibres: the normal number of eggs is two; they are white with claret-coloured and purplish spots."

A nest of this species taken at Coonoor on the 14th March, 1869, by Mr. Carter, to whom I owe this and many other nests from the Nilghiris, reminds one much of those of the Red-cheeked Bulbuls. A wisp of dry grass and dead leaves, with the dead leaves greatly predominating exteriorly, twisted into a shallow cup, some 4½ inches in diameter externally, and with a shallow depression tolerably neatly lined with finer grass-stems measuring some 3 inches across and perhaps an inch in depth. The bottom of the nest is almost exclusively composed of dead leaves; while even in the sides, externally, little but these are visible, only a few grass-stems crossing in and out, here and there, sufficiently to keep the leaves in their places.

Mr. Wait remarks, writing from Coonoor:—" Our Black Bulbul breeds from March to June. It builds a cup-shaped nest neatly and firmly made. Outside, the nest is chiefly composed, as a rule, of green moss, grass-stalks, and fibres, while inside it is lined with fine stalks and hair. The cavity is from 2·5 to 3 inches in diameter and about half that depth. Two is certainly the normal number of eggs; indeed, I have never found more."

Mr. Rhodes W. Morgan, writing from South India, says in 'The Ibis':—" It breeds in lofty trees in the Nilghiris, building a shallow cup-shaped nest, from 20 to 60 feet from the ground. The nest is constructed of the dried stems of the wild forget-me-not,

and lined with a moss much resembling black horsehair.   The
eggs, which are two in number, are pretty thickly spotted with
pale lilac and claret on a light pink ground-colour.   I found these
birds migrating in vast flights, numbering several thousands, in the
Bolumputty valley in July.   They were flying westwards towards
Malabar."

Mr. Darling, Junior, writes :—" I have taken the eggs of this
Black Bulbul every year from 1863 to 1870 during March, April,
May, and part of June, all over the Nilghiris.   The nests were all
made of moss, dry leaves, and roots, lined with roots and fibres.
I have only once found three eggs (the normal number being two):
in this case the eggs are very much smaller than usual, and more
blotched with the reddish spots.   I have found them at all heights
from the ground up to 30 feet, and mostly in rhododendron trees.
I found two nests in S. Wynaad, at an elevation of about 4000 feet,
both with young, in June 1873."

Mr. C. J. W. Taylor informs us that he procured the nest of this
bird with three fresh eggs at Manzeerabad in Mysore on the 7th April.

Colonel Legge tells us that this Bulbul breeds in Ceylon from
January till March.

That the Nilghiris bird should lay usually only *two* eggs, and this
seems a well ascertained fact, while our very closely allied Himalayan
form lays, as I can personally certify, regularly *four*, is certainly
very strange.

The eggs of this species, sent me from the Nilghiris by Messrs.
Carter and Davison, very closely resemble those of *H. psaroides*
from the Himalayas.   The eggs are of course of the Bulbul type,
but in form are typically much more elongated and conical than
the true Bulbuls.   The ground-colour varies from white to a delicate
pink.   The markings consist of different shades of deep red and
pale washed-out purple.   In some the markings are bold, large,
and blotchy, in others minute and speckly; and in both forms
there is a tendency to confluence towards the large end, where
there is commonly a more or less perfect, but irregular, zone.   The
eggs though smooth and satiny have commonly little or no gloss,
and, considering their size, are very delicate and fragile.

In length they vary from 1·0 to 1·17, and in breadth from 0·7 to 0·8.

275. Hemixus macclellandi (Horsf.).   *The Rufous-bellied Bulbul.*

Hypsipetes mclellandi, *Horsf., Jerd. B. Ind.* ii, p. 79.
Hypsipetes m'clellandii, *Horsf., Hume, Rough Draft N. & E.* no. 447.

The Rufous-bellied Bulbul, according to Mr. Hodgson's notes,
breeds in the central region of Nepal, and low down nearly to the
Terai, from April to June.   Its nest is a shallow saucer suspended
between a slender horizontal fork, to the twigs of which it is firmly
bound like an Oriole's with vegetable fibres and roots.   It is composed of roots and dry leaves bound together with fibres, and
lined with fine grass or moss-roots.   The bird is said to lay four
eggs, but these are neither figured nor described.

Dr. Scully writes from Nepal :—"This Bulbul is common

throughout the year on the hills round the valley of Nepal, but never tenants the central woods. It is generally found in bushes and bush trees, not in high tree-forest; and is commonly seen in pairs. The breeding-season appears to be May and June. A nest was taken on the 6th June, which contained two fresh eggs. The nest was somewhat oval in shape, measuring 3·35 inches in length and 2·5 across; the egg-cavity was about 1 inch deep in the centre, and the bottom of the nest 1·25 thick. It was attached to a slender fork of a tree, and was composed externally of ferns, dry leaves, roots, grass, and a little moss, bound together with fine black hair-like fibres, which were wound round the prongs of the fork so as regularly to suspend the nest like an Oriole's. There was a regular lining, distinct from the body of the nest, composed of fine long yellowish grass-stems, and a little cobweb was spread here and there over the branches of the fork and the outside of the nest. The eggs are rather long ovals, smaller at one end, and fairly glossy; they measure 1·0 by 0·7, and 0·97 by 0·7. The ground-colour is pure pinkish white, abundantly speckled and finely spotted with reddish purple; the spots closely crowded together at the large end, but not confluent, forming in one egg a broadish zone, and in the other a cap; in the latter egg there are a few faint underlying stains of purplish inky at the large end."

Two eggs sent me by Mr. Mandelli from Darjeeling, said to belong to this species, are elongated ovals, much pointed towards the small end. The shell is fine and fairly glossy; the ground-colour a dull salmon-pink, and they are profusely and minutely freckled, speckled, and streaked (so densely at the large end that the markings there are almost confluent) with dull reddish purple.

The eggs measure 1·06 and 1·11 by 0·67.

277. **Alcurus striatus** (Bl.). *The Striated Green Bulbul.*

Alcurus striatus (*Bl.*), Jerd. B. Ind. ii, p. 81.

Mr. Mandelli sent me a nest of this species which was found, he said, on the 8th May about 4 feet from the ground amongst the foliage of a kind of prickly bamboo growing out of the crevices of a patch of large stones near Lebong (elevation 5000 feet), and contained two eggs nearly ready to hatch. The nest is a shallow cup, about 3·75 inches in diameter and 1·5 in height externally, composed entirely of fine brown fibrous roots, a little bound together outside with wool and the silk of cocoons and with two or three little bits of moss stuck about it, and sparingly lined with hair-like grass. It is altogether a light brown nest, no dark material being used in it at all. The cavity is 2·75 inches in diameter and about 1 deep.

278. **Molpastes hæmorrhous** (Gm.). *The Madras Red-vented Bulbul.*

Pycnonotus hæmorrhous (*Gm.*), Jerd. B. Ind. ii, p. 94.
Molpastes pusillus (*Bl.*), Hume, Rough Draft N. & E. no. 462.

The Madras Red-vented Bulbul, which by the way extends

northwards throughout the Central Provinces, Chota-Nagpoor, Rajpootana (the eastern portions), the plains of the North-Western Provinces, Oudh, Behar, and Western Bengal, breeds in the plains country chiefly in June and July, although a few eggs *may* also be found in April, May, and August. In the Nilghiris the breeding-season is from February to April, both months included.

Elsewhere I have recorded the following notes on the nidification of this species in the neighbourhood of Bareilly :—

" Close to the tank is a thick clump of sâl-trees (*Shorea robusta*), the great building-timber of Northern India, whose natural home is in that vast sub-Himalayan belt of forest which passes only 30 miles to the north of Bareilly.

" In one of these a Common Madras Bulbul had made its home. The nest was compact and rather massive, built in a fork, on and round a small twig. Externally it was composed of the stems (with the leaves and flowers still on them) of a tiny groundsel-like (*Senecio*) asteraceous plant, amongst which were mingled a number of quite dead and skeleton leaves and a few blades of dry grass : inside, rather coarse grass was tightly woven into a lining for the cavity, which was deep, being about 2 inches in depth by 3 inches in diameter.

" This is the common type of nest; but half an hour later, and scarcely 100 yards further on, we took another nest of this same species. This one was built in a mango-tree, towards the extremity of one of the branches, where it divided into four upright twigs, between which the Bulbul had firmly planted his dwelling. Externally it was as usual chiefly composed of the withered stems of the little asteraceous plant, interwoven with a few jhow-shoots (*Tamarix dioica*) and a little tow-like fibre of the putsan (*Hibiscus cannabinus*), while a good deal of cobweb was applied externally here and there. The interior was lined with excessively fine stems of some herbaceous exogenous plant, and there did not appear to be a single dead leaf or a single particle of grass in the whole nest.

" The eggs, however, in both nests, three in each, closely resembled each other, being of a delicate pink ground, with reddish-brown and purplish-grey spots and blotches nearly equally distributed over the whole surface of the egg, the reddish brown in places becoming almost a maroon-red. Two eggs, however, that we took out of a nest, similar to the first in structure but situated like the second in a mango-tree, were of a somewhat different character and very different in tint. The ground was dingy reddish pink, and the whole of the egg was thickly mottled all over with very deep blood-red, the mottlings being so thick at the large end as to form an almost perfectly confluent cap. Altogether the colouring of these two eggs reminded one of richly coloured types of *Neophron's* eggs. Some of the Bulbuls' eggs that we have taken earlier in the season were much feebler coloured than any of those obtained to-day, and presented a very different appearance, with a pinkish-white ground, and only moderately thickly but very uniformly speckled all over with small spots of light purplish grey, light reddish brown, and very dark brown. These eggs scarcely seem to belong to the same bird as the boldly blotched and richly mottled specimens that we have taken to-day."

Writing from the neighbourhood of Delhi, Mr. F. R. Blewitt says : " This Bulbul breeds from the middle of May to about the middle of August. Its selection of a tree for its nest is arbitrary, as I have found the latter on almost every variety of bush and tree. The nest is neatly cup-shaped, generally fragile in structure, though I have seen many a nest strong and compact. The outer diameter of the nest varies from 3 to nearly 4 inches, and the inner diameter from 2 to almost 3 inches.

" The chief material of the nest is, on the outside, coarse grass, with fine *khus* or fine grass for the lining. Very frequently horse-hair is likewise used for lining the interior of the cavity.

" I have seen some nests bound round on the outside with hemp, other kinds of vegetable fibres, and even spider's web.

" The regular number of the eggs is four."

Mr. W. Theobald found the present species breeding in Mon-ghyr in the fourth week of June.

Mr. Nunn remarks :— " I took a nest of this species at Hoshun-gabad on 26th June, 1868, which contained four eggs; it was placed in a lime-tree, was composed of very small twigs, and lined inside with fine grass-roots; it was cup-shaped, and measured internally 2·25 inches in breadth by 1·75 in depth."

The late Mr. A. Anderson wrote from Futtehgurh:—" On the 30th April last (1874) I took a very beautifully and curiously constructed nest of our Common Bulbul. In shape and size it resembled the ordinary nest, but the curious part of it was that the upper por-tion of the nest for an inch all round was composed entirely of *green twigs* of the neem tree on which it was built, and the under surface (below) was felted with fresh blossoms belonging to the same tree. The green twigs had evidently been broken off by the birds, but the flowers were picked up from off the ground, where they were lying thick."

Colonel Butler says :—" The Madras Red-vented Bulbul breeds in the neighbourhood of Deesa all through the hot weather and in the monsoon. I found a nest at Mount Aboo in a garden on the 15th of April in the middle of a pot of sweet peas, containing three fresh eggs. I found other nests in Deesa, from the 11th May to 20th August, each containing three eggs.

" The nest is usually built of dry grass-stems, lined with fine roots and a few horsehairs neatly woven together. One nest I found was in a very remarkable situation, viz. inside an uninhabited bungalow upon the top of a door leading out of a sitting-room; the door was open and the bolt at the top had been forced back, and it was between the top of the door and the top of the bolt that the nest rested. The old bird entered the building by passing first of all through the lattice-work of the verandah and then through a broken window-pane into the room where the nest was built."

Mr. R. M. Adam informs us that this bird breeds at Sambhur during June and July.

Lieut. H. E. Barnes, speaking of Rajputana in general, states that this Bulbul breeds from April to September. Nests are occasionally found even earlier than this, but they are exceptions to the general rule.

Major C. T. Bingham writes :—" The first nest I have a note of taking was at Allahabad on the 2nd April. At Delhi it breeds from the end of April to the end of July; I have, however, found most nests in May. All have been firmly made little cups of slender twigs, sometimes dry stems of some herbaceous plant, and lined with fine grass-roots. Five is the usual number of eggs laid."

Mr. G. W. Vidal, writing of the South Konkan, says :—" Abundant everywhere. Breeds in April, and again in September."

Dr. Jerdon, whose experience of this species had been gained mainly in Madras, states that " it breeds from June to September, according to the locality. The nest is rather neat, cup-shaped, made of roots and grass, lined with hair, fibres, and spiders' webs *, placed at no great height in a shrub or hedge. The eggs are pale pinkish, with spots of darker lake-red, most crowded at the thick end. Burgess describes them as a rich madder colour, spotted and blotched with grey and madder-brown : Layard as pale cream, with darker markings."

Mr. Benjamin Aitken writes :—" The Common Bulbul lays at Khandalla in May, but I never found a nest in the plains till after the rains had set in. I have found one nest in Bombay, one in Poona, and two in Berar, as late as October ; and my brother found a nest in Berar in September, with three eggs which were duly hatched."

Writing from the Nilghiris, Miss Cockburn says that " the nests, which in shape closely resemble those of the Southern Red-whiskered Bulbul, are composed chiefly of grass. The eggs are three in number, and may occasionally be found in any month of the year, though most plentiful during February, March, and April."

In shape the eggs are typically rather long ovals, slightly compressed or pointed towards the small end. Some are a good deal pointed and elongated ; a few are tolerably perfect broad ovals, and abnormal shapes are not very uncommon. The ground is universally pinkish or reddish white (in old eggs which have been kept a long time a sort of dull French white), of which more or less is seen according to the extent of the markings. These markings take almost every conceivable form, defined and undefined—specks, spots, blotches, streaks, smudges, and clouds ; their combinations are as varied as their colours, which embrace every shade of red, brownish, and purplish red. As a rule, besides the primary markings, feeble secondary markings of pale inky purple are exhibited, often only perceptible when the egg is closely examined, sometimes so numerous as to give the ground-colour of the egg a

---

* This is some *lapsus pennæ*. Spiders' webs are sometimes used exteriorly, never as a lining.

universal purple tint. In about half the eggs there is a tendency to exhibit, more or less, an irregular zone or cap at the large end, but solitary eggs occur in which there is a cap at the small end. Three pretty well marked types may be separately described. First, an egg thickly mottled and streaked all over with deep blood-red, which is entirely confluent over one third of the surface, namely at the large end, and leaves less than a third of the ground-colour visible as a paler mottling over the rest of the surface. Then there is another type with a very delicate pure pink ground, and with a few large, bold, deep red blotches, chiefly at the large end, where they are intermingled with a few small pale inky-purple clouds, and with only a few spots and specks of the former colour scattered over the rest of the surface. Lastly, there is a pale dingy pink ground, speckled almost uniformly, but only moderately thickly, over the whole surface, with minute specks and spots of blood-red and pale inky purple.

The dimensions are excessively variable. In length the eggs vary from 0·7 to 1·02, and in breadth from 0·6 to 0·75, but the average of sixty eggs measured was 0·89 by 0·65.

279. **Molpastes burmanicus** (Sharpe). *The Burmese Red-vented Bulbul.*

The Burmese Red-vented Bulbul occurs from Manipur down to Rangoon. Writing from Upper Pegu, Mr. Oates says :—" On the 29th July I found a nest in the extremity of a bamboo-frond forming one of a large clump near my house at Boulay. It was circular, the internal diameter about 2·5 and the external 4 inches ; the depth inside 1·5, and the total height 2·5. Foundation of dead leaves, the bulk of the nest coarse grass and small roots, and the interior of much finer grass carefully curved to shape. Altogether the nest was a very pretty structure. Two eggs measured 0·9 by 0·62 and 0·65. Another nest found at the same time was placed in a small shrub about 4 feet from the ground. It was very similar in construction and size to the above and contained three eggs."

Subsequently writing from Lower Pegu, he says :—" Breeds abundantly from May to September, and has no particular prefer-ence for any one month."

281. **Molpastes atricapillus** (Vieill.). *The Chinese Red-vented Bulbul.*

Molpastes atricapillus (*V.*), *Hume, Cat.* no. 462 ter.

Mr. J. Darling, Jr., found a nest of the Chinese Red-vented Bulbul in Tenasserim with three fresh eggs on the 16th March. It was built in a bush little more than a foot above the ground on a hill-side.

Except that they seem to run smaller, these eggs are not dis-tinguishable from those of the other species of this genus, and there is really nothing to add to the description already given of the eggs of *M. hæmorrhous.* The three eggs measured 0·79 by 0·6.

282. Molpastes bengalensis (Blyth). *The Bengal Red-vented Bulbul.*

Pycnonotus pygæus (*Hodgs.*), *Jerd. B. Ind.* ii, p. 93.
Molpastes pygmæus (*Hodgs.*), *Hume, Rough Draft N. & E.* no. 401.

I have taken many nests of the Bengal Red-vented Bulbul in many localities, and while the birds vary, getting less typical as you go westwards, the nests are all pretty much the same, though the eastern birds go in rather more for dead leaves than the western. Sikhim birds are very typical, and I will therefore confine myself to quoting a note I made there.

Several nests taken at Darjeeling in June, at elevations of from 2000 to 4000 feet, each contained three or four, more or less incubated, eggs. The nests were mostly very compact and rather deep cups about $3\frac{1}{2}$ inches in diameter and 2 inches in height, very firmly woven of moss and grass-roots, but with a certain quantity of dry and dead leaves, and here and there a little cobweb worked into the outer surface. Sometimes a little fine grass was used as a lining; but generally there was no lining, only the roots that were used in finishing off the interior of the nests were rather finer than those employed elsewhere. The egg-cavity is very large for the size of the nest, the sides, though very firm and compact, being scarcely above half an inch in thickness. The nests differ very much in appearance, owing to the fact that in some all the roots used are black, in others pale brown.

Mr. Gammie says :—" I took two or three nests of this species in the latter half of May at Mongpho, in Sikhim, at elevations of 3500 feet or thereabouts. They contained three eggs each, hard-set. The nests were in trees, at a moderate height, and rather flimsy structures; shallow cups, composed externally of fine twigs and vegetable fibre, and generally some dead leaves intermingled, especially towards their basal portions, and lined with the fine hair-like stem portion of the flowering tops of grass. One nest measured internally $2\frac{1}{2}$ inches in diameter by nearly $1\frac{1}{2}$ inch in depth; externally it was nearly 4 inches in diameter and 2 inches in height. The eggs were of the usual type."

Mr. J. R. Cripps, writing from Fureedpore, Eastern Bengal, says :—" Excessively common and a permanent resident; commits great havoc in gardens amongst tomatoes and chillies, the red colour of which seems to attract them. Builds its nest in very exposed places and at all heights from two to thirty feet off the ground, in bushes and trees. One nest I saw containing two young ones, on the 28th June, was built on a small date-tree which stood on the side of a road along which people were passing all day, and within six feet of them. The nest was only five feet from the ground, but the materials of which it was made and the colour of the bird assimilated so perfectly with the bark of the tree that detection was difficult. I have found the nests with eggs from the 3rd of April to the end of June; dead leaves and cobwebs were incorporated with the twigs and grasses in all nests which I have seen in Dacca. The natives keep these birds for fighting purposes; large sums are lost at times on these combats."

Writing from Nepal, Dr. Scully remarks :—" It breeds in May and June in the Residency grounds, the nests being very commonly placed in small pine-trees (*Pinus longifolia*). Three is the usual number of eggs found, and a clutch taken on the 29th May measured in length from 0·85 to 0·93, and in breadth from 0·64 to 0·65."

I have fully described the leading types of the eggs of these Bulbuls under *Molpastes hæmorrhous*. I shall therefore only here say that the eggs of this species in shape and colour exactly resemble those of its congener, but that as a body they are larger in size; every variety observable in the eggs of the one is, as far as I know, to be met with amongst those of the other. Taking only the eggs of typical birds from Lower Bengal and Sikhim, they vary from 0·88 to 1·05 in length and from 0·67 to 0·75 in breadth.

283. Molpastes intermedius (A. Hay). *The Punjab Red-vented Bulbul.*

All my specimens from the Salt Range belong to this species, and not to *M. bengalensis*, so that Mr. W. Theobald's remarks in regard to the Common Bulbul's nidification about Pind Dadan Khan and the Salt Range must refer to this species. He says :—

" Lay in May, June, and July; eggs, four; shape, blunt ovato-pyriform; size, 0·87 by 0·62; colour, deep pink, blotched with deep claret-red; nest, a neat cup of vegetable fibres in bushes."

From Murree, Colonel C. H. T. Marshall writes :—" This Bulbul breeds in large numbers on the lower hills."

From Mussoorie, Captain Hutton remarked :—" This is more properly a Dhoon species, as although it does ascend the hills, it is represented there to a great extent by *M. leucogenys*. It breeds in April, May, and June, constructing its nest in some thick bush. On the 12th May one nest contained three eggs of a rosy-white, thickly irrorated and blotched with purple or deep claret colour, and at the larger end confluently stained with dull purple, appearing as if beneath the shell. The nest is small and cup-shaped, composed of fine roots, dry grasses, flower-stalks chiefly of forget-me-not, and a few dead leaves occasionally interwoven; in some the outside is also smeared over here and there with cobwebs and silky seed-down; the lining is usually of very fine roots. Some nests have four eggs, which are liable to great variation both in the intensity of colouring and in the size and number of spots."

284. Molpastes leucogenys (Gr.). *The White-cheeked Bulbul.*

Otocompsa leucogenys (*Gray*), *Jerd. B. Ind.* ii, p. 90; *Hume, Rough Draft N. & E.* no. 458.

The White-cheeked Bulbul breeds throughout the Himalayas, from Afghanistan to Bhootan, from April to July, and at all heights from 3000 to 7000 feet. The nest is a loose, slender fabric, externally composed of fine stems of some herbaceous plant and a few blades of grass, and internally lined with very fine hair-like grass. The nests may measure externally, at most, 4 inches in diameter; but the egg-cavity, which is in proportion very large and deep, is fully $2\frac{1}{4}$ inches across by $1\frac{3}{4}$ inch deep. As I before said, the nest is usually very slightly and loosely put together, so that it is difficult to remove it without injury; but sometimes they are more substantial, and occasionally the cup is much shallower and wider than I have above described. Four is the full complement of eggs.

Captain Unwin says:—"I found a nest containing three fresh eggs near the village of Jaskote, in the Agrore Valley, on the 24th April, 1870. The nest was placed about 5 feet from the ground in a small wild bēr-tree in a watercourse. On the 7th May I found another nest placed in a small thick cheer-tree in the same valley, which contained four eggs."

From Murree, Colonel C. H. T. Marshall tells us that this species "breeds in the valleys, at about 4000 or 5000 feet up, in the end of June. Lays four eggs with a white ground, very thickly blotched with claret-red; nest roughly made of grass and roots, in low bushes."

About Simla and the valleys of the Sutlej and Beas I have found it common, and my experience of its nidification in these localities has been above recorded.

From Mussoorie, Captain Hutton wrote that it is "common in the Dhoon throughout the year, and in the hills during the summer. It breeds in April and May. The nest is neat and cup-shaped, placed in the forks of bushes or pollard trees, and is composed externally of the dried stalks of forget-me-not, lined with fine grass-stalks. Eggs three or four, rosy or faint purplish white, thickly sprinkled with specks and spots of darker rufescent purple or claret colour. Sometimes the outside of the nest is composed of fine dried stalks of woody plants, whose roughness causes them to adhere together."

Mr. W. E. Brooks remarks:—"I found this bird common at Almorah, and procured several nests. They were placed in a bush or small tree, and were slightly composed of fine grass, roots, and fibres: eggs three; ground-colour purplish white, speckled all over, most densely at the larger end, with spots and blotches of purple-brown and purplish grey: laying in Kumaon from the beginning of May to June."

Dr. Scully states that in Nepal this Bulbul "breeds in May and June, principally at elevations of from 5000 to 6000 feet. Its nests were secured on the 2nd, 5th, 6th, 14th, and 28th June; the usual number of eggs laid seems to be three."

Colonel G. F. L. Marshall writes:—"This species breeds both at Naini Tal (7000 feet) and at Bheem Tal (4000 feet). In Kumaon the eggs seem to be laid in the first half of June; the earliest date I have taken them was a single fresh egg on the 23rd May, and the latest, four eggs on the 25th June: the nest is seldom more than six feet from the ground, and is placed either in a thick bush or in the outer twigs of a low bough of a tree."

The eggs are of the regular Bulbul type, as exemplified in those of *Molpastes hæmorrhous*, and vary much in colour, size, and shape. Typically they are rather a long oval, somewhat pointed at one end, have a pinkish or reddish-white ground with little or no gloss, and are thickly speckled, freckled, streaked, or blotched, as the case may be, with blood-, brownish-, or purplish-red, &c., and here and there, chiefly towards the large end, exhibit, besides these primary markings, tiny underlying spots and clouds of pale inky purple. Some eggs have a pretty well-marked zone or irregular cap at the large end, but this is not very common. In size they average somewhat larger than those of *Molpastes leucotis* and *Otocompsa emeria*, both of which they closely resemble; but they are smaller and as a body less richly coloured than those of *O. fuscicaudata*. They vary in length from 0·82 to 0·95, and from 0·58 to 0·7 in breadth; but the average of fifty-seven specimens measured was 0·88 by 0·65.

285. Molpastes leucotis (Gould). *The White-eared Bulbul.*

Otocompsa leucotis (*Gould*), *Jerd. B. Ind.* ii, p. 91; *Hume, Rough Draft N. & E.* no. 459.

The White-eared Bulbul is, so far as my experience goes, entirely a Western Indian form. In the cold weather it may be met with at Agra, Cawnpoor, and even Jhansi, Saugor, and Hoshungabad; but during the summer months I only know of its occurring in Cutch, Katywar, Sindh, Rajpootana, and the Punjab. In all these localities it breeds, laying for the most part in July and August in the Punjab, but somewhat earlier in Sindh. I have, even in Rajpootana, seen eggs towards the end of May, but this is the exception.

The nests are usually in dense and thorny bushes—acacias, catechu, and jhand (*Prosopis spicigera*)—and are placed at heights of from 4 to 6 feet from the ground. The Customs hedge is a great place for their nests, but I have noticed that they are partial to bushes in the immediate neighbourhood of water; and at Hansie, whence he sent me many nests and eggs, Mr. W. Blewitt always found them either in the fort ditch or along the banks of the canal.

The nests, which very much resemble those of *Molpastes hæmorrhous*, are usually composed of very fine dry twigs of some herbaceous plant, intermingled with vegetable fibre resembling tow, and scantily lined with very fine grass-roots. They are rather slender structures, shallow cups measuring internally from 2½ to 3 inches in diameter, and a little more than 1 inch in depth. Three was the largest number of eggs I ever found in any nest, and several sets were fully incubated.

Mr. W. Theobald makes the following note on the nidification

of this bird in the neighbourhood of Pind Dadan Khan and Katas
in the Salt Range:—"Lay in May, June, and July: eggs four;
shape ovato-pyriform; size 0·91 inch by 0·64 inch; colour white,
much dotted with claret-red; nest a neat cup of vegetable fibres
in bushes."

Mr. S. Doig informs us that this bird breeds on the Eastern
Narra in Sind from May to August.

Colonel Butler writes:—"I found a nest of the White-eared
Bulbul at Deesa on the 5th August containing three fresh eggs.
It was placed in the fork of a low Bēr tree about 4 feet from the
ground, and in structure closely resembled the nest of *M. hæmorrhous*.

"On the 17th August I found another nest built by the same
pair of birds in an exactly similar situation, about 60 yards from
the first nest, containing three more fresh eggs."

The eggs, which I need not here describe in detail, are precisely
similar to, but as a body slightly smaller than, those of *Molpastes
leucoyenys*. The only point of difference that I seem to notice,
and this might disappear with a larger series before me, is that
there is a rather greater tendency in the eggs of this species to
exhibit a zone or cap. In length they vary from 0·75 to 0·9, and
in breadth from 0·52 to 0·68; but the average of twenty-three eggs
measured was 0·83 barely, by 0·64.

288. **Otocompsa emeria** (Linn.).   *The Bengal Red-whiskered Bulbul.*

Otocompsa jocosa (*L.*), *Jerd. B. Ind.* ii, p. 92 (part.).
Otocompsa emeria (*Shaw*), *Hume, Rough Draft N. & E.* no. 460.

The Bengal Red-whiskered Bulbul breeds from March to the end
of May. Its nest is placed, according to my experience in Lower
Bengal, in any thick bush, clump of grass, or knot of creepers;
sometimes in the immediate proximity of native villages or in the
gardens of Europeans, and sometimes quite away in the jungle.
It is a typical Bulbul nest, a broad shallow saucer, compactly put
together with twigs of herbaceous plants, amongst which, especially towards the base, a few dry leaves are incorporated, and
lined with roots or fine grass. Exteriorly a little cobweb is wound
round to keep twigs and leaves firm and in their places. All the nests
I have seen were tolerably near the ground, at heights
ranging from 3 to 5 feet.

Three is the normal number of the eggs, but only the other day
I obtained one containing four.

Mr. R. M. Adam says:—"This bird is very common in Oudh.
It affects gardens and low scrub-jungle, flying about with a jerky
flight from bush to bush. They are very fond of the fruit of the
peepot-tree (*F. indica*), and may be seen in great numbers about
these trees when the fruit is ripe. Their note is something like
that of the common Bulbul, but livelier and louder. I have seen a

number of this year's young birds well grown, but as yet without the red cheek-tuft.

"They build in clumps of moong-grass about 2 to 3 feet from the ground. One I found in the tendrils of a creeper about 20 feet from the ground. The nest is well fixed in the grass and fastened to it by the intertwining of some of the fibres of which it is composed. It is cup-shaped, and measures 4 inches in diameter, about 0·75 in thickness, with an egg-cavity 2·75 in diameter and 1·5 deep.

"The nest is formed of roots, twigs, and grass loosely worked together, and over the exterior, with the view of binding the mass together, dried or skeleton leaves, pieces of cloth, broad pieces of grass, and plaintain-bark are fastened carelessly on by means of cobwebs and the silk from cocoons. The egg-cavity is lined with fine roots.

"I never have found more than three eggs; on several occasions only two."

I do not think it possible to separate the Andaman bird. Of its nidification in those islands Mr. Davison says:—"I found a nest of this species in April near Port Blair, in a low mangrove-bush growing quite at the edge of the water; it (the nest) was cup-shaped and composed of roots, dried leaves, and small pieces of bark, lined with fine roots and cocoa-nut fibres; it contained three eggs, with a pinkish-white ground thickly mottled and blotched with purplish red, the spots coalescing at the thicker end to form a zone."

Mr. J. R. Cripps writes from Eastern Bengal:—"Very common and a permanent resident; it freely enters gardens and orchards. In my garden there was a kaminee-tree (*Murraya exotica*), in which I found a nest of this species on the 27th March in course of construction; and on looking at it on the 12th April found two young that had just been hatched. Cane-brakes are favourite places for them to nest in. On the 6th May I found a nest in one of these about 4 feet off the ground, and containing three partly incubated eggs. This species does not, as a rule, build in such exposed situations as *M. bengalensis*; it eats the fruit of jungly trees, *Ficus*, &c., as well as insects."

On the breeding of this Bulbul in Pegu Mr. Oates remarks:— "This bird breeds as early as February, on the 27th of which month I procured a nest with two eggs nearly hatched. It stops nesting, I think, at the beginning of the rains."

Mr. W. Davison informs us that he "took a nest of this bird at Bankasoon, in Southern Tenasserim, on the 15th March. It was placed in a small bush growing in an old garden about 4 feet above the ground. The nest was of the usual type, a compactly-woven cup, composed externally of dry twigs, leaves, &c., the egg-cavity lined with fibres. It contained three nearly fresh eggs."

The eggs in size, colour, and shape closely resemble those of *Molpastes leucotis*. All that I have said in regard to these latter is applicable to those of the present species, and, so far as varieties of

12*

coloration go, the description of the eggs of *Molpastes leucogenys*
is equally applicable to those of the present species. If any
distinction can be drawn, it is that, as a body, bold blotches of
rich red and pale purple are more commonly exhibited in the eggs
of this species than in those of either of the preceding ones.

In length the eggs vary from 0·8 to 0·9, and in breadth from
0·85 to 0·7, but the average of twenty-seven eggs was 0·83 nearly,
by 0·63 barely.

289. Otocompsa fuscicaudata, Gould. *The Southern Red-whiskered Bulbul.*

Otocompsa fuscicaudata, *Gould, Hume, Rough Draft N. & E.* no. 460
bis.

The Southern Red-whiskered Bulbul is found throughout the
more hilly and more or less elevated tracts of the peninsula,
from Cape Comorin northwards as far as Mount Aboo on the west,
and the Eastern Ghâts, above Nellore, on the east. How far
northwards it extends in the centre of the peninsula I am not
certain, but I have seen a specimen from the Satpooras.

They breed any time from the beginning of February to the
end of May. Their nests are usually placed at no great height
from the ground (say at from 2 to 6 feet) in some thick bush.

The nests of this species that I procured at Mount Aboo, and
which have been sent me by Mr. Carter both from Coonoor and
Salem, and by other friends from other parts of the Nilghiris,
where the bird is excessively common, very much resemble those
of *O. emeria*, but they are somewhat neater and more substantial
in structure. They differ a good deal in size and shape, as the
nests of Bulbuls are wont to do. Some are rather broad and
shallow, with egg-cavities measuring 3¼ inches across, and perhaps
1 inch in depth; while others are deeper and more cup-shaped, the
cavity measuring only 2½ inches across and fully 1½ inch in depth.
They are composed in some cases almost wholly of grass-roots, in
others of very fine twigs of the furash (*Tamarix furas*), in others
again of rather fine grass, and all have a quantity of dead leaves
or dry ferns worked into the bottom, and all are lined with either
very fine grass or very fine grass-roots. The external diameter
averages about 4½ inches, but some stand fully 3 inches high,
while others are not above 2 inches in height. As might be
expected, the White-cheeked and White-eared and the two Red-
whiskered Bulbuls' types of architecture differ considerably; *inter
se*, the nests of *M. leucotis* and *M. leucogenys* differ just sufficiently
to render it generally possible to separate them, and the same may
be said of the nests of *O. emeria* and *O. fuscicaudata*. But there
is a very wide difference between the nests of the two former and
the two latter species, so that it would be scarcely possible to
mistake a nest belonging to the one group for that of the other.
The incorporation of a quantity of dead leaves in the body of the
nests, reminding one much of those of the English Nightingale, is

characteristic of the Red-whiskered Bulbul, and is scarcely to be met with in those of the White-cheeked or White-eared ones.

Mr. H. R. P. Carter says:—"At Coonoor on the Nilghiris I have found the nests from the 13th March to the 22nd April, but I believe they commence laying in February. They are generally placed in coffee-bushes and low shrubs, as a rule in a fork, but I have frequently found them suspended between the twigs of a bush which had no fork. I have also found the nest of this bird in the thatch of the eaves of a deserted bungalow, and in tufts of grass on the edge of a cutting overhanging the public road.

"The nest is cup-shaped, rather loosely constructed outside, but closely and neatly finished inside. The outside is nearly always fern-leaves at the bottom, coarse grass and fibres above, and lined inside either with fine fibres or fine grass.

"I have never found more than two eggs, and I have taken great numbers of nests; but I am told that three in a nest is not uncommon."

Writing from Kotagherry, Miss Cockburn says:—"Our Red-whiskered Bulbul builds a cup-shaped nest in any thick bush. The foundation is generally laid with pieces of dry leaves and fern, after which small sticks are added, and the whole neatly finished with a lining of fine grass. They lay two (sometimes three) very prettily spotted eggs of different shades of red and white, which are found in February, March, and April."

Mr. Wait remarks:—"This bird breeds at Coonoor from February to June. It builds usually in isolated bushes and shrubs, in gardens and open jungle. The nest is cup-shaped, loosely but strongly built of grass-bents, rooty fibres, and thin stalks, and is lined with finer grass-stems and roots. I think the internal diameter averages about 2½ inches, and about an inch in depth; but they vary a good deal in size. They lay two or three eggs, rarely four; and the eggs vary a good deal in shape and size, being sometimes very round and sometimes comparatively long ovals. The birds swarm on our coffee estates, and breed freely in the coffee-bushes."

Dr. Jerdon says:—"I have frequently had its nest and eggs brought me on the Nilghiris. The nest was very neatly made, deep, cup-shaped, of moss, lichens, and small roots, lined with hair and down. The eggs are barely distinguishable from those of the next bird [*M. bengalensis*], being reddish white with spots of purplish or lake-red all over, larger at the thick end."

But Dr. Jerdon rarely took nests with his own hand, and in this case clearly wrong nests must have been brought to him.

From Trevandrum Mr. F. Bourdillon says:—"It lays three or four eggs of a pale pink colour, with purple spots, in a nest of roots, lined with finer roots and interwoven with the leaves of a jungle-shrub gathered green. The nest, 3 inches in diameter and 2 inches deep, is generally situated in a bush 4 to 5 feet from the ground."

Mr. J. Davidson remarks:—"This bird simply swarms along

the Western Ghâts from Mahabuleshwur down the Koina and Werna valleys, and seems to have a very extended breeding-time. Last year (1873) I took its nests in March and May on several occasions, and this year I found three nests in March and April in the Werna valley ; and the Hill people, who seem intelligent and fairly trustworthy, stated that this species breeds there throughout the Rains, a season when, owing to the tremendous rainfall, no European can remain. If this be true they must breed at least twice a year. All the nests I saw were placed in bushes from 2 to 4 feet high, some of them most carefully concealed amongst thorns. Out of, I think, nine nests, all taken by myself personally, I never found more than two eggs in any ; and on two occasions last year I obtained single eggs nearly fully incubated."

Messrs Davidson and Wenden, writing of the Deccan, remark :— " Commonish in wooded localities. D. took several nests in the Satara Hills in March and the two following months."

Captain Butler writes :—" The Red-whiskered Bulbul is common at Mount Aboo and breeds in March, April, and May. The nest is usually placed in low bushes from 4 to 8 feet from the ground, and is a neat cup-shaped structure composed externally of fibrous roots and dry grass-stems, and lined with fine grass, horsehair, &c. Round the edge and woven into the outside I have generally found small spiders' nests looking like lumps of wool. The eggs, usually two but sometimes three in number, are of a pinkish-white colour, covered all over with spots and blotches and streaks of purplish or lake-red, forming a dense confluent cap at the large end. A nest I examined on the 24th April contained two nestlings almost ready to fly.

"On the 3rd May, 1875, I took a nest in a low carinda bush, containing two fresh eggs."

Mr. C. J. W. Taylor, writing from Manzeerabad, Mysore, says :— " Most abundant in the wooded district. Common everywhere. Eggs taken March and April. On the 5th July, 1883, I procured a nest of this species with three pure white eggs. I found it in a coffee-bush the day before leaving, so snared parent bird to make sure it was O. fuscicaudata, or otherwise should have left a couple of the eggs to see if young would turn out true to parents."

Captain Horace Terry states that on the Pulney hills this species is "a most common bird, found wherever there are bushes. In the small bushes along the banks of the streams is a very favourite place. I found several nests with usually two, but sometimes three eggs."

Mr. Benjamin Aitken tells us :—" I never saw this bird in the plains, but it is, perhaps without exception, the commonest bird at Matheran, Khandalla, and other hill-stations in the Bombay Presidency. I have found the nests, always with eggs in May, placed from four to seven feet from the ground, and often in the most exposed situations. It is not unusual to find only two eggs in a nest. The bird is not in the least shy, and sets up no clatter, like the Common Bulbul, when its nest is disturbed."

Finally, Mr. J. Darling, Junior, remarks :—"I really wonder if anyone down south does not know the Red-whiskered Bulbul and its nest. On the Nilghiris and in the Wynaad I can safely say it is the commonest nest to be met with, built in all sorts of places, sometimes high up. They generally lay two, but very often three, eggs. In a friend's bungalow in the Wynaad there were three nests built on the wall-plate of the verandah and two eggs laid in each nest. The young were safely hatched.

"This year the nests have been rebuilt and contain eggs. As I am writing, there are two pairs building in a rose-bush about 3 yards from me. They breed from 15th February to 15th May."

The numerous eggs of this species that I possess, though truly Bulbul-like in character, all belong to one single type of that form. Almost all have a dull pinkish or reddish-white ground, very thickly freckled, mottled, and streaked all over with a rich red ; in most blood-red, in others brick-red, underneath which, when closely looked into, a small number of pale inky-purple spots are visible. In half the number of eggs the markings are much densest at the large end : these eggs are one and all more brightly and intensely coloured than any of those that I possess of *M. leucotis, M. leucogenys,* and *O. emeria*; they are, moreover, larger than any of these.

In length they vary from 0·82 to 0·97, and in breadth from 0·63 to 0·71 ; but the average of thirty-six eggs measured was 0·9 by 0·66.

290. Otocompsa flaviventris (Tick.). *The Black-crested Yellow Bulbul.*

Rubigula flaviventris (*Tick.*), *Jerd. B. Ind.* ii, p. 88.
Pycnonotus flaviventris (*Tick.*), *Hume, Rough Draft N. & E.* no. 456.

The Black-crested Yellow Bulbul is another very common species of which I have as yet seen very few eggs. The first notice of its nidification I am acquainted with is contained in the following brief note by Captain Bulger, which appeared in 'The Ibis.' He says:—"I obtained several specimens, chiefly from the vicinity of the Great Rungeet River. From a thicket on the bank, near the cane-bridge, a nest was brought to me on the 16th May, of the ordinary cup-shape, made of fibres and leaves, and containing three eggs, which my *shikaree* said belonged to this species. The eggs were of a dull pinkish hue, very thickly marked with small specks and blotches of brownish crimson."

Major C. T. Bingham, writing of this Bulbul in Tenasserim, says :—"Common enough in the Thoungyeen forests, affecting chiefly the neighbourhood of villages and clearings. The following is a note of finding a nest and eggs I recorded in 1878:—On the 14th April I happened to be putting up for the day in one of the abandoned Karen houses of the old village of Podeesakai at the foot of the Warmailoo toung, a spur from the east watershed range of the Meplay river. Having to wait for guides, I had nothing particular to do that day, a very rare event in my forest work ; I

devoted it to a fruitless search for bears.   I had returned tired and
rather dispirited, and was moving about among the ruined houses,
between and among which a lot of jungle was already springing up,
when, just as I passed a low bush about 3 feet high, out went one
of the above-mentioned birds; of course the bush contained a nest,
a remarkably neat cup-shaped affair, below and outside of fine twigs,
then a layer of roots, above which was a lining of the stems of
the flower of the 'theckay' grass.   It contained three eggs on
the point of hatching, out of which I was only able to save one.
It is one of the loveliest eggs I have seen; in colour I can liken it
only to a peculiar pink granite that is so common at home in Ireland.
Its ground-colour I should say was white, but it is so thickly spotted
with pink and claret that it is hard to describe.   It measured 0·85
× 0·61 inch."

Captain Wardlaw Ramsay writes in 'The Ibis':—"I found· a
nest containing two eggs in April at the foot of the Karen hills
in Burma."

I have seen too few eggs of this species to say much about them.
What I have seen were rather elongated ovals pretty markedly
pointed towards the small end.   The shell fine, but with only a
slight gloss; the ground a pinky creamy white, everywhere very
finely freckled over with red, varying from brownish to maroon,
and again still more thickly with pale purple or purplish grey,
this latter colour being almost confluent over a broad zone round
the large end.

292. **Spizixus canifrons**, Blyth.   *The Finch-billed Bulbul.*

Spizixus canifrons, *Bl., Hume, Cat.* no. 453 bis.

Colonel Godwin-Austen says :—"*Spizixos canifrons* breeds in the
neighbourhood of Shillong, in May.   Young birds are seen in
June." *

---

* Trachycomus ochrocephalus (Gm.).   *The Yellow-crowned Bulbul.*

Trachycomus ochrocephalus (*Gm.*), *Hume, Cat.* no. 449 bis.

As this bird occurs in Tenasserim, the following description of the nest and
eggs found a short distance outside our limits will prove interesting.

Mr. J. Darling, Junior, writes :—"I found the nest of this bird on the 2nd
July at Kossoom.   The nest was of the ordinary Bulbul type, but much larger,
and like a very shallow saucer.   The foundation was a single piece of some
creeping orchid, 3 feet long, coiled round; then a lot of coils of fern, grass, and
moss-roots.   The nest was 4 inches in diameter on the inside, the walls ¼ inch
thick, and the cavity 1 inch deep.   It was built 10 feet from the ground, in a
bush in a very exposed position, and exactly where any ordinary Bulbul would
have built."

The eggs of this species are of the ordinary Bulbul type, rather broad at the
large end, compressed and slightly pyriform, or more or less pointed, towards
the small end.   The shell fine and smooth, but with only a moderate amount of
gloss.   The ground-colour varies from very pale pinky white to a rich warm
salmon-pink.   The markings are two colours: first, a red varying from a dull
brownish to almost crimson; the second, a paler colour varying from neutral tint
through purplish grey to a full though pale purple.   The first may be called the

295. Iole icterica (Strickl.). *The Yellow-browed Bulbul.*

Criniger ictericus, *Strickl., Jerd. B. Ind.* ii, p. 82; *Hume, Rough Draft N. & E.* no 450.

The Yellow-browed Bulbul breeds apparently throughout the hilly regions of Ceylon and the southern portion of the Peninsula of India. I have never taken the nests myself, and I have only detailed information of their nidification on the Nilghiris, which they ascend to an elevation of from 6000 to 6500 feet, and where they lay from March to May.

A nest of this species, taken by Mr. Wait near Coonoor on the 20th of March, is a small shallow cup hung between two twigs, measuring some 3½ inches across and ¾ inch in depth. It is composed of excessively fine twigs and lined with still finer hair-like grass, is attached to the twigs by cobwebs, and has a few dead leaves attached by the same means to its lower surface. It is a slight structure, nowhere I should think above ¼ inch in thickness, and apparently carelessly put together; but for all that, owing to the fineness of the materials used, it is a pretty firm and compact nest. It is not easy to express it in words; but still this nest differs very considerably in appearance from the nests of any of the true Bulbuls with which I am acquainted, and more approaches those of *Hypsipetes*.

Mr. Wait sends me the following note :—

" This bird, although very common on the Nilghiris at elevations of from 4000 to 5000 feet, is a very shy nester, and its nest, which is not easily found, is, as far as my experience goes, invariably placed in the top of young thin saplings at heights of from 6 to 10 feet from the ground. The saplings chosen are almost always in thick cover near the edge of dry water-courses. They generally lay during May, but I have found nests in March. In shape the nest is a moderately deep cup, nearly hemispherical, with an internal diameter of from 2·5 to 3 inches—a true Bulbul's nest, composed of grass and bents and lined with finer grasses. The nest is always suspended by the outer rim between two lateral branches, and never, I believe, built in a fork as is so common in the case of many other Bulbuls. They lay only two eggs, and never, I believe, more. The eggs are longish ovals, rather pointed at one end, a dull white or reddish white, more or less thickly speckled and spotted or clouded with pale yellowish or reddish brown; occasionally the eggs exhibit a few very fine black lines."

primary markings; the others, which seem to be somewhat beneath the surface of the shell, the secondary ones. Varying as both do in *different* eggs, all the primary markings of any one egg are almost precisely the same shade; and the same is the case with the secondary ones, and there is always a distinct harmony between both these and the ground tint. As for the markings, they are generally much the most dense, in a more or less confluent mottled cap, round one end, generally the largest, and are usually more or less thinly set elsewhere. In some eggs all the markings are rather coarse and sparse, in others fine and more thickly set. Two eggs measured 1·06 by 0·76 and 1·03 by 0·73.

Miss Cockburn, writing from Kotagherry, says :—" The Yellow-browed Bulbul is common on the less elevated slopes of the Nilghiris, where it is often seen feeding upon guavas, loquots, pears, peaches, &c. They lay generally in April and May.

" Their nests are constructed very much like those of the common Bulbuls, except that, instead of being placed in the forked branches of trees, they are suspended between two twigs, and fastened to them by cobwebs, the inside being neatly lined with fine grass. Two nests of this bird were found, each containing two fresh eggs, of a pretty pinkish salmon colour, with a dark ring at the thick end ; but another nest had three nearly *white* eggs! The whole structure of the nests was slight and thin, and the eggs could be plainly seen through. The notes of the Yellow-browed Bulbul are loud and repeated often."

Writing on the birds of Ceylon, Colonel Legge remarks :—" I once found the nest of this bird in the Pasdun-Korale forests in August ; little or nothing, however, is known of its breeding-habits in Ceylon, so that it most likely commences earlier than that month to rear its brood. My nest was placed in the fork of a thin sapling about 8 feet from the ground. It was of large size for such a bird, the foundation being bulky and composed of small twigs, moss, and dead leaves, supporting a cup of about 2½ inches in diameter, which was constructed of moss, lined with fine roots ; the upper edge of the body of the nest was woven round the supporting branches. . . . . . . The bottom of the nest was in the fork."

The eggs of this species sent to me by Mr. Wait from Coonoor are totally unlike any other egg of this family with which I am acquainted. They remind one more of the eggs of *Stoparola melanops* or one of the *Niltavas* than anything else. The eggs are moderately long and rather perfect ovals, almost devoid of gloss, and with a dull white or pinkish-white ground, speckled more or less thickly over the whole surface with rather pale brownish red or pink. The specklings becoming confluent at the large end, where they form a dull irregular mottled cap. Other specimens received from Miss Cockburn from Kotagherry exhibit the same general characters ; but the majority of them are considerably elongated eggs, approaching, so far as shape is concerned, the *Hypsipetes* type. In some eggs only the faintest trace of pale pinkish mottling towards the large end is observable ; in others, the whole surface of the egg is thickly freckled and mottled all over, but most densely at the large end, with salmon-pink or pale pinkish brown.

In length the eggs vary from 0·9 to 1·03, and in breadth from 0·64 to 0·7.*

---

* PYCNONOTUS ANALIS (Horsf.). *The Yellow-vented Bulbul.*

Otocompsa analis (*Horsf.*), *Hume, Cat.* no. 452 sex.

Mr. J. Darling, Junior, writes :—" I found the nest of this Bulbul at Salang, in the Malay peninsula, on the 14th February. The nest was built in a bush in secondary jungle, with a few trees scattered about. It was in a fork 6 feet from the ground. The foundation was of dried leaves, then fine twigs, and lined

299. **Pycnonotus finlaysoni**, Strickl. *Finlayson's Stripe-throated Bulbul.*

Ixus finlaysoni (*Strickl.*), *Hume, Cat.* no. 452 ter.

Major C. T. Bingham says :—" On the 22nd May, 1877, while wandering about collecting in the jungles below the Circuit-house at Maulmain, I came across a neat, though thinly made, cup-shaped nest in the fork of a tall sapling, some 12 feet above the ground. Coming closer, I perceived it contained eggs, which were plainly visible through the frail structure of the sides. On looking about to find the owner, I saw a couple of *Pycnonotus finlaysoni* flitting about uneasily in a tree close at hand ; so I hid myself a few yards off, and was almost immediately rewarded by seeing one of them (it turned out to be the female) fly down on to the nest, and seat herself on the eggs. Approaching cautiously, I managed to shoot her as she slipped off ; but, on taking down the nest, I found I had fired too soon, as one of the eggs (there were but two) was smashed by a pellet of shot. The nest was rather a deep cup, and, notwithstanding its flimsy sides, strongly made of grass-roots, lined with very fine black roots of fern. The one unbroken egg was rather roundish in shape, of a dull whitish and claret colour, mixed and spotted and clouded with deeper vinous red, chiefly at the larger end."

Mr. J. Darling, Junior, found the nest of this Bulbul on more than one occasion at Taroar in the Malay peninsula. He writes :—" I shot this bird off a nest with two eggs on the 8th February ; the nest was in a bush 5 feet from the ground ; the foundation was of leaves and fine grass, lined with fine grass and a few cocoanut fibres. The nest was 3 inches in diameter and 2 inches deep. The eggs were too hard-set to blow.

" On the 10th February I took another nest of *Pycnonotus finlaysoni* at Taroar. The nest was built in a small shrub 3 feet from the ground, in a fork ; foundation of dead leaves, built of fine twigs and fibrous bark ; lined with fine grass-bents and moss-roots. Egg-cavity 2¼ inches in diameter, 1¾ deep ; walls ¼ inch thick, bottom ¾ inch.

with fine grass-bents. There was a good deal of cobweb in the construction. It was an exact facsimile of many nests of *Otocompsa fuscicaudata* from the Nilgherry Hills. The egg-cavity was 3 inches in diameter and 2½ inches deep ; the walls were ½ inch thick, the bottom 1 inch."

The eggs are of the usual variable Bulbul type, some broader and more regular, some more elongated, some more or less pyriform. The shell as in others, and apparently rarely showing any very perceptible gloss. The ground-colour pinky white to a warm pink ; the markings, specks, and spots, or, when three or four of these latter have coalesced, occasionally small blotches of a rich maroon-red intermixed with spots and specks and clouds of pale purple. The markings always apparently pretty thickly set everywhere, but almost invariably most densely in a zone about the larger end, where they become at times more or less confluent. Of course as in others of the genus, in some eggs all the markings are very fine and speckly, while in others they are somewhat bolder. In some the red greatly predominates ; in others, again, the grey underlying clouds are very widely extended, and form by far the most conspicuous part of the markings, giving a grey tinge to the entire egg. The eggs vary from 0·82 to 0·91 in length and from 0·61 to 0·65 in breadth.

"Found a nest of *Pycnonotus finlaysoni*, with two fresh eggs, on the 16th March. The nest was built in a thin small sapling, 5½ feet from ground, on the top of a thinly wooded hill; the nest was of the ordinary Bulbul type, but better put together and neater. The foundation was of broad fibrous bark and twigs, lined with fine grass-stalks."

The eggs vary in shape from broad ovals a good deal pointed towards one end, to pyriform and elongated shaped, very obtuse even at the small end. The shell is fine and compact, in some has a fine gloss, in others it is rather dull. The ground-colour is a beautiful pink, sometimes with a creamy tinge, and the markings are bold blotches, spots, and streaks of a maroon of varying degrees in richness, and of a subsurface-looking purple, varying to almost inky grey. In some eggs the maroon, in some the purple or grey seems to predominate ; in some eggs the markings seem pretty equally distributed over the egg ; in others they form a more or less conspicuous zone about the larger end. The eggs measure from 0·85 to 0·92 in length by 0·6 to 0·7 in breadth.

300. Pycnonotus davisoni (Hume).   *Davison's Stripe-throated Bulbul.*

Ixus davisoni, *Hume* ; *Hume, Cat.* no. 452 quat.

Mr. Oates writes from Kyeikpadein in Pegu :—" A nest of this bird was found on the 1st June, and another on 6th of the same month, each containing two fresh eggs. The females, which were shot off the nest, showed, however, no signs on dissection of being about to lay more.

" The nest is a flimsy structure, built of the stems of small weeds and lined with grass. A few fine black tree-roots are twisted round the inside of the egg-chamber. The outside and inside diameters measure 4 and 3 inches, and the depths are similarly 3 and 1¼. Both nests were placed low down about 4 feet from the ground— one in a bush, and the other in a creeper.

" The eggs vary much in size. One pair measure ·92 and ·88 by ·60 and ·65, and the other ·83 and ·82 by ·65 and ·61 respectively ; the ground-colour of all is a pinkish white. In one pair the shell-blotches of washed-out purple are spread over the whole egg, and the surface-spots and dashes of carneous red are also equally spread over the whole shell. In the other pair the shell-marks are grouped round the larger end to form a broad ring, and the whole egg is thickly speckled and spotted with bright reddish. The eggs are very slightly glossy."

301. Pycnonotus melanicterus (Gm.).   *The Black-capped Bulbul.*

Rubigula melanictera (*Gm.*), *Hume, Cat.* no. 455 bis.

Colonel Legge writes :—" In April 1873 I received from a friend in Ceylon three eggs of this bird ; but I was unable to identify them until lately, when I had an opportunity of comparing them with a clutch taken last year in the Western Province, and about which there was no doubt. In the latter case the nest was fixed on the top of a small stump, and was a loose structure of grass and bents; in shape rather a deep cup; and contained two eggs of a reddish-white ground-colour, profusely speckled with reddish brown (in one example confluent round the obtuse end, in the other distributed over the whole surface) over freckles of bluish grey. Dimensions : 0·79 by 0·58, 0·78 by 0·57. The other nest was made of grass on a foundation of dry leaves and herbaceous stalks, loosely lined with fine hair-like tendrils of creepers. The eggs were of a reddish-white ground, thickly covered throughout with brownish-red and dusky red spots, becoming somewhat confluent round the obtuse end. In form they are regular ovals, and measure 0·78 by 0·6, 0·79 by 0·58."

305. **Pycnonotus luteolus** (Less.). *The White-browed Bulbul.*

Ixos luteolus (*Less.*), *Jerd. B. Ind.* ii, p. 84 ; *Hume, Rough Draft N. & E.* no. 452.

Common as is the White-browed Bulbul in Midnapoor, throughout the Tributary Mehals, along the Eastern Ghâts, and again, it appears, in Bombay, only two of my correspondents appear as yet to have procured the nest or eggs.

Mr. Benjamin Aitken, writing from Bombay under date the 11th June, says :—"I now send you a nest of *Pycnonotus luteolus* with two eggs. I took it this morning from a thickly foliaged tree in a garden. It was placed on the top of the main stem of the tree, which had been abruptly cut off about 5 feet from the ground, where the stem was about 3 inches thick. The nest was begun this day week, Thursday, and the first egg was laid the day before yesterday (Tuesday). The bird is a very common one in gardens in Bombay, though I never saw it in Berar nor even in Poona. They build in situations similar to, but perhaps rather more sheltered than, those chosen by the Common Bulbul; but I remember finding one nest placed at a height of only 2 feet from the ground.

"This present nest was begun, as already mentioned, last Thursday, just two days after the first severe thunder-shower preliminary to the monsoon, now fairly on us.

"I draw your attention to the manner in which the nest has been tied at *one* place to a twig to prevent its being blown off its very (apparently) insecure site. I was obliged to take the nest, as I was leaving at once, otherwise one or perhaps two more eggs would have been laid."

The nest is a rather loose straggling structure, exteriorly composed of fine twigs. The cavity, hemispherical in shape, is carefully lined with fine grass-stems. Outside it is very irregularly shaped, and many of the twigs used are much too long and hang down several inches from the nest; but on one side the outer framework has been firmly tied with wool and a little cobweb to a live twig to which the leaves, now withered, are still attached. No roots or hair have entered into the composition of this nest.

Mr. E. Aitken writes :—" I once found a nest in Bombay, not many feet above the level of the sea of course.

"The first egg was laid on 14th September. The nest was built in a bush on the edge of an inundated field, but in our garden. It was fixed to a thin waving branch underneath the bush, which completely overshadowed it. It was only 2 feet from the ground, a cup just large enough to hold the body of the bird, whose head and tail always projected over the edge ; and it was made of thin twigs and neatly lined with *coir*. The bird laid two eggs and then deserted the nest. One of these, which I took, was thicker and rounder than a Bulbul's, and thickly spotted with claret-coloured spots, which gathered into a ring at the larger end.

"The eggs were laid on successive days. I think the birds had already had one brood (in another nest), for I saw apparently the same pair followed by a young one not long before."

Dr. Jerdon says :—" I found the nest in my garden at Nellore. It was rather loosely made with roots, grass, and hair, placed in a hedge, and the eggs, four in number, were reddish white, with darker lake-red spots, exceedingly like those of the Common Bulbul."

Colonel Legge, in his 'Birds of Ceylon,' tells us that this Bulbul breeds in the west and south-west of Ceylon from December to June, the months of April and May, however, appearing to be the favourite time. On the eastern side of the island it breeds during the north-east rains.

The eggs answer well enough to Dr. Jerdon's description, but to an oologist's eye they are excessively *un-like* those of the Common Bulbul; shape, tone of colour, and character of markings alike differ.

In shape they are decidedly elongated ovals. The shell is very fine and smooth, and moderately glossy. The ground is reddish white, and this is profusely speckled and blotched (the blotches being chiefly confined, however, to a broad irregular zone round the broader end) with a deep but certainly, I should say, *not* lake-red, but much nearer what one would get by mixing brown with vermilion. Besides these red markings sundry clouds and spots of a pale greyish lilac are intermingled in a zone, and one or two spots of the same colour may be traced elsewhere.

The eggs measure 0·92 by 0·62, and 0·97 by 0·63.

306. Pycnonotus blanfordi (Jerd.). *Blanford's Bulbul.*

Ixus blanfordi (*Jerd.*), *Hume, Cat.* no. 452 quint.

Mr. Oates writes from Pegu :—" Nest in a small tree, well concealed by leaves, about 7 feet from the ground, near Pegu. A very neat cup measuring 3 inches diameter externally and 2¼ internally. The depth 1¾ inch outside and 1¼ inside. The sides of the nest, though very strongly woven, can be seen through. The materials consist of small fine branchlets of weeds, and the inside is neatly lined with grass. One or two dead leaves, or rather fragments, are used in the exterior walling.

" The nest was found on the 25th May, and contained three eggs slightly incubated. The ground-colour is a fresh pink, but with little gloss. The whole egg is covered with a profusion of dark purplish-red spots, more thickly disposed at the thick end, but everywhere frequent. In addition there are some underlying and much paler smears. The three eggs measured respectively ·75, ·78, and ·77 in length, by ·63, ·62, and ·61 in breadth.

" Subsequently I found five other nests, from the 1st April to the 20th June, all similar to the one described. Eggs invariably three. Average size of twelve eggs ·82 by ·6."

The nests of this species that I have seen have been very slight flimsy structures, nearly hemispherical cups, composed of fine twigs and the leaf-stalks of pennated leaves a little bound together with cobwebs and thinly lined with fine hair-like grass. In some cases a leaf or two has been attached to the outer surface to aid the concealment of the nest. The nest is very loosely woven just like a sieve, as a rule nowhere more than 0·25 inch thick, and with a truly hemispherical cavity, diameter about 2·5, depth about 1·25.

The eggs are of the ordinary Bulbul type, but not amongst the more richly-coloured examples of these; in shape and size they vary a good deal, but typically they seem to be moderately broad ovals slightly compressed towards the small end. The shell is fine and smooth, but has scarcely any appreciable gloss; the ground is pale pink or pinky white. At the large end the markings are dense, forming in some eggs an almost confluent zone, in others a mottled cap; they consist of irregular-shaped spots and specks of deep red and pale subsurface-looking greyish purple; over the rest of the surface of the egg outside the zone or cap the markings are much smaller in size and much more thinly scattered, and it is observable that the secondary purple markings are to a great extent confined to the zone or cap, as the case may be, and its immediate neighbourhood.

Occasionally the markings, which seem always to be small and speckly, are very sparsely set, leaving comparatively large portions of the surface unmarked; and occasionally eggs are met with in which the primary markings are wholly wanting, and there is nothing but a pale reddish-purple cloudy mottling over the greater portion of the surface of the egg.*

---

* PYCNONOTUS PLUMOSUS, Bl. *The Large Olive Bulbul.*

Ixus plumosus (*Bl.*), *Hume, Cat.* no. 452 sept.

Mr. W. Davison writes:—" I found one nest of this Bulbul at Kossoom : it was of the ordinary Bulbul type and placed in a small but dense clump of cane, about 18 inches from the ground. The parent birds were very vociferous when the nest was approached."

The eggs of all these Bulbuls, though they are separable when individually compared, follow so closely the same type of colouring that it is almost impossible to make their distinctions apparent by any verbal descriptions.

The eggs of the present species are like those of so many others, moderately broad ovals, obtuse at the large end, somewhat compressed towards the small

Family SITTIDÆ.

315. Sitta himalayensis, Jard. & Selby. *The White-tailed Nuthatch.*

Sitta himalayensis, *J. & S.*, Jerd. *B. Ind.* i, p. 385; *Hume, Rough Draft N. & E.* no. 248.

According to Mr. Hodgson's notes and drawings this species begins to lay in April, constructing a shallow saucer-like nest of moss lined with moss-roots, in holes of trees at no great elevation from the ground. One such nest, the measurements of which are recorded, was 3·25 inches in diameter and 2 in height externally ; the cavity was 2·25 inches in diameter and 1·25 inch in depth. They lay three or four pure white eggs slightly speckled with red, which measure about 0·72 inch in length by 0·55 inch in width. They breed once a year, and both sexes assist in incubating the eggs and rearing the young.

Mr. R. Thompson says :—" In Kumaon the White-tailed Nuthatch breeds in May and June, laying five or six eggs, in holes in trees, especially in oaks."

Colonel G. F. L. Marshall writes :—" This bird is an early breeder in Naini Tal ; a nest found on the 25th April contained half-fledged young. It was in a natural hollow of a tree about 10 feet from

---

end, at times slightly pyriform. The shell very fine, smooth and thin, but strong, and generally with an appreciable though not at all conspicuous gloss.

The ground-colour is pink or pinky white, and they are very thickly speckled and spotted everywhere, but extremely densely so, and there blotched also in a broad irregular zone, round the large end with rich reddish maroon and dull greyish or inky purple—the rich colour predominating in some eggs, the dull colour in others ; and in some the markings being all extremely fine and speckly, while in others they are rather bolder. Two eggs measure 0·9 by 0·66.

PYCNONOTUS SIMPLEX, Less. *Moore's Olive Bulbul.*

Ixus brunneus (*Bl.*), *Hume, Cat.* no. 452 oct.

Mr. W. Davison says:—" I took a nest of *P. simplex* in some rather thick jungle at Klang. The nest, of the ordinary Bulbul type (in fact it might easily have passed for a nest of *Otocompsa*), was placed in the fork of a small sapling about 6 feet from the ground. The nest contained two eggs. The female was shot from the nest."

The eggs are moderately elongated, rather regular ovals, some specimens having a slight pyriform tendency. The shell is fine and compact, and seems to have generally an appreciable but not striking gloss. The ground-colour appears to have been creamy pink, and it is very thickly freckled and speckled all over with a rich maroon, in amongst which tiny clouds of pale purple may be faintly discerned ; dense as are the markings everywhere, they are generally most so in a zone round the large end. Very possibly this species will be found to exhibit somewhat different types of coloration, as the eggs of all Bulbuls vary very much ; but certainly typically the markings of this species are much more speckly than in most of the others, forming a universal stippling over the entire surface. The two eggs measure 0·9 and 0·88 in length by 0·62 in breadth.

the ground in a thick trunk; the hole was closed up with a kind
of stiff gummy substance, leaving only a circular entrance about an
inch in diameter, just as I have seen in nests of *Sitta europœa.*
The old birds were busily engaged in feeding the young. Another
nest containing young was found on the 28th April in an oak tree
at about 7000 feet elevation; both birds were feeding the young,
and the nest was similar to the last except that in this case it was
so low down in the trunk that, sitting on the ground, I could put
my ear against the hole. From a third nest, found on the 2nd
May, the young had apparently just fled. My experience bears
out Mr. Hodgson's observations: I have often been up here in May
and June searching closely and never found a nest; this year I
came up for the first time in April, and within a few days find
three nests with young. I may add that after the 10th May all
the Nuthatches I have seen were in small parties, apparently
parents with their young."

316. **Sitta cinnamomeiventris,** Blyth. *The Cinnamon-bellied Nuthatch.*

Sitta cinnamomeoventris, *Bl., Jerd. B. Ind.* i, p. 387.

Writing from Sikhim, Mr. Gammie says:—"I lately took the
nest of *Sitta cinnamomeiventris* at 2000 feet. It was 20 feet from
the ground in a soft decaying bamboo on the edge of large jungle.
The birds had made a small hole just below an internode, and
from the next internode below had filled up the hollow of the
bamboo with alternate layers of green moss and pieces of tree-
bark of about an inch or more square to within a few inches of the
entrance-hole. Each layer of moss was about an inch thick, but
the bark layer not more than a quarter of an inch, the thickness of
the bark itself. On the top of this pile, which was a foot high,
was a pad three inches wide by two in depth, of fine moss, fur,
a feather or two, and a few insects' wings intermixed, for the
eggs to rest on. The fur looks like that of a rat. There were four
hard-set eggs, which, unfortunately, got broken in the taking.
One of them only was measurable, and it was 0·65 inch by 0·5.
I send the shell-fragments to show the coloration."

317. **Sitta neglecta,** Walden. *The Burmese Nuthatch.*

Sitta neglecta, *Wald., Hume, Cat.* no. 250 bis.

The Burmese Nuthatch probably breeds throughout Pegu and
Tenasserim. Of its nidification in the latter division Major C.
T. Bingham writes:—"On the 21st March, wandering about in a
deserted clearing, I saw a couple of Nuthatches (*Sitta neglecta*)
flying to and from a tree, carrying food apparently. Watching
them closely with a pair of binoculars, I saw them disappear near
a knot in a branch. The tree was a dead dry one and rather
difficult to climb, but a peon of mine went up and reported five
young ones unfledged, the nest-hole being 6 inches deep, and the

opening, which was originally a large one, and probably caused by water wearing into the site of a broken branch, narrowed by an edging of clay. The young lay on a layer of broken leaves. As they were featherless, blind little things I left them alone, and was delighted to see the parents continuing to feed them."

321. Sitta castaneiventris, Frankl. *The Chestnut-bellied Nuthatch.*

Sitta castaneoventris, *Frankl., Jerd. B. Ind.* i, p. 386.

The late Captain Cock furnished me with the following note a long time ago regarding the breeding of this Nuthatch :—" A very common bird at Sitapur in Oudh, every mango-tope containing one or more pairs. They pair early and commence making their nests in February, laying their eggs in March. The nests are in cavities of trees, at no great height from the ground, and unless observed in course of construction are difficult to find—the bird filling the whole cavity up with mud consolidated with some viscid seed of a parasitical plant, and merely leaving a small round hole for entrance. This composition hardens like pucca masonry in a very short time, and secures the nest from all marauders except the oologist. The nest consists of a few dry leaves at the bottom of the cavity at no great depth, and upon this four eggs are laid. The birds sit close and do not easily desert their nests, as the following instance will show. In 1873 I found a *Sitta's* nest in a mango-tree, and after watching the birds for some days, when the eggs had been laid I took the nest, placing my handkerchief in the nest to prevent bits of mud falling in on the eggs. I opened out the cavity, cleaning away the mud, and putting in my hand I caught the female bird. I looked at her and let her go. In 1874 curiosity induced me to look at the place again, and to my surprise I saw the cavity had been built up again. I caught a bird on the nest and took four eggs ; it may have been a different bird, but there was only one pair in that tope of trees, and was probably the same bird I caught in 1873. I found another nest in my garden about 2 feet from the ground, and I often used to flash the sunlight from a small hand-mirror, that I use out birds' nesting, on to the hen bird while she sat on her eggs. Our collection contains a large series of these eggs, the produce of some five-and-twenty nests taken by myself at Sitapur."

Major C. T. Bingham writes :—"At Allahabad I found two nests of this little Nuthatch, one in July and one in September. I regret to say neither contained any eggs, though the birds were going in and out constantly The nests were in tiny holes in mango-trees, the entrances being still more contracted by earth being plastered round."

Colonel C. H. T. Marshall observes :—"A nest of the Chestnut-bellied Nuthatch was pointed out to me at Umballa in the next garden to mine. It was about 12 feet above the ground in an old mango-tree; the locality chosen was the stump of a branch which had been cut off and had rotted down. Outside there was a grea deal of masonry work as hard and firm as that on white-ant hills, in the middle of which was a neat circular hole just large enough for the passage of the bird. The masonry continued down inside the hole as far as I could see; I did not break it open, as there were nearly fledged young ones inside. I knew this because the parent birds had been seen for some days carrying in food. I did not see the nest till the end of May. The following spring I found another nest at Kurnal in a bokain tree; it was constructed after the same fashion; the nest itself, which consisted only of dead leaves, was not very far down. I was unfortunately this time (March 15th) too early for the eggs. The holes are not easy to see from the ground, as they are most skilfully concealed from view."

The eggs of this species are very regular, slightly elongated ovals, scarcely compressed or pointed towards the small end at all. The shell is fragile, and is either entirely glossless or has only a trace of gloss. The ground-colour is white, with at times a faint pinkish tinge, and the markings consist of specks, spots, and splashes (always most numerous at the large end, where they usually form a more or less conspicuous though irregular cap) of dull or bright brick-red, more or less intermingled in most specimens with dull reddish lilac. The arrangement and size of the markings are very variable. In some eggs they are all mere specks, forming a small speckly cap at the large end, and elsewhere very thinly scattered about the surface; in others many of the spots are (for the size of the egg) large, the majority are well-marked spots and not mere specks, and the whole surface of the egg is pretty thickly studded with them, while the broad end exhibits a large blotched and mottled cap. The majority of the eggs are intermediate between these two extremes.

In length the eggs vary from 0·61 to 0·72 and in breadth from 0·5 to 0·54, but the average of numerous specimens is 0·67 by 0·52.*

---

* SITTA TEPHRONOTA, Sharpe. *The Eastern Rock-Nuthatch.*

Sitta neumayeri, *Mich., Hume, Cat.* no. 248 quint.

The Eastern Rock-Nuthatch is abundant in Baluchistan, and without doubt breeds there. The following note by Lieut. H. E. Barnes will therefore be interesting. He writes from Afghanistan:—"This Nuthatch is very common on the hills. It appears to choose very different localities to build in. In some instances a hole in the face of a rock is selected, and this it lines with agglutinated mud and resin, continuing the lining-case until it projects in the shape of a cone to fully 8 inches. It seems fond of decorating its little palace with feathers to a distance of 2 or even 3 feet, and it is thus a conspicuous object; but most nests are found in holes in trees, and even here feathers are stuck into crevices all around. They are usually well lined with camel-hair.

"They breed in March and April. The eggs are usually four in number (I have sometimes found five), oval in shape, more or less glossy white, and more or less densely or sparsely (generally most densely towards the large end) spotted and blotched with varying shades of chestnut to reddish brown, more or less intermingled with pale purple and occasionally purplish grey. Some eggs are very richly marked. Some are almost pure white. They average 0·87 by 0·57."

The eggs of this species are typically moderately broad ovals, slightly pointed

13*

323. Sitta leucopsis, Gould. *The White-cheeked Nuthatch.*

Sitta leucopsis, *Gould, Jerd. B. Ind.* i, p. 385; *Hume, Rough Draft N. & E. no.* 249.

Captain Cock took the eggs of the White-cheeked Nuthatch late in May and early in June (1871) in Kashmir at Sonamurg.

Captain Wardlaw Ramsay says, writing of Afghanistan :—" I observed it hanging about a nest-hole on the 21st May, but on returning to take the eggs some days later was unable to find the tree;" and he adds, "On the 21st of June I shot a young bird just fledged near the Peiwar Kotul."

The eggs of this species vary somewhat in size. In shape some are moderately elongated, some are somewhat broad ovals, and all are, more or less, compressed towards the smaller end, which, however, is obtuse and not at all pointed. The ground is white and has a slight gloss. The markings consist of small spots and minute specks, some eggs exhibiting only the latter. In all cases the markings are most dense towards the large end, where they generally form an irregular and ill-defined mottled cap or zone. In colour the markings are red and pale purple, the red varying from bright brickdust-red to brownish and even purplish red, and the purple being sometimes lilac and sometimes grey, and here and there in a single speck, almost black. In length the eggs vary from 0·67 to 0·75 inch, and in breadth from 0·54 to 0·55 inch.

325. Sitta frontalis, Horsf. *The Velvet-fronted Blue Nuthatch.*

Dendrophila frontalis (*Horsf.*), *Jerd. B. Ind.* i, p. 388; *Hume, Rough Draft N. & E. no.* 253.

The Velvet-fronted Nuthatch lays from the middle of February to the end of May. It breeds in the forest-tracts of the Sub-Himalayan ranges, in the Central Indian forests, the Ghâts of Southern India, and the well-wooded slopes of the Nilghiris, Palnis, &c.

It builds a compact little nest of moss and feathers in a tiny hole in a tree, selecting, I believe, generally a natural cavity, but certainly trimming the entrance and interior itself.

Mr. R. Thompson says :—" This species is common in all the low densely wooded valleys of the Sub-Himalayan ranges of Kumaon,

---

towards the small end, but elongated and more or less blunt-ended pyriform examples occur. The shell is extremely fine and smooth, but has only a moderate amount of gloss in any specimen that I have seen, and in some specimens has only a trace of this. The ground-colour is pure white, and the eggs are generally thinly speckled, spotted, or blotched, about the broad end only, with a pale red; occasionally a few greyish-purple spots and blotches are intermingled with the other markings, and specks and tiny spots of both red and grey sometimes extend to the smaller end of the egg also. I have seen no such examples myself, but very probably in some eggs the principal markings may be at the small end. Eighteen eggs vary from 0·81 to 0·91 in length by 0·61 to 0·69 in breadth.

at an elevation of from 1500 to 2500 feet. It breeds in May and June in hollows of trees. Any small hole suits for a nest, and it lays four or five eggs, for I have seen it with as many young, though I never took the trouble of getting out the eggs themselves."

Mr. Davison says :—" This Nuthatch breeds on the Nilghiris as high up as Ootacamund, nesting in holes of trees, and laying three or four eggs, spotted with chestnut, pinkish red, or reddish brown. The nest is composed of moss, moss-roots, &c., and lined with feathers. I am not quite certain how long the breeding-season lasts, but I think that it is from the middle of April to the early part of May."

Miss Cockburn, of Kotagherry, sends me the following account of the first nest she took of this species :—

" After having wished for some years to obtain the eggs of this bird, I was delighted to hear from my brother that he had seen a Nuthatch go into a *small* hole in a tree, and that, on looking into it, he had seen something like a nest. I went prepared with a chisel and hammer, but wished first to ascertain fully who the owner of the nest was. After watching at a respectful distance for a long time, an Indian Grey Tit flew to the hole and peeped in. My first thought was one of great disappointment at having ridden many miles with such high expectations to find only a Common Titmouse's nest; but it did not last long; the inquisitive Grey Tit found the hole too small for him, and flew off just as happily as he had flown to it. I continued to watch, and was quite repaid by seeing a Velvet-fronted Nuthatch fly to the top of the tree containing the nest, and descend rapidly down the trunk (which was about 12 or 13 feet high), as if it knew where the wee hole was, and disappear into it. This was sufficient proof as to the proprietor of the nest; I walked quietly up to the tree, and when within a foot of it out flew the bird. My handkerchief was stuffed into the hole to prevent any chips breaking the eggs, should there be any ; and making use of the chisel and hammer, I soon made the hole large enough to admit my hand. The nest contained three eggs, which I most carefully extracted one by one. The nest was then brought out, and consisted of a quantity of beautiful green moss, feathers (many of which belong to the bird), some soft fine hair, and a few pieces of lichen. This nest was discovered on the 10th February. The tree it was found in grew nearly alone, at the side of a road not much frequented.

" The eggs were quite fresh, and most probably the bird would have laid at least one more ; but these were sufficient to show the colour of the eggs, which were pure white, with dark and light red spots and blotches, chiefly at the thick end, besides a circle of spots like a Flycatcher's eggs."

Mr. Rhodes W. Morgan, writing of South India, says, in ' The Ibis' :—" It breeds in holes of trees, preferring the deserted ones excavated by *Megalæma caniceps*. The nest is built of moss, and lined with the fluff of hares and soft feathers. The eggs are always four in number, spotted with pinkish red on a white ground, the spots being more numerous towards the larger end. They breed in March. Dimensions, 0·71 inch long by 0·57 broad."

Mr. Mandelli sent me a small pad-like nest of this species found on the 4th May in Native Sikhim. It was placed in a hollow of a trunk of a large tree about 3 feet from the ground. It is composed of very fine moss felted together with a little fine vegetable fibre, and the upper surface coated with a little fine short silky fur, probably that of a rat.

Major Bingham, writing from Tenasserim, says :—" Fairly common in the Thoungyeen valley. On the 18th February I found a nest in a hole in a branch of a pynkado tree (*Xylia dolabriformis*), but I was too early for eggs."

One egg of this very beautiful species was sent me by Miss Cockburn. It is intermediate in size and colour between those of the European Creeper and Nuthatch, while at the same time it strongly recalls the eggs of *Parus atriceps*. In shape the egg is a broad oval (not quite so broad, however, as those of the European Nuthatch are), slightly compressed towards one end. The ground-colour is white, and the egg is blotched, speckled, and spotted, chiefly, however, in a sort of irregular zone round the large end, with brickdust-red and somewhat pale purple. The shell is fine and compact, but devoid of gloss. The egg measures 0·68 by 0·55 inch.

Three other eggs from the Sikhim Terai measure 0·68 by 0·51.

Family DICRURIDÆ.

327. Dicrurus ater (Hermann).   *The Black Drongo.*

Dicrurus macrocercus (*V.*), *Jerd. B. Ind.* i, p. 427.
Buchanga albirictus, *Hodgs., Hume, Rough Draft N. & E.* no. 278.

The Black Drongo or Common King-Crow lays throughout India, at any rate in the plain country ; it does not appear to breed either in the Himalayas or the Nilghiris at any height exceeding 5000 feet.

A few eggs may be found towards the close of April, and again during the first week of August, but May, June, and July are *the* months.

It builds usually pretty high up in tall trees, in some fork not quite at the outside, constructing a broad shallow cup, and lays normally four eggs, although I *have* found five. Elsewhere I have recorded the following in regard to its nidification :—

" Close at our own gate is a pretty neem tree, the ' *Melia azadirachta,*' a species now naturalized in Provence and other parts of the south of France. High up in a fork a small nest was visible, and projecting over it on one side a black forked tail that could belong to nothing but the King-Crow. Of this bird we have already taken during the last six weeks at least fifty nests, and in many cases where we had left the empty nest in *statu quo*, we found it a week later with a fresh batch of eggs laid therein. Many birds will never return to a nest which has once been robbed, but others, like the King-Crow and the Little Shrike (*Lanius vittatus*), will continue laying even after the nest has been *twice* robbed. The very day after the nest has been cleared of perhaps four slightly incubated eggs, a fresh one that otherwise would assuredly never have seen the light is laid, and that, too, a fertile egg, which, if not meddled with, will be hatched off in due course. It might be supposed that immediately on discovering their loss, nature urged the birds to new intercourse, the result of which was the fertile egg, and this, in some cases, is probably really the case; Martins and others of the Swallow kind being often to be seen busy with 'love's pleasing labour' before their eggs have been well stowed away by the collector. But this will not account for instances that I have observed of birds in confinement, who separated from the male before they had laid their full number, and then later, just when they began to sit deprived of their eggs, straightway laid a second set, neither so large nor so well coloured as the first, but still fertile eggs that were duly hatched. But for the removal of the first set, these subsequent eggs would never have been developed or laid. Now, the theory has always been that the contact of the sperm- and germ-cells causes the development and fertilization of the latter. In these cases no fresh accession of sperm-cells was possible, and hence it would seem as if in some birds the female organs were able to store up living sperm-cells, which only work to fertilize and develop ova in the event of some accident rendering it necessary, and which otherwise ultimately lose vitality and pass away without action.

"The nest of the King-Crow that we took was of the ordinary type; in fact I have noticed scarcely any difference in the shape or materials of all the numerous nests of this common bird that I have yet seen. They are all composed of tiny twigs and fine grass-stems, and the roots of the khus-khus grass, as a rule, neatly and tightly woven together, and exteriorly bound round with a good deal of cobweb, in which a few feathers are sometimes entangled. The cavity is broad and shallow, and at times lined with horsehair or fine grass, but most commonly only with khus. The bottom of the nest is very thin, but the sides or rim rather firm and thick; in this case the cavity was 4 inches in diameter, and about $1\frac{1}{2}$ in depth, and contained three pure white glossless eggs. In the very next tree, however (a mango, and this is perhaps their favourite tree), was another similar nest, containing four eggs, slightly glossy, with a salmon-pink tinge throughout, and numerous well-marked brownish-red specks and spots, most numerous towards the large end, looking vastly like Brobdingnagian specimens of the Rocket-bird's eggs. The variation in this bird's eggs is remarkable; out of more than one hundred eggs nearly one third have been pure white, and between the dead glossless purely white egg and a somewhat glossy, warm pinky-grounded one, with numerous well-marked spots and specks of maroon colour, dull-red, and red-brown or even dusky, every possible gradation is found. Each set of eggs, however, seems to be invariably of the same type, and we have never yet found a quite white and a well coloured and marked egg in the same nest.

"These birds are very jealous of the approach of other birds even of their own species to a nest in which they have eggs, and many a little family would this year have been safely reared, and their ovate cradles have escaped the plundering hands of my shikaries, had not attention been invariably called to the whereabouts of the nest by the pertinacious and vicious rushes of one or other of the parents from near their nest at every feathered thing that passed them by."

Captain Hutton says :—" This species, which appears to be generally diffused throughout India, is not uncommon in the Dehra Doon, but does not ascend the hills ; it breeds in June, laying four eggs of somewhat variable size. They are pure white, thus differing widely from those of the supposed *D. longicaudatus* of Mussoorie.

"It is evident likewise that the eggs which Captain Tickell assigns to this species do not belong to it. (*Vide* Journal As. Soc. vol. xvii. p. 304.)

"The nest differs from that of our hill species, being larger and far less neatly made ; it is placed in the bifurcation of the smaller branches of a tall tree, and is composed exteriorly of the hard semi-woody stalks of various plants, plastered over with cobwebs. Another one was constructed entirely of fine roots, like the khus-khus used for tatties, and plastered over like the former with cobwebs. It is flattened or saucer-shaped, and about 5 inches in diameter."

Mr. F. R. Blewitt remarks :—" It breeds from the middle of May well into August. I do not think it has two broods in the year, at least close observation has not proved the fact. Trees of various sizes are chosen indiscriminately for the nest, from the lofty mango and tamarind to the low-growing roonji, &c.

"The nest is a peculiarly slight-formed structure (occasionally I have seen it otherwise, but this is the exception), always neatly made. The exterior of the nest is composed of small fine twigs, roots, and grass, with generally a good deal of spider's web round the outer surface. The average exterior diameter of the nest is about 5·5 inches. The cavity is frequently lined with horsehair. On three or four occasions I have seen very fine khus substituted for the hair. The average inner diameter of the nest is about 3·4 inches.

"The regular number of eggs is four; in colour they are a light reddish white, with a few spots or blotches, here and there, of a purplish red or red-brown. The eggs often differ much in size.

"I happened to find in one nest two eggs, one of the usual size, the other only about one third of the size. What is more surprising, it was perfectly formed, as regards the white and yolk.

The instance of sagacity related by Mr. Phillips, and quoted by Jerdon, was related to him by the late Mr. Davis, my old Collector of Customs.

" I have on two or three occasions myself witnessed similar instances of sagacity. This bird, during the breeding-season, is pugnacious to a degree, fearlessly attacking every bird that approaches the tree on which the nest may be."

Writing from the Sambhur Lake, Mr. R. M. Adam says :— " Very common here. The King-Crow breeds here in June and July. The eggs vary much with regard to colouring ; some are pure white without spots, some have dark brown spots on the white ground, whilst others have a pale rufous ground darker at the broader end, with spots of deep rust-colour and lilac."

Colonel G. F. L. Marshall writes :—" At Bheem Tal, fully 4000 feet above the sea, I found two nests of this species on the 24th May, one contained four eggs, and the other three ; the eggs varied much in size, and out of the seven, six were pure white, almost like Barbet's eggs, and the seventh had only a faint sprinkling of tiny dark spots at one end. The birds, all four of which I shot, were typical *D. ater*, with the white spot well developed. On the same day, and in the same place, I found eggs of *D. longicaudatus*. I record this, as it is not usual to find *D. ater* breeding at this elevation. It may be noticed that the eggs of this species found by Hutton in the Doon were all pure white, while in the plains I think white is more exceptional."

Dr. Scully says :—" In Nepal it breeds freely at elevations of from 4000 to 5000 feet. Three nests were taken in the valley, in May and June ; these contained each three or four pure white eggs."

Major C. T. Bingham remarks :—" I have found many nests of the King-Crow both at Allahabad and Delhi. In both places they begin laying towards the end of May, and I got fresh eggs at Allahabad as late as the 13th August. The nests and eggs have been nearly always of the same type. The former, a shallow, but well-made saucer, rather small sometimes for the size of the bird, of grass-roots and twigs, and absolutely without lining ; the latter white, when fresh with a pink tinge, spotted, chiefly at the larger end, rather scantily with claret-colour and dark brown. I have never found a pure white egg."

Lieut. H. E. Barnes, writing of Rajputana in general, tells us :— " The King-Crow breeds during May and June. A few nests may be found in July, but by far the greater number are to be found during the latter part of May and the commencement of June."

Colonel Butler informs us that " The Common King-Crow breeds in the neighbourhood of Deesa during the rains. I have taken nests on the following dates :—

" June 6, 1875. A nest containing 4 fresh eggs.
" June 7, 1875. ,, ,, 4 fresh eggs.
" June 9, 1875. ,, ,, 2 fresh eggs.
" ,, ,, ,, ,, 4 young birds.
" June 10, 1875. ,, ,, 4 fresh eggs.

"June 11, 1875. A nest containing 4 fresh eggs.
"June 13, 1875.   „   „   3 fresh eggs.
      „   „   „   „   4 fresh eggs.
"July 8, 1875.   „   „   4 fresh eggs.
"July 12, 1875.   „   „   4 fresh eggs.

"The nest consists of a broad shallow saucer about 3½ inches in diameter measured from the inside, composed of dry twigs and fine roots, and is invariably fixed in the fork of a tree. The bottom of the nest, though strongly woven, is often so thin that the eggs are visible from below. The eggs, usually four in number, are of the Oriole type, being white or creamy buff, sparingly spotted and speckled with deep chocolate or rusty brown, with, occasionally, markings of inky purple. The markings of the eggs of this species, like those of the Oriole, are apt to run if washed."

Messrs. Davidson and Wenden, writing from the Deccan, say:— "Common and breeds."

Mr. Vidal remarks of this bird in the South Konkan:— "Abundant. Breeds in May."

Mr. Rhodes W. Morgan, writing from South India, says in 'The Ibis':—"Breeds from March to the end of May, constructing a slight cup-shaped nest in a tree. The nest is composed of fine twigs bound together with cobwebs, and is rather a flimsy concern, the eggs often being visible from below. It is generally placed in the fork of a branch, at from 10 to 30 feet from the ground. The eggs are three in number, occasionally only two, and vary very greatly in colour, some being almost of a pure white, whilst others again are spotted and blotched, especially at the larger end, with claret and light purple on a rich salmon-coloured ground. The birds are very noisy in the breeding-season, keeping all intruders off, not hesitating to attack Kites and Crows. They seem to have an especial antipathy to the latter."

Mr. Benjamin Aitken states that in Madras "the King-Crow, so conspicuous on the backs of cattle, telegraph-wires, &c., all through the cold and hot seasons, is conspicuous by its absence during the breeding-season. Many of them retire to woods and gardens to breed, but even when they do not, they keep very quiet while they have their nests. Last June there was a nest in a tree in the Thieves' bazaar at Madras, but the birds hardly ever showed themselves out of the tree."

Mr. J. Inglis informs us that in Cachar "this King-Crow is extremely common. It breeds all through the summer. It lays four or five pure white eggs on the top of a few grasses placed in the fork of a tree. It is very pugnacious, and attacks birds of all sizes if they approach it."

There are two very distinct types of this bird's eggs. The one pure white and spotless, the other a pale salmon-colour, spotted with a rich brownish red. These eggs unquestionably both belong to the same species, as I have taken them times without number myself and can positively certify to their parentage; moreover, connecting links are not wanting in a large series. I have one egg perfectly white, with the exception of three or four blackish-brown spots, another with more of these spots, another with almost as many as the ordinary spotted eggs have, the ground-colour in all these being still pure white, and the spots being blackish or very deep reddish brown. Then I have others similar to those just described, but showing a faint salmon-coloured halo round one or two of the largest spots, others in which the halo is further developed, and others again with the entire ground-colour an excessively pale salmon throughout, and so on a complete series gradually increasing in intensity of colour till we get the pure rich salmon-buff which is at the other end of the scale. I am particular in this description, because the eggs of this bird have been a subject of almost as many contradictions between Indian naturalists as the chameleon of pious memory. In shape the eggs are typically a rather long oval, somewhat pointed towards one end. Very much elongated varieties are common, recalling in this respect the eggs of *Chibia hottentotta*. Spherical varieties, if they occur, must be very rare, the enormous series I possess containing no example. In the colour of the ground, as above remarked, there is every possible variety of shade between pure white and a very rich salmon-colour. In the intensity and number of the markings there is an equally great variety. The markings, always spots and specks, the largest never exceeding 0·1 inch in diameter, are invariably most numerous towards the large end, where they are sometimes, though rarely, slightly confluent. They vary from only two or three to a number too large to count, and in colour through many shades of reddish, blackish, and purplish brown, the latter being rare and abnormal.

The eggs are entirely devoid of gloss, as a rule, though here and there a slight trace of it is observable. It is this want of gloss alone that distinguishes some of the larger white, black-spotted varieties from the eggs of the common Oriole, which they occasionally exactly resemble not only in shape, colour, and character of marking, but even (though generally smaller) in size.

In length they vary from 0·87 to 1·15 inch, and in breadth from 0·7 to 0·85, but the average of 152 eggs measured is 1·01 by 0·75 inch. I have two dwarf eggs of this species not included in the above average which I myself obtained in different nests, measuring only 0·78 by 0·5 inch, and 0·87 by 0·62 inch.

328. Dicrurus longicaudatus, A. Hay. *The Indian Ashy Drongo.*

Dicrurus longicaudatus, *A. Hay, Jerd. B. Ind.* i, p. 430.
Buchanga longicaudata (*A. Hay*), *Hume, Rough Draft N. & E.* no. 280.

The Indian Ashy Drongo, a species that, with the really large series before me from all parts of India, I find it impossible to subdivide into two or more species, breeds alike in the plains, in well-watered and wooded districts, and in the Himalayas up to an elevation of 6000 to 7000 feet, and lays during the months of May and June.

They build generally in large trees, at a considerable height from the ground, placing their somewhat shallow cup-shaped nests in some slender fork towards the summit or exterior of the tree. The nest is neatly and firmly built, of fine grass-stems, slender twigs, and grass-roots, closely interwoven, and externally bound together with cobwebs, in which, as in the body of the nest, lichens of several species are much intermingled. Exteriorly the nests are from 4 to 5 inches in diameter, and from 2 to 2½ in height. Interiorly they are lined with moss, roots, hairs, and fine grass ; the cavity measuring from 3 to 3·5 inches in breadth, and from 1·1 to 1·4 inch in depth. The normal number of the eggs is four.

Mr. Brooks says:—"The nest is usually fixed on the upper surface of a thin branch about 15 to 20 feet from the ground, and at its junction with another branch, the nest being partly embedded in the fork of two *horizontal* branches. It is composed of grass, fibres, and roots, and lined with finer grasses and a few hairs. The nest is broader and much shallower than that of *D. ater* ; outside it is covered with spiders' webs and small bits of lichen.

" The eggs are four in number, sometimes only three, and vary much in size, shape, and colour ; size 1·0 by 0·7 inch : some are buff, blotched with light reddish brown and pale purple-grey ; others are lighter buff, almost white in fact, spotted and marked more sparingly than the first described with the same two colours, but each of a darker tint ; others are white, marked sparingly with spots and blotches of dark purple-brown and reddish brown, and intermixed with larger blotches of deep purple-grey, the markings principally forming a zone at the larger end. Others, again, are pale purplish white, spotted with dark and light purple-brown, and intermixed with spots and blotches of purple-grey. The shape of the egg varies as much as the colouring, some being of a fine oval form, while others are quite pyriform. Laying in Kumaon from the middle to end of May."

As I shall notice further on, I think that Mr. Brooks is mistaken about some of his eggs.

Captain Hutton remarks :—" This species, the only one that visits Mussoorie, arrives from the Doon about the middle of March, and retires again about September. It is abundant during the summer months, and breeds from the latter end of April till the middle of June, making a very neat nest, which is placed in the bifurcation of a horizontal branch of some tall tree, usually an oak tree ; it is constructed of grey lichens gathered from the trees, and fine seed-stalks of grasses, firmly and neatly interwoven ; with the latter it is also usually lined, although sometimes a black fibrous lichen is used ; externally the materials are kept compactly together by being plastered over with spiders' webs. It is altogether a light and elegant nest. The shape is circular, somewhat shallow ; internal diameter 3 inches. The eggs are three or four, generally the latter number, and so variable in colour and distribution of spots that until I had got several specimens and compared them narrowly, I was inclined to think we had more than one species of *Dicrurus* here. I am, however, now fully convinced that these variable eggs belong to the same species. Sometimes they are dull white with brick-red spots openly disposed in form of a rude ring at the larger end; at other times the spots are rufescent claret, with duller indistinct ones appearing through the shell; others are of a deep carneous hue, clouded and coarsely blotched with deep rufescent claret; while again some are faint carneous with large irregular blotches of rufous clay with duller ones beneath the shell."

Some of Captain Hutton's eggs which he sent me were clearly those of *Hypsipetes psaroides* (of which also he sent me specimens), and the fact is that in thick foliage where the Red-bill is not seen nothing is easier than to mistake this bird for *D. longicaudatus*. I have taken a great many of these nests, and I never found eggs other than of the two types to be below described.

Colonel G. F. L. Marshall writes:—"In Kumaon this species breeds from 4000 to 5000 feet above the sea; the eggs are laid in the last week of May. I have never seen a nest at Naini Tal itself (6000 to 7000 feet), but at Bheem Tal (4000 feet) I found numerous nests within three days, in the first week of June; all without exception had young. The next season I visited the place in the last week of May, and found the eggs just laid.

"The nests were of the usual *Dicrurus* type, wedged in a fork at heights varying from fifteen to fifty feet from the ground, but as far as my experience goes always in conspicuous places and generally on trees almost or quite bare of leaves. The nests are usually only to be obtained by sawing off the bough they are built on."

Long ago Captain Cock, writing from Dhurmsala, said:—"I took a nest on the 8th of May, containing four eggs. The eggs are regular, roundish ovals, somewhat pointed towards one end. The ground-colour is white, here and there suffused with a faint pinkish tinge, and it is spotted and blotched with purplish red and pale lilac, most of the spots being gathered into an irregular zone about the large end."

Colonel C. H. T. Marshall, writing from Murree, says:— "Breeds in May, in almost inaccessible places, about 7000 feet up, choosing a thin fork at the outermost end of a bough about 50 or 60 feet from the ground, and always on trees that have no lower branches. The nest is almost invisible from below, as it is very neatly built on the top of the fork; and when the female sits on it, she places her tail down the bough so as entirely to hide herself. The eggs are only to be obtained either by climbing higher up the tree than the nest is, and extracting the eggs by means of a small muslin bag at the end of a long stick, or else by lashing the bough on which the nest is to an upper bough as the climber goes along so as to make it strong enough to support him. The nest is much neater than that of *D. ater*; the eggs are light salmon-coloured, with brick-red blotches sparsely scattered over them, and are ·95 by ·7 inch."

Dr. Scully records the following note from Nepal:—"This species lays in the valley in May and June, the nest being placed high up in trees, often in *Pinus longifolia*. The eggs are usually four in number, fairly glossy, in shape moderate ovals, smaller at one end. The ground-colour is pinkish white, with a tinge of buff, sparingly spotted and blotched with brownish red, chiefly at the large end, where the marks tend to coalesce, so as to form an irregular incomplete ring. Four eggs taken on the 28th May measured 1·09 to 1·12 in length, and 0·75 to 0·76 in breadth. The race which I identify with *D. himalayanus* was found, in very small numbers, on the summit of Sheopuri, at an elevation of about 7500 feet, and was breeding at the time I shot my specimen, viz. the 20th May."

Mr. Gammie found a nest at Mongpho, near Darjeeling, at an elevation of about 3500 feet on the 13th May. It was placed on an outer branch of a tall tree and contained only one partially incubated egg. The nest was a beautifully compact, but shallow cup, placed on the upper surface of the bough, composed externally of roots and coated with a little lichen and a great deal of cobweb. Interiorly lined with the finest grass and moss-roots. The cavity measured about 3 inches in diameter and scarcely more than 1 inch in depth. At the bottom, where it rested on the bough, the nest was not above ¼ inch thick, and consisted only of the lining materials. Laterally it was about ¾ inch thick.

The egg was a broad oval, slightly compressed towards one end, but not at all pointed. The shell very fine and with a slight gloss, the ground-colour a delicate salmon-pink, and with a broad ring of deep brownish-pink spots and blotches intermingled with pale purple subsurface-looking clouds and spots round the large end. The rest of the egg with some half-dozen similar spots.

He subsequently sent me the following note:—"This species is common in the Darjeeling district up to 4000 feet or so. It rather affects the neighbourhood of bungalows, and is a very lively neighbour, especially in the mornings and evenings. These birds are continually quarrelling among themselves, sallying after insects, or making their best attempts at singing. They are *dead* on Kites, Crows, and such-like depredators. For several days an Owl (*Bulaca newarensis*) was flying about near the Cinchona Bungalow at Mongpho, and being a stupid creature at the best, and doubly so during daylight when it had no business to be abroad, was evidently considered fair game by the Long-tailed Drongo and Swallow-Shrikes, and so awfully 'sat upon' by them, that its life must have become a burden to it until it left the place in despair of ever getting either peace or comfort about Mongpho.

"They lay in April and May, and have but one brood in the year. The nest is generally either built against a tall bamboo, well up, supported on the branch of twigs at a node, or near the extremity of a branch of a tree, sometimes on quite slender branches of young trees, which get so tremendously wafted about by the wind as to make the retention of the eggs or young in the nest appear almost miraculous. When anyone meddles with the nest, the owners make bold dashes at the head of the robber. The Darjeeling birds are not so knowing as their fellows of Murree, the females of whom are said to sit on the nests with their tails along the boughs so as to entirely conceal themselves. I have seen dozens of the nests here, and never once saw the female in this position, but always with her tail *across* the bough. The nest is a compact shallow cup, measuring externally 4·5 inches across by 1·75 in height, while the cavity is 3 inches in diameter by about 1·2 in depth. It is made of twigs bound up with cobwebs, among which a few lichens are intermingled. The lining is a mixture of straw-coloured root-fibres and fine branchlets of the same coloured grass-panicles."

Mr. Mandelli sent me nests of this species, which were taken at Ging, near Darjeeling, on the 26th April and on the 22nd May, the one contained one fresh egg, the other three. They were both placed on branches of large trees at heights of about 20 feet from the ground. They are broad shallow cups, from 4 to 5 inches in diameter, about 2 in height, compactly composed of fine twigs and grass-stems, bound together with cobwebs and with many pieces of lichen and some tiny dry leaves worked in on the outer surface. Interiorly, they are lined with very fine hair-like grass-stems. The saucer-like cavities are about 3 inches in diameter and about 1¼ in depth.

Dr. Jerdon says :—" I found its nest on one occasion, in April, in Lower Malabar. It was shallow and loosely made with roots, and lined with hair, about 20 feet from the ground, on the fork of a tree ; and it contained three eggs of a pinkish-white colour, with some longish rusty or brick-red spots."

There are two very strongly marked types of this bird's eggs. The eggs of both types are moderately broad, or, at most, some-what elongated ovals, and comparatively devoid of gloss. The first, in its colouring, exactly resembles the eggs of *Caprimulgus indicus* ; a pinkish salmon-coloured ground, streaked, blotched, and clouded, but nowhere densely (except towards the large end, where there is a tendency to form a cap or zone), with reddish pink, not differing widely in hue from, though deeper in shade than, the ground-colour. Here and there, where the markings are thickest, under-clouds of very faint purple occur, but these are too feeble to attract attention, unless the egg is looked into closely. In the other type of egg, the ground-colour is pale pinkish white, pretty boldly blotched and spotted almost exclusively towards the large end, where there is a broad irregular imperfect zone, with brownish red, intermingled with blotches of very faint inky purple. My description possibly fails to make this as apparent as it should be, but no two eggs can, to a casual observer, appear more distinct than these two types. There is yet, according to Mr. Brooks, a third type of this bird's eggs ; of this he has given me a single

example. In shape it is excessively long and narrow, of the type
of the eggs of *Chibia hottentotta*, but its coloration and character
of markings are unlike those of any Shrike or Drongo with which
I am acquainted, and exactly resemble those of many types of the
eggs of the several Bulbuls. The ground-colour is pinkish white,
and is thickly speckled and spotted throughout with primary markings of rich brownish red, and feeble secondary ones of excessively
pale inky purple. This egg, moreover, possesses a degree of gloss
never observable in those of the *Dicruri*, and therefore, well
assured though Mr. Brooks is of the parentage of this egg which
he took with his own hands, I feel confident, having since obtained
many eggs of *Hypsipetes psaroides* which are exactly similar to
this last described egg, that in, perhaps, indifferent light he mistook this bird for a *Dicrurus*. I may add that the first described
type, of which I have procured numerous specimens from different
parts of the Himalayas, taking *several* nests with my own hands, is
most characteristic of this species.

In the type with the pinky-white ground, large or small spots
often occur about the large end of a deep purple colour, so deep
as to be almost black, and but for the absence of gloss some of
these paler eggs are very close to those of some of the Orioles.
Intermediate varieties between the two types above described occur,
but in not one of more than sixty specimens that I have examined
has there been any perceptible gloss.

The eggs vary in length from 0·85 to 1·01 inch, and in breadth
from 0·7 to 0·75 inch, but the average of fifty-one eggs is 0·95
by 0·74 inch.

329. Dicrurus nigrescens, Oates. *The Tenasserim Ashy Drongo.*

Dicrurus nigrescens, *Oates*; *Oates, B. I.* i, p. 315.

Mr. Oates found the nest of this Drongo in Pegu. He says:—
" I found one nest on the 27th April at Kyeikpadein, near the
town of Pegu, on a small sapling near the summit. It contained
four eggs *; they are without gloss; the ground-colour in all is
white. In three eggs the whole shell is marked with spots of pale
purple; these are perhaps more numerous at the thick end, but
not conspicuously so. The fourth egg is blotched, not spotted,
with the same colour.

"The nest is composed of fine twigs and the dry branches of
weeds; it is lined very firmly and neatly with grass. Exterior
diameter 5 inches and depth 2 ; egg-chamber 3½ inches across and
1¼ deep. The outside of the nest is profusely covered with lichens
and cobwebs. The eggs measure from ·83 to ·95 in length, and
·68 to ·71 in width."

---

* I recorded the nest and eggs of this bird under the name of *Buchanga
intermedia* (S. F. v, p. 149). The parent birds of these eggs are fortunately still
in the British Museum, and I am able to identify them with this species, which
occurs generally throughout Tenasserim and many parts of Lower Pegu.—ED.

330. **Dicrurus cærulescens** (Linn.). *The White-bellied Drongo.*

Dicrurus cærulescens (*L.*), *Jerd. B. Ind.* i, p. 432.
Dicrurus cæruleus (*Müll.*), *Hume, Rough Draft N. & E.* no. 281.

I have never seen a nest of the White-bellied Drongo. Mr. R. Thompson says:—"This bird's breeding-habitat is from 2500 to 6000 feet in the Himalayas. It is common on the south-eastern slopes of Nyneetal. It lays in May and June, placing its shallow cup-shaped nest in some little fork near the top of a moderate-sized oak-tree, if breeding on a mountain-side, but of some tall *Alnus nipalensis, Acacia elata,* or *Acer oblongum,* if nesting in deep dells or valleys The nest appeared to be exactly like that of *D. ater*; but I can say nothing very positive about it or the eggs, as, though continually seeing them, I never, I think, took the trouble of getting one down."

Colonel G. F. L. Marshall, commenting on Mr. Thompson's remark that this Drongo is common near Naini Tal, says:—"My experience on this point is negative; I have carefully searched the south-eastern slopes of Naini Tal for four years without even seeing the bird, so that I do not think it can be classed as a common breeder here."

Mr. J. Davidson informs us that on the 16th July he saw a brood of *Dicrurus cærulescens* on the Kondabhari Ghât, just able to fly. Referring to Western Khandeish, he tells us that he saw only two nests. They were on adjoining trees in the Akrani; they were largish nests, not like those of *D. ater*, but more resembling those of *D. longicaudatus* described in 'Nests and Eggs.' One nest contained three young ones, the other was only building; and nothing could have been more plucky than the way the old ones defended their nest.

331. **Dicrurus leucopygialis**, Blyth. *The White-vented Drongo.*

Buchanga leucopygialis (*Bl.*), *Hume, Rough Draft N. & E.* no. 281 bis.

Colonel Legge gives us the following account of the breeding of this Drongo, which is confined to Ceylon :—" The breeding-season of this Drongo is from March until May ; and the nest is almost invariably built at the horizontal fork of the branch of a large tree, at a considerable height from the ground, sometimes as much as 40 feet. It is a shallow cup, measuring about $2\frac{1}{4}$ inches in diameter by 1 in depth, and is compactly put together, well finished round the top, but sometimes rather loose on the exterior, which is composed of fine grass-stalks and bark-fibres, the lining being of fine grass or tendrils of creepers. The number of eggs varies from two to four, three being the most common. They vary much in shape, and also in the depth of their ground-tint ; some are regular ovals, others

are stumpy at the small end, while now and then very spherical
eggs are laid.   They are either reddish white, 'fleshy,' or pure
white, in some cases marked with small and large blotches of faded
red, confluent at the obtuse end, and openly dispersed over the rest
of the surface, overlying blots of faint lilac-grey ; others have a
conspicuous zone round the large end, with a few scanty blotches
of light red and bluish grey on the remainder; in others, again,
the markings are confined to a few very large roundish blotches of
the above colours at one end, or, again, several still larger clouds of
brick-red at the obtuse end, with a few blotches of the same at the
other.   Dimensions from 1·0 to 0·86 inch in length, by 0·72 to 0·68
in breadth.   I once observed a pair in the north of Ceylon very
cleverly forming their nest on a horizontal fork by first constructing
the side furthest from the angle, thus forming an arch, which was
then joined to the fork by the formation of the bottom of the
structure.

   "The parent birds in this species display great courage, vigour-
ously sweeping down on any intruder who may threaten to molest
their young."

334. Chaptia ænea (Vieill.).   *The Bronzed Drongo.*

Chaptia ænea (*V.*), *Jerd. B. Ind.* i, p. 433 ; *Hume, Rough Draft N.
   & E.* no. 282.

   The Bronzed Drongo breeds, according to Mr. Hodgson's notes,
in the central hills of Nepal, or rather in the plains near to these
hills, rarely quitting large woods.   They begin to lay in March,
and build a broad somewhat saucer-shaped nest some 4 or 5 inches
in width and 2 to 3 in depth externally.   The nest is placed in
some slender horizontal fork, to one at least of the twigs of which
it is firmly attached by vegetable fibres ; it is composed of fine
twigs and grass, and bound round with cobwebs in which pieces of
lichen and small cocoons are often intermingled.   Mr. Hodgson
specially notes :—" *June 6th, valley.*   Female, nest and eggs ; nest
on fork of upper branch of large tree, 4·5 inches wide by 2·25 deep,
cup-shaped, made of fibres of grass bound with cobweb, lining
none ; three eggs, obtusely oval, the ground fawn tinged white,
blotched (especially at larger end) with fawn or reddish brown."

   It appears that four is the maximum number of eggs laid ; both
sexes participate in the work of incubation and rearing the young,
but they are very jealous of the approach of any birds when they
have eggs or young, driving all such intruders away with the
utmost bravery.   The eggs measure from 0·88 to 0·95 inch by
0·65.

   From Sikhim Mr. Gammie writes :—" I have found the Bronzed
Drongo breeding from April to June in the low hot valleys at
about 2000 feet above the sea.   It suspends its nest in a slender
horizontal fork at 10 feet or more from the ground, and appears,
like its frequent neighbour *Dicrurus longicaudatus*, to prefer a

bamboo-clump to breed in. The nest is a compact cup, neatly made of fine grass-stalks, with an outer coating of dry bamboo-leaves plastered over with cobwebs; it is fastened to the supporting branches by cobwebs. Externally it measures 3·5 inches wide by 2 inches deep, internally 2·5 by 1·5.

" The usual number of eggs is three."

Major M. Forbes Coussmaker, writing from Bangalore, tells us:— " I took the nest of this bird on 6th April in the Shemagah District, Mysore. It was built on the fork of a bare branch about 20 feet from the ground in big tree-jungle, and was composed of fine grass, fibre, and a few dry bamboo-leaves woven together with cobwebs, making a small compact cup-like nest which measured 3 inches in diameter externally, 2·5 internally, and 1·4 deep.

" From where I stood I saw the bird come and sit on the nest and fly off again a dozen times at least. The eggs, three in number, measured ·9 by ·65, and were pinkish white with darker pink and light purple blotches and spots all over, principally at the larger end."

Mr. J. R. Cripps informs us that at Furreedpore, in Eastern Bengal, this species is "rather common; generally to be found perching on the dead branches of high trees overlooking water, especially whenever there is a dense undergrowth of jungle. On the 1st June, 1878, I secured a nest with three fresh eggs; it was built on a slender twig on the outer side of a mango-tree which was standing near a ryot's house, and was about 15 feet off the ground. External diameter 3¼ inches, depth 2; internal diameter 2⅓, depth 1⅛. Saucer-shaped; the outside consisted of plaintain-leaves torn up into slips, all of which were firmly bound together by fibres of the plaintain-leaf and jute, which were wound round the twigs and secured the nest. Inside lining was made of very fine pieces of 'sone' grass. The pair were very pugnacious, attacking any birds coming near their nest. These birds have a clear mellow ringing whistle."

Mr. Oates writes from Pegu:—"I procured one nest on the 23rd April. It was placed at the tip of an outer branch of a jack tree, and attention was drawn to it by the vigorous attacks the parents made on passing birds. The nest was suspended in a fork; the outside diameter is 4 inches and inside 3, total depth 2½, and the egg-cup is about 1½ deep. The nest is composed of fine grass, strips of plaintain-bark, and other vegetable fibres closely woven together; the edges and the interior are chiefly of delicate branchlets of the finer weeds and grasses. It is overlaid at the edges, where it is attached to the branches, with cobwebs, and a few fragments of moss are stuck on at various points.

" There were two fresh eggs; the ground-colour is a pale salmon-fawn, and the shell is covered with darker spots and marks of the same. They are only very slightly glossy. The two eggs measure 0·85 by 0·62."

Major C. T. Bingham writes from Tenasserim :—"On the 10th March, 1880, being encamped at the head-waters of the Queebaw-

14*

choung, a feeder of the Meplay, and having an hour to spare, I
took my gun and climbed up a steep hill to the very sources of the
Queebaw. Here, hanging over the trickling stream, was a nest of
Chaptia ænea firmly woven and tied on to a fork in the branch of
a little tree, at a height of about 10 feet from the ground. The
nest was of roots and grass lined by soft fine black roots, and held
three eggs, of a rich salmon-pink, obscurely spotted darker at the
large end; they measure 0·83 by 0·61, 0·82 by 0·61, and 0·80 by
0·61 respectively.

"On the 15th March, 1880, in the fork of a branch of a small
zimbun-tree (*Dillenia pentagyna*), hanging over a pathway along
the bank of the Meplay stream, I found a nest of the above species.
A neat strongly-made little cup of vegetable fibres and cobwebs,
containing two fresh eggs; ground-colour dull salmon, obscurely
spotted with brownish pink. They measure 0·86 by 0·64 and
0·88 by 0·65."

Mr. J. L. Darling, Jun., records the following notes :—

"26th March. Found a nest of *Chaptia ænea*, building, when on
the march from Tavoy to Nwalabo, some seven miles east of Tavoy,
in the fork of a bamboo-branch 12 feet from ground.

"29th March. Took two fresh eggs of *Chaptia ænea*, and shot
the bird off nest, about twenty-three miles east of Tavoy, in open
bamboo-land, very low elevation. The nest was built in the fork
of an overhanging branch of a bamboo some 50 feet from the
ground.

"13th April. Found a nest of *Chaptia ænea* with two large young
ones. Nest built in a tree some 40 feet from ground, in open
forest about twenty miles east of Tavoy.

"22nd April. Found a nest of *Chaptia ænea* with two large
young ones. Nest built at the end of a bough about 30 feet from
ground, near Tavoy."

The nests of this species are quite of the Oriole type, more or
less deep cups suspended between the forks of small branches or
twigs of some bamboo-clump or tree. Exteriorly they are composed of dry flags of grass, bits of bamboo-spathes, or coarse grass,
bound together with vegetable fibres and often with a good deal of
cobweb worked over them; sometimes a tiny bit or two of moss
may be found added, and often the fine thread-like flower-stems of
grass. Interiorly they are generally lined with excessively fine
grass. In one or two nests very fine black fern-roots are intermingled with the grass lining. The nests vary a good deal in size,
but are all extremely compact, and while some are decidedly massive, nearly an inch thick at bottom, others are scarcely a quarter
of this in thickness beneath. In one the cavity is 2·5 inches broad
by 3 long, and fully 2 deep; in another it is about 2·5 inches in
diameter by scarcely 1·25 inches in depth. In one nest four fresh
eggs were found; in another three fully incubated ones. The
nests were suspended at heights of from 10 to 30 feet from the
ground.

The eggs sent by Mr. Gammie very much recall the eggs of

Niltava and others of the Flycatchers. They are moderately elongated ovals, in some cases slightly pyriform, in others somewhat pointed towards the small end. The shell is fine and compact, smooth and silky to the touch, but they have but little gloss. The ground-colour varies from a pale pinkish fawn to a pale salmon-pink, and they exhibit round the large end a feeble more or less imperfect and irregular zone of darker-coloured cloudy spots, in some cases reddish, in some rather inclining to purple, which zone is more or less involved in a haze of the same colour, but slightly darker than the rest of the ground-colour of the egg.

The eggs vary in length from 0·76 to 0·88, and in breadth from 0·6 to 0·64. The average of fifteen eggs is 0·82 by 0·61.

335. Chibia hottentotta (Linn.). *The Hair-crested Drongo.*

Chibia hottentota (*L.*), *Jerd. B. Ind.* i, p. 439; *Hume, Rough Draft N. & E.* no. 286.

Mr. R. Thompson says :—" The Hair-crested Drongo is extremely common as a breeder in all our hot valleys (Kumaon and Gurwhal). It lays in May and June, building in forks of branches of small leafy trees situated in warm valleys having an elevation of from 2000 to 2500 feet. The nest is circular, about 5 inches in diameter, rather deep and hollow ; it is composed of fine roots and fibres bound together with cobwebs, and it is lined with hairs and fine roots. They lay from three to four much elongated, purplish-white eggs, spotted with pink or claret colour."

Dr. Jerdon remarks :—" The Lepchas at Darjeeling brought me the nest, which was said to have been placed high up in a large tree ; it was composed of twigs and roots and a few bits of grass, and contained two eggs, livid white, with purplish and claret spots, and of a very elongated form."

The Jobraj, according to Mr. Hodgson's notes and figures, begins to lay in Nepal in April. It builds a large shallow nest, 8 or 9 inches in diameter externally, with the cavity of about half that diameter, attached, as a rule, to the slender branches of some horizontal fork, between which it is suspended much like that of an Oriole, though much shallower than this latter ; it is composed of small twigs, fine roots, and grass-stems bound together, and it is attached to the branches by vegetable fibre, and more or less coated with cobwebs ; little pieces of lichen and moss are also blended in the nest. It lays three or four eggs, rather pyriform in shape, measuring 1·25 by 0·86 inch, with a whitish or pinky-whitish ground, speckled and spotted pretty well all over, but most densely towards the large end, with reddish pink.

From Sikhim Mr. Gammie writes :—" I took two nests of the Hair-crested Drongo this year in June, both at about an elevation of 1500 feet in wooded valleys, placed well up in the outer branches of tall, slender trees ; they are of a broad saucer-shape, openly but firmly made of roots and stems of slender climbers, and desti-

tute of lining. There is a good deal of cobweb on the outsides of
the nests, and they were attached to the supporting branches by
the same material. One was fixed in among several upright sprays,
the other suspended in a slender fork after the manner of an
Oriole. They measured about 6 inches broad by 2¼ deep externally,
internally 4 by 1¾. One nest contained four fresh eggs, the other
three partially-incubated eggs."

Mr. Oates, writing from Pegu, says :—" In the first week of
May I took several nests of this bird, but in all cases the nests
were situated in such dangerous places that most of the eggs got
broken ; there were three in each nest. The position of the nest
and the nest itself are very much like those of *D. paradiseus*.
Comparing many nests of both species together, the only difference
appears to be that the nests of the Hair-crested Drongo are slightly
larger on the whole.

" The only two eggs saved measure 1·10 by ·8 and 1·11 by ·81 ;
they are slightly glossy, dull white, minutely and thickly freckled
and spotted with reddish brown and pale underlying marks of
neutral tint.

" I may add that at the commencement of May all the eggs were
much incubated."

Major C. T. Bingham remarks :—" During the breeding-season
in the end of March and in April I saw a great number of nests
round and about Meeawuddy in Tenasserim, but all inaccessible, as
they were invariably built out at the very end of the thinnest
branches of eng, teak, thingan (*Hopea odorata*), and other trees.

" Except during those two months, I have not seen the bird
plentiful anywhere."

Mr. J. R. Cripps has written the following valuable notes
regarding the breeding of the Hair-crested Drongo in the Dibrugarh district in Assam :—

" 17th May, 1879. Nest with three fresh eggs, attached to a
fork in one of the outer branches of an otinga (*Dillenia pentagyna*)
tree, and about 15 feet off the ground.

" 15th May, 1880. Three fresh eggs in a nest 20 feet off the
ground, and a few yards from my bungalow, in an oorian (*Bischoffia
javanica*, Bl.).

" 5th June, 1880. Nest with three partly-incubated eggs, in one
of the outer branches of a jack (*Artocarpus integrifolia*) tree, and
about 15 feet off the ground.

" 27th May, 1881. Three fresh eggs in a nest on a soom (*Ma-
chilus odoratissima*) tree at the edge of the forest bordering the
tea. The nests are deep saucers, 3½ inches in diameter, internally
1½ deep, with the sides about ¼ thick ; but the bottom is so
flimsy that the eggs are easily seen from below, the materials being
grass, roots, and fine tendrils of creepers, especially if these are
thorny, when they are used as a lining. The nest is always situated
in the fork of a branch."

The nests are large, shallow, King-Crow-like structures, often
suspended between forks, sometimes placed between four or five

upright shoots, at times resting on a horizontal bough against and attached to some more or less upright shoots. They are composed mainly of roots thinly but firmly twisted together, have sometimes a good deal of cobweb twisted round their outer surface, often a good deal of vegetable fibre used for the same purpose, and, though they have no lining, are always composed interiorly of finer material than that used for the outer portion of the structure. Exteriorly the diameter varies from 6 to nearly 7 inches, the height from nearly 2 to 2½ ; the cavity is usually about 4 inches in diameter and 1·5 to 1·75 in depth. I have taken the nests in May and June alike in small and large trees, at elevations of from 10 to 30 feet from the ground.

Typically the eggs are rather broad ovals, a good deal pointed towards the small end, but they vary a great deal both in size and shape, are occasionally very much elongated, and again, at times, exhibit the characteristic pointing but feebly. The ground-colour varies from greyish white to a delicate pale pink ; as a rule the markings are small and inconspicuous frecklings and specklings of pale purple reddish where the ground is pink, greyish where it is white, tolerably thickly set about the large end and somewhat sparsely elsewhere ; but in some eggs these markings are everywhere almost obsolete. In many there is a dull pale purplish cloud underlying the primary markings, extending over the greater part of the large end of the egg. Not uncommonly a few specks and spots of yellowish brown are scattered here and there about the egg. In one egg before me the markings are larger, more decided, and fewer in number—distinct spots, some of them one tenth of an inch in diameter ; and in this egg the spots are decidedly brownish red, while intermixed with them are a few specks and clouds of inky purple. The ground in this case is a pale pinky white.

As a rule the eggs are entirely devoid of gloss, but one or two have a very faint gloss.

The eggs measure from 1·01 to 1·21 in length, and from 0·79 to 0·86 in breadth ; but the average of twenty-nine eggs is 1·12 by 0·81.

338. **Dissemurulus lophorhinus (Vieill.).** *The Ceylon Black Drongo.*

Dissemuroides lophorhinus (*V.*), *Hume, Cat.* no. 283 quat.

Colonel Legge says, in his 'Birds of Ceylon' :—"This species breeds in the south of Ceylon in the beginning of April. I have seen the young just able to fly in the Opaté forests at the end of this month ; but I have not succeeded in getting any information concerning its nest or eggs."

upright shoots, at times resting on a horizontal bough against and attached to some more or less upright shoots. They are composed mainly of roots thinly but firmly twisted together, have sometimes a good deal of cobweb twisted round their outer surface, often a good deal of vegetable fibre used for the same purpose, and, though they have no lining, are always composed interiorly of finer material than that used for the outer portion of the structure. Exteriorly the diameter varies from 6 to nearly 7 inches, the height from nearly 2 to 2½; the cavity is usually about 4 inches in diameter and 1·5 to 1·75 in depth. I have taken the nests in May and June alike in small and large trees, at elevations of from 10 to 30 feet from the ground.

Typically the eggs are rather broad ovals, a good deal pointed towards the small end, but they vary a great deal both in size and shape, are occasionally very much elongated, and again, at times, exhibit the characteristic pointing but feebly. The ground-colour varies from greyish white to a delicate pale pink; as a rule the markings are small and inconspicuous frecklings and specklings of pale purple reddish where the ground is pink, greyish where it is white, tolerably thickly set about the large end and somewhat sparsely elsewhere; but in some eggs these markings are everywhere almost obsolete. In many there is a dull pale purplish cloud underlying the primary markings, extending over the greater part of the large end of the egg. Not uncommonly a few specks and spots of yellowish brown are scattered here and there about the egg. In one egg before me the markings are larger, more decided, and fewer in number—distinct spots, some of them one tenth of an inch in diameter; and in this egg the spots are decidedly brownish red, while intermixed with them are a few specks and clouds of inky purple. The ground in this case is a pale pinky white.

As a rule the eggs are entirely devoid of gloss, but one or two have a very faint gloss.

The eggs measure from 1·01 to 1·21 in length, and from 0·79 to 0·86 in breadth; but the average of twenty-nine eggs is 1·12 by 0·81.

338. Dissemurulus lophorhinus (Vieill.). *The Ceylon Black Drongo.*

Dissemuroides lophorhinus (*V.*), *Hume, Cat.* no. 283 quat.

Colonel Legge says, in his 'Birds of Ceylon':—"This species breeds in the south of Ceylon in the beginning of April. I have seen the young just able to fly in the Opaté forests at the end of this month; but I have not succeeded in getting any information concerning its nest or eggs."

339. Bhringa remifer (Temm.). *The Lesser Racket-tailed Drongo.*

Bhringa remifer (*Temm.*), *Jerd. B. Ind.* i, p. 434.
Bhringa tenuirostris, *Hodgs., Hume, Rough Draft N. & E.* no. 283.

Of the Lesser Racket-tailed Drongo Mr. R. Thompson says:—
"This elegant Drongo is somewhat common in our lower Kumaon
ranges. Its lively clear and ringing notes are one of the greatest
charms of the spring season in our forests. It breeds in May and
June, and builds upon lofty trees in dense forests, usually in some
deep damp valley. The nest from below looks just like that of a
common King-Crow—a broad shallow cup; but I never closely
examined either nest or eggs."

Dr. Jerdon remarks:—"A nest with eggs were brought to me
in June, said to be of this species. The nest was loosely made of
sticks and roots, and contained three eggs, reddish white, with a
very few reddish-brown blotches."

From Sikhim, Mr. Gammie writes:—"I have taken but one
nest of this Drongo. It was suspended between two small horizontal forking branches of a tall tree, some 20 feet from ground.
It is a neat, saucer-shaped structure, somewhat triangular, to fit
well up to the fork, built of fibry roots, and firmly bound to the
branches by spiders' webs. The sides and bottom are strong, but
so thin that they can everywhere be seen through. Externally it
measures 4·5 inches across by 1·9 in height; internally 3·5 by 1·3.
It was taken on the 15th May at 2500 feet, and contained three
partially incubated eggs."

A nest of this species taken by Mr. Gammie at Rishap (elevation
4800) in Sikhim, on the 20th May, is a very broad shallow saucer,
composed almost entirely of moderately fine dark brown roots,
but with a few slender herbaceous twigs intermingled. It is suspended in the fork of two widely diverging twigs, to which either
margin is attached, chiefly by cobwebs, though on one side at one
place part of the substance of the nest is wound round the twig:
the cavity, which is not lined, is oval, and measures 3·25 inches by
2·75, by barely 0·75 in depth. The female seated on the nest had
long tail-feathers, so this species does not drop these for convenience in incubating.

Several nests of this species obtained in Sikhim by Messrs.
Gammie, Mandelli, &c. are all precisely similar—broad saucers,
suspended Oriole-like between the fork of a small branch. Exteriorly composed of moderately fine brown roots, more or less
bound together, especially those portions of them that are bound
round the twigs of the fork with cobwebs, and lined interiorly with
fine black horsehair-like roots. They seem to be always right up in
the angle of the fork, whereas in *Chaptia* they are often some inches
down the fork, and consequently the cavity is triangular on the
one side, and semicircular on the other. The cavities measure from
3 to nearly 4 inches in their greatest diameters, and vary from 1
to 1½ inch in depth; though strong and firm, and fully ¼ of an

inch thick at bottom, the materials are so put together that, held up against the light, they look like a fine network.

The eggs of this species obtained by Mr. Gammie, though more elongated in shape and somewhat larger, very closely resemble in coloration the more ordinary type of the eggs of *Dicrurus longicaudatus*. In shape they are elongated ovals, a good deal compressed towards the smaller end. The shell is fine, but has scarcely any gloss. The ground-colour is a moderately warm salmon-pink. It is spotted, streaked, and blotched thickly about the large end (where there is a tendency to form a cap or zone), thinly elsewhere, with somewhat brownish red, or in some merely a darker shade of the ground-colour; where the markings are thickest about the large end, in some only one or two, in others numerous blotches and clouds of a dull inky purple are intermingled, and a few specks and spots of the same colour often occur elsewhere about the egg.

Two eggs measure 1·09 by 0·75, and a third measures 0·98 by 0·75.

340. Dissemurus paradiseus (Linn.). *The Larger Racket-tailed Drongo.*

Edolius paradiseus (*L.*), *Jerd. B. Ind.* i, p. 435.
Edolius malabaricus (*Scop.*), *Jerd. t. c.* p. 437.
Dissemurus malabaroides (*Hodgs.*), *Hume, Rough Draft N. & E.* no. 284.

Of the Larger Racket-tailed Drongo Dr. Jerdon tells us that he has "had its nest brought him several times at Darjeeling; rather a large structure of twigs and roots; and the eggs, usually three in number, pinkish white, with claret-coloured or purple spots, but they vary a great deal in size, form, and colouring. They breed in April and May."

The solitary egg that I possess of this species, given me by Dr. Jerdon, is probably an exceptionally small one. It is a broad oval, tapering a good deal towards one end, a good deal smaller than the eggs of *Chibia hottentotta*, and not very much larger than some eggs of *D. ater*. Its coloration, however, resembles that of *Chibia hottentotta*, and differs conspicuously, *when compared with them* (though it may be difficult to make this apparent by description), from those of the true *Dicruri*. The ground-colour is a dead white, and it is very thinly speckled all over, a little more thickly towards the large end, with minute dots and spots, chiefly of a very pale inky purple, a very few only of the spots being a dark inky purple. The texture of the egg is fine and close, but it is devoid of gloss. This egg measures 1·1 by 0·87 inch.

Mr. Iver Macpherson writes from Mysore:—

"*Kakencotte State Forest, Mysore District.*—I send you six eggs, specimens from three different nests.

"This bird is very common in the heavy forests of the Mysore District, but the only nest I have ever found myself was on the 2nd May, 1880, and contained two or three young birds. I could not distinctly see how many. The nest was fixed towards the end of a branch of a tree, at a considerable height from the ground, and was almost impossible to get at. Had there been eggs in it I could not have taken them.

"The breeding-season I should say was from the beginning of April to the end of May.

"Three nests, each containing three eggs, were brought to me this season on the 10th and 26th April, and 9th May, 1880, by Cooroobahs (the jungle-tribes in these forests); and although the eggs in each nest vary considerably from one another, there is no doubt in my mind that the eggs belong to one and the same species of bird.

"It is a bird so well known in these forests that it would be impossible to mistake it for any other.

"In one case only was the nest brought to me, and this, which unfortunately I did not keep, was loosely made of twigs and roots."

Professor H. Littledale, quoting Mr. J. Davidson, informs us that this species breeds in the east of Godhra, and therefore probably throughout the Panch Mehals.

Mr. J. Inglis, writing from Cachar, says:—"The Bhimraj is very common, frequenting thick jungle; it often goes in company with other birds, which it mimics to perfection. It lays about four eggs in a shallow nest made of grass similar to the above; it is very easily tamed. The hill-tribes use the long tail-feathers for ornamenting their head-dresses."

Mr. Oates writes from Pegu:—"I have taken the eggs of this species on all dates, from the 30th April to the 16th June.

"The nest is placed in forks of the outer branches of trees at all heights from 20 to 70 feet, and in all cases they are very difficult to take without breaking the eggs.

"The nest is a cradle, and the whole of it lies below the fork to which it is attached. It is made entirely of small branches of weeds and creepers, finer as they approach the interior. The egg-cup is generally, but not always, lined with dry grass.

"The outside dimensions are 6 inches in diameter and 3 deep. The interior measures 4 inches by 2. In one nest the sides are bound to the fork by cotton thread in addition to the usual weeds and creepers.

"The eggs have very little or no gloss, and differ among themselves a good deal in colour. In one clutch the ground-colour is white, spotted and blotched, not very thickly, with neutral tint and inky purple, chiefly at the larger end. Other eggs are pinkish salmon, and the shell is more or less thickly or thinly covered with pale greyish purple or neutral tint, and brownish-yellow or orange-brown spots and dashes.

"They vary in size from 1·2 to 1·06 in length, and ·85 to ·8 in breadth."

Major C. T. Bingham has the following note:—"About five miles below the large village of Meplay, in the district of that name, the main stream of the Meplay river is joined by a tributary, the Theedoquee. On the 4th April I was wading across the mouth of the latter, when my attention was attracted by seeing a pair of the above birds dart from a small tree growing at the very point of the fork where the streams met, and sweep down at my dog, not actually striking him, but nearly doing so. Of course, I made for the tree, and sure enough there, about 15 feet from the ground, in a fork, was a large mass of twigs, above which was placed a neatly made cup-shaped nest, lined with fine black roots, and containing three fresh eggs, densely spotted, chiefly at the larger end, with yellowish brown and sepia, on a ground-colour of dull greenish white. The whole time the peon I had sent up was climbing up and getting the nest, the two birds kept sweeping round and round with harsh cries. I secured them both for the identification of the eggs."

The eggs of this species are typically rather long ovals, generally a good deal pointed towards the small end. They are dull eggs, and never seem to have any perceptible gloss. The ground-colour varies from white to a rich warm pink. The markings are of all sizes and shapes, from large blotches to the tiniest specks, and they vary in every egg, being thickly set in some, thinly in others, but as a rule the largest and most conspicuous markings are about the large end. Again, in colour the markings vary very much: they are red, purplish red, reddish brown, pale purple, and inky grey; generally the eggs exhibit both coloured markings reddish and lilac, but sometimes the white-grounded eggs have only these latter. Some of the pink eggs are strikingly handsome, and remind one of those of some of the Bulbuls. Others are dull eggs with only a few irregular grey clouds about the large end, thinly interspersed with brownish-red spots, usually darker about the centre, and elsewhere excessively minutely and thinly speckled with spots too small to render it possible to say what colour they are.

An egg I received from Darjeeling measures 1·1 by 0·87; others received from Mynall from Mr. Bourdillon, and the Kakencotte Forest, Mysore, from Mr. I. Macpherson, vary in length from 1·16 to 1·1, and in breadth from 0·84 to 0·75. Three eggs, taken in Pegu by Mr. Oates, measure from 1·1 to 1·05 in length, by 0·83 to 0·81 in breadth, and are smaller than those the dimensions of which he himself records above.

Family CERTHIIDÆ.

341. Certhia himalayana, Vigors. *The Himalayan Tree-Creeper.*

Certhia himalayana, *Vig., Jerd. B. Ind.* i, p. 380; *Hume, Rough Draft N. & E.* no. 243.

Writing from Murree of the Himalayan Tree-Creeper, Colonel C. H. T. Marshall says:—"This is a most difficult nest to find, as the little bird always chooses crevices where the bark has been broken or bulged out, some 40 or 50 feet from the ground, and generally on tall oak-trees which have no branches within 40 feet of their roots. There were young in the few nests we found. Captain Cock secured the eggs in Kashmir; they are very small, being only 0·6 by 0·45; the ground is white, with numerous red spots. The nests we found were in the highest part of Murree, about 7200 feet."

Two eggs of this species which I possess measure 0·69 and 0·68 respectively in length, by 0·5 in breadth.

342. Certhia hodgsoni, Brooks. *Hodgson's Tree-Creeper.*

Certhia hodgsoni, *Brooks, Hume, Rough Draft N. & E.* no. 243 bis.

Hodgson's Tree-Creeper is the supposed *C. familiaris* obtained by Dr. Jerdon in Cashmir, of which he gave me two specimens.

Mr. Brooks says:—"It was seen at Gulmurg and also at Sonamurg, where Captain Cock took a few nests. The egg is much more densely spotted than that of the English Creeper, so as almost to hide the reddish-white ground-colour. Size 0·59 to 0·65 inch long by 0·48 inch broad; time of laying, the *first* week in June."

The egg is of smooth texture, without gloss, of a purplish-white ground-colour, and fully spotted all over with light brownish red, especially at the larger end. Numerous spots of reddish grey or pale inky purple are intermingled with red ones.

In shape the egg varies from a somewhat elongated oval, more or less compressed towards the smaller end, to a comparatively broad oval, also slightly compressed towards the latter end. In all the eggs that I have seen, the markings were more or less confluent towards the large end. Their dimensions are correctly recorded by Mr. Brooks.

347. Salpornis spilonota (Frankl.). *The Spotted-Grey Creeper.*

Salpornis spilonota (*Frankl.*), *Jerd. B. I.* i, p. 382.

Mr. Cleveland found a nest of this species at Hattin, in the Gurgaon district, on the 16th April. The nest was placed on a large ber-tree in a patch of preserved jungle, at a height of about 10 feet from the ground. It was cup-shaped, placed on the upper surface of a horizontal bough at the angle formed between this and a vertical shoot, to which it was attached on one side, the other three sides being free. The nest itself is unlike any other that I have seen. It is composed entirely of bits of leaf-stalks, tiny bits of leaves, chips of bark, the dung of caterpillars, all cemented together everywhere with cobwebs, so that the whole nest is a firm but yet soft and elastic mass. The nest is cup-shaped, but oval and not circular; its exterior diameters are 4 and 3 inches respectively; its greatest height 2 inches; the cavity measures 2·6 by 2·2, and 1·1 in depth.

The texture of the nest, as I have already said, is extremely peculiar; it is extremely strong, and though pulled off the bough on which it rested and the off-shoot to which it was attached, is as perfect apparently as the day it was found, bearing on the lower surface an exact cast of the inequalities of the bark on which it rested; but it is soft, yielding, and flabby in the hand, almost as much so as if it was jelly. The nest contained two almost full-grown nestlings and one addled egg.

This egg is a very regular oval, slightly broader at one end, the shell fine and fairly glossy; the ground-colour is pale greenish white; round the large end there is an irregular imperfect zone of blackish-brown specks and tiny spots, and round about these is more or less of a brown nimbus, and over the rest of the egg a very few specks and spots of blackish, dusky, and pale brown are scattered. It measures 0·68 by 0·53.

Another nest was found about 15 feet up a tree. It was partly seated on and partly wedged in between the fork of two thick oblique branches, to the rough bark of which the bottom only was firmly cemented with cobwebs, the sides, as in the case of the first nest, being quite free and detached from its surroundings. As regards dimensions and composition, the latter nest was an exact counterpart of that first taken. It contained two partially fledged nestlings.

352. Anorthura neglecta (Brooks). *The Cashmir Wren.*

Troglodytes neglecta, *Brooks, Hume, Cat.* no. 333 bis.
Troglodytes nipalensis, *Hodgs., Hume, Rough Draft N. & E.* no. 333.

The Cashmir Wren breeds in Cashmir in May and June at elevations of from 6000 to nearly 10,000 feet. I have never seen the nest, though I possess eggs taken by Captain Cock and Mr. Brooks in Cashmir. The latter says:—" Only two nests of this bird were found (both at Gulmurg), one having four eggs and the other three. In the latter case the full number was not laid, as the nest, when first found, was empty; on three successive mornings an egg was laid and then they were taken.

" In shape they vary as much as do those of the English Wren, and like them they are white, sometimes minutely freckled with pale red and purple-grey specks, which are principally confined to the large end, with a tendency to form a zone. Other eggs are plain white, without the slightest sign of a spot; but these, I think, must be the exception, for the egg of the English Wren is usually spotted. The egg has very little gloss, and the ground-colour is pure white."

The eggs are very large for the size of the bird. There appear to be two types. The one somewhat elongated ovals, slightly compressed towards the lesser end; the others broad short ovals, decidedly pointed at one end. Some eggs are perfectly pure unspotted white; others have a dull white ground, with a faint zone of minute specks of brownish red and tiny spots of greyish purple towards the large end, and a very few markings of a similar character scattered about the rest of the surface. All the eggs of the latter type vary in the amount and size of markings; these latter are always sparse and very minute. The pure white eggs appear to be less common. The eggs have always a slight gloss, the pure white ones at times a very decided, though never at all a brilliant gloss.

In length they vary from 0·61 to 0·7 inch, and in breadth from 0·5 to 0·52 inch.

Mr. Brooks subsequently wrote:—" The Cashmir Wren is not uncommon in the pine-woods of Cashmir, and in habits and manners resembles its European congener. Its song is very similar and quite as pretty. It is a shy, active little bird, and very difficult to shoot. I found two nests. One was placed in the roots of a large upturned pine, and was globular with entrance at the side. It was profusely lined with feathers and composed of moss and fibres. The eggs were white, sparingly and minutely spotted with red, rather oval in shape; measuring 0·66 by 0·5. A second nest was placed in the thick foliage of a moss-grown fir-tree, and was about 7 feet above the ground. It was similarly composed to the other nest, but the eggs were rounder and plain white, without any spots."

355. Urocichla caudata (Blyth).   *The Tailed Wren.*

Pnoepyga caudata (*Blyth*), *Jerd. B. Ind.* i, p. 490; *Hume, Rough Draft N. & E.* no. 331.

The Tailed Wren, according to Mr. Hodgson's notes, lays in April and May, building a deep cup-shaped nest about the roots of trees or in a hole of fallen timber; the nest is a dense mass of moss and moss-roots, lined with the latter. One measured was 3·5 inches in diameter and 3 in height; internally, the cavity was 1·6 inch in diameter and about 1 inch deep. They lay four or five spotless whitish eggs, which are figured as broad ovals, rather pointed towards one end, and measuring 0·75 by 0·54 inch.

356. Pnoepyga albiventris (Hodgs.). *The Scaly-breasted Wren.*

Pnoepyga squamata (*Gould*), Jerd. B. Ind. i, p. 488.

From Sikhim, Mr. Gammie writes :—" I found two nests of the Scaly-breasted Wren this year within a few yards of each other. They were in a small moist ravine in the Rishap forest, at 5000 feet above sea-level. One was deserted before being quite finished, and the other was taken a few days after three eggs had been laid. The two nests were alike, and both were built among the moss growing on the trunks of large trees, within a yard of the ground. The only carried material was very fine roots, which were firmly interwoven, and the ends worked in with the natural moss. These fine roots were worked into the shape of a half-egg, cut lengthways, and placed with its open side against the trunk, which thus formed one side of the nest. Near the top one side was not quite close to the trunk, and by this irregular opening the bird entered. Internally the nest measured 3 inches deep by 2 in width. I killed the female off the eggs; she had eaten a caterpillar, spiders, and other insects."

Mr. Mandelli found a nest of this species at Pattabong, elevation 5000 feet, near Darjeeling, on the 19th May, containing three fresh eggs. The nest was placed amongst some small bushes projecting out of a crevice of a rock about three feet from the ground. It was completely sheltered above, but was not hooded or domed; it was, for the size of the bird, a rather large cup, composed of green moss rather closely felted together and lined with fine blackish-brown roots. The cavity measured about 2 inches in diameter and 1 in depth.

The eggs of this species seem large for the size of the bird; they are rather broad at the large end, considerably pointed towards the small end. They are pure white, almost entirely devoid of gloss, and with very delicate and fragile shells.

The eggs varied from 0·72 to 0·78 in length, and from 0·54 to 0·57 in breadth.

Family REGULIDÆ.

358. Regulus cristatus, Koch. *The Goldcrest.*

Regulus himalayensis, *Blyth*, Jerd. B. Ind. ii, p. 206 ; *Hume, Rough Draft N. & E.* no. 580.

All I know of the nidification of this species is that Sir E. C. Buck, C.S., found a nest at Rogee, in the Sutlej Valley, on the 8th June, on the end of a deodar branch 8 feet from the ground and partly suspended. It contained seven young birds fully fledged; no crest or signs of a crest were observable in the young. Both the parent birds and the nest were kindly sent to me.

The nest is a deep pouch suspended from several twigs, with the entrance at the top, and composed entirely of fine lichens woven or interwined into a thick, soft, flexible tissue of from three eighths to half an inch in thickness. Externally the nest was about 3½ to 4 inches in depth, and about 3 inches in diameter.

Family SYLVIIDÆ.

363. **Acrocephalus stentoreus** (H. & E.). *The Indian Great Reed-Warbler.*

Acrocephalus brunnescens (*Jerd.*), *Jerd. B. Ind.* ii, p. 154.
Calamodyta stentorea (*H. & E.*), *Hume, Rough Draft N. & E.* no. 515.

Both Mr. Brooks and Captain Cock succeeded in securing the nests and eggs of the Indian Great Reed-Warbler in Cashmere. Common as it is, my own collectors failed to get eggs, though they brought plenty of nests.

The nest is a very deep massive cup hung to the sides of reeds. A nest before me, taken in Cashmere on the 10th June, is an inverted and slightly truncated cone. Externally it has a diameter of 3¼ inches and a depth of nearly 6 inches. It is massive, but by no means neat; composed of coarse water-grass, mingled with a few dead leaves and fibrous roots of water-plants. The egg-cavity is lined with finer and more compactly woven grass, and measures about 1¾ inch in diameter and 2¼ inches in depth.

It breeds in May and June; at the beginning of July all the nests either contained young or were empty. Four is the full complement of eggs.

Mr. Brooks noted *in epist.* :—" *Srinuggur*, 10*th June.* I went out early this morning on the lake here to look for eggs of *Acrocephalus stentoreus*, but it came on to rain so heavily that I only partially succeeded. I took three nests, two with three eggs each, and one with four young ones, the latter half-hatched. The eggs very much resemble large and boldly-marked Sparrows' eggs. They are smaller than the eggs of *A. arundinaceus*, but very similar. The latter have larger clear spaces without spots than those of our bird. I neither saw nor heard any other aquatic warbler."

Later, in a paper on the eggs and nests he had obtained in Cashmere, he stated that this species " breeds abundantly in the Cashmere lakes. The nest is supported, about 18 inches above the water, by three or four reeds, and is a deep cup composed of grasses and fibres. The eggs are four, very like those of *A. arundinaceus*, but the markings are more plentiful and smaller."

Captain Cock writes to me that " the Large Reed-Warbler is very common in the reeds that fringe all the lakes in Cashmere. It breeds in June, builds a largish nest of dry sedge, woven round five or six reeds, of a deep cup form, which it places about 2 feet above the water. It lays four or five eggs, rather blunt ovals, equally blunt at both ends, blotched with olive and dusky grey on a dirty-white ground."

Mr. S. B. Doig, who found this bird breeding in the Eastern Narra in Sind, writes :—" On the 4th August, while my man was poling along in a canoe in a large swamp on the lookout for eggs, he passed a small bunch of reeds and in them spotted a nest with a bird on it. The nest contained three beautiful fresh eggs. A few days later I joined him, and on asking about these eggs he described the bird and said he had found several other nests of the same species, but all of them contained young ones nearly fledged. I made him show me some of these nests, all of which were situated in clumps of reed, in the middle of the swamp, and in these same reeds I found and shot the young ones which, though fledged, were not able to fly. These I sent with one of the eggs to Mr. Hume, who has identified them as belonging to this species. The nests were composed of frayed pieces of reed-grass and fine sedge, the latter being principally towards the inside, thus forming a kind of lining. The nests were loosely put together, were about 3 inches inner diameter, 1¼ inch deep, the outer diameter being 6 inches. They were situated about a foot over water-line in the tops of reeds growing in the water."

Colonel Legge says :—" This species breeds in Ceylon during June and July. Its nest was procured by me in the former month at the Tamara-Kulam, and was a very interesting structure, built into the fork of one of the tall seed-stalks of the rush growing there ; the walls rested exteriorly against three of the branches of the fork, but were worked round some of the stems of the flower itself which sprung from the base of the fork. It was composed of various fine grasses, with a few rush-blades among them, and was lined with the fine stalks of the flower divested, by the bird I conclude, of the seed-matter growing on them. In form it was a tolerably deep cup, well shaped, measuring 2½ inches in internal diameter by 2 in depth. The single egg which it contained at the time of my finding it was a broad oval in shape, pale green, boldly blotched with blackish over spots of olive and olivaceous brown, mingled with linear markings of the same, under which there were small clouds and blotches of bluish grey. The black markings were longitudinal and thickest at the obtuse end. It measured 0·89 by 0·67 inch."

The eggs of this species, as might have been expected, greatly resemble those of *A. arundinaceus*. In shape they are moderately elongated ovals, in some cases almost absolutely perfect, but generally slightly compressed towards one end. The shell, though fine, is entirely devoid of gloss.

The ground-colour varies much, but the two commonest types are pale green or greenish white and a pale somewhat creamy stone-colour. Occasionally the ground-colour has a bluish tinge.

The markings vary even more than the ground-colour. In one type the ground is everywhere minutely, but not densely, stippled

with minute specks, too minute for one to be able to say of what
colour; over this are pretty thickly scattered fairly bold and well-
marked spots and blotches of greyish black, inky purple, olive-
brown, yellowish olive, and reddish-umber brown; here and there
pale inky clouds underlay the more distinct markings. In other
eggs the stippling is altogether wanting, and the markings are
smaller and less well-defined. In some eggs one or more of the
colours predominate greatly, and in some several are almost entirely
wanting. In most eggs the markings are densest towards the
large end, where they sometimes form more or less of a mottled,
irregular, ill-defined cap.

In length the eggs vary from 0·8 to 0·97, and in breadth from
0·58 to 0·63; but the average of the only nine eggs that I measured
was 0·89, nearly, by rather more than 0·61.

366. Acrocephalus dumetorum, Blyth. *Blyth's Reed-Warbler.*

Acrocephalus dumetorum, *Bl., Jerd. B. Ind.* ii, p. 155.
Calamodyta dumetorum (*Bl.*), *Hume, Rough Draft N. & E.* no. 516.

Blyth's Reed-Warbler breeds, I believe, for the most part along
the course of the streams of the lower Himalayan and sub-Himalayan ranges, and in suitable localities on and about these ranges;
such at least is my present idea. They are with us in the plains
up to quite the end of March, and are back again by the last day
of August, and during May at any rate they may be heard and
seen everywhere in the valleys south of the first snowy range.

Mr. Brooks remarks that " this species was excessively common
on the Hindoostan side of the Pir-pinjal Range, but I have never
seen it in Cashmere. I think it breeds in the low valleys by the
river-sides, for it was in very vigorous song there at the end of
May." This is my experience also, and probably while many may
go north to Central Asia to breed, a good many remain in the
localities indicated.

Captain Hutton says :—" This species arrives in the hills up to
7000 feet at least, in April, when it is very common, and appears
in pairs with something of the manner of a *Phylloscopus.* The
note is a sharp *tchick, tchick,* resembling the sound emitted by a
flint and steel.

" It disappears by the end of May, in which month they breed;
but, owing to the high winds and strong weather experienced in
that month in 1848, many nests were left incomplete, and the
birds must have departed without breeding.

" One nest, which I took on the 6th May, was a round ball with
a lateral entrance; it was placed in a thick barberry-bush growing
at the side of a deep and sheltered ditch; it was composed of
coarse dry grasses externally and lined with finer grass. Eggs
three and pearl-white, with minute scattered specks of rufous,
chiefly at the larger end. Diameter 0·62 by 0·5."

The late Mr. A. Anderson wrote the following note :—" On the

fifth day after leaving Naini Tal—ever mindful of my friend
Mr. Brooks's parting advice to me (in reference to the part of the
country which required to be investigated), 'avoid the lower hills
as the plague'—I reached Takula, which is the first march beyond
Almora on the road to the Pindari glacier, late on the evening of
the 10th of May. It rained heavily all that night, so that I was
obliged to halt the next day, my tents being far too wet to be
struck, and the distance to the next halting-place necessitating a
start the first thing in the morning.

"Takula is at an elevation between 5000 and 6000 feet; it is
beautifully wooded, with a small mountain-stream flowing right
under the camping-ground, and the climate is delightful. All
things considered, I was not sorry at having an opportunity of
exploring such productive-looking ground; and before it was fairly
daylight the next morning operations were commenced in right
earnest. To each of my collectors I apportioned off a well-wooded
mountain-slope, reserving for my own hunting-ground (as I had
not yet got my *hill-legs*) the water-courses and ravines in the immediate vicinity of my camp.

"Not more than 20 yards from where my tent stood, there is a
deep ravine clothed on both banks with a dense jungle of the larger
kind of nettle (*Girardinia heterophylla*: such nettles too!), the hill-
dock (*Rumea nepalensis*), and wild-rose trees. Wending my way
through this dark, damp, and muggy nullah to the best of my
ability, I came upon the nest of this interesting little bird; it was
placed in the centre of a rose-bush, at an elevation of some two
feet above the bank and about four feet from where I stood, but
yet in a most tantalizing situation, inasmuch as it was necessary to
remove several thorny branches before an examination of the nest
was possible.

"The act of cutting away the branches alarmed my sombre little
friend (I knew that the nest was tenanted, as the bill and head
were distinctly visible through the lateral entrance), and out she
darted with such a '*whir*' that anything like satisfactory identification for a bird of this sort was utterly hopeless. The nest contained four beautiful little eggs, so that to bag the parent bird was
a matter of the first importance; all my attempts, however, first
to capture her on the nest and next to shoot her as she flew off,
were equally futile, her movements being as rapid and erratic as
forked lightning. And here let me give a word of advice to my
brother ornithologists : Never attempt to shoot a *wary little bird
in the act of leaving its nest*, as you only run the risk, and mortifi-
cation I may add, of wounding perhaps an unknown bird, in which
case she will never again return to her nest; but *lie in ambush* for
her with outlying scants, *and make certain of her as she is returning
to her nest.* She will first alight on a neighbouring tree, then on
one closer, coming nearer and nearer each time; finally, she will
perch on the very tree or bush in which the nest is built, and
while taking a look round to see that all is well before making a
final ascent, you have yourself to blame if you fail to bag her. All

15*

this sounds very cruel; but if a bird must be shot for scientific purposes, it is surely preferable to kill it outright than to let it die a lingering death. Thus it was that I eventually succeeded, even at the expense of being devoured alive by midges and mosquitoes; but then had I not the satisfaction of knowing that to become the happy possessor of *authentic* eggs of *Acrocephalus dumetorum* was in itself sufficient to repay me for my hill excursion!

"I cannot, however, pretend to lay claim to originality in the discovery of the breeding-habits of this bird, for Hutton's description of the nest and eggs taken by him so fully accords with my own experience, that it is but fair to conclude he was correct in his identification. I would add, however, with reference to his remarks, that the nest above alluded to was *more elliptical* than *spherical*, being about the size and shape of an Ostrich's egg, that it was constructed throughout of the *largest* and *coarsest* blades of various kinds of dry grass, the egg-cavity being lined with grass-bents of a finer quality, and that it was domed over, having a lateral entrance about the middle of the nest. The whole structure was so loosely put together as to fall to pieces immediately it was removed.

"The eggs, four in number, are pure white, beautifully glossed, and well covered with rufous or reddish-brown specks, most numerous at the obtuse end. Owing to its similarity to a number of eggs, particularly to those of the Titmouse group, it is just one of those that I would never feel comfortable in accepting on trust.

"It was a remarkable coincidence that the very day I took this nest my post brought me part iv. of the P. Z. S. for 1874, containing Mr. Dresser's interesting paper on the nidification of the *Hypolais* and *Acrocephalus* groups; and if I understand him rightly, he is certainly correct in his surmise as to the eggs of *Acrocephalus dumetorum* approaching those of the *Hypolais* group.

"My good luck, as regards Blyth's Reed-Warbler, did not end here, for on the following day, at Bagesur, at an elevation of only 3000 feet, I again encountered a pair of these birds, finding their nest on the banks of the Surjoo. The position, shape, and architecture of this nest were identical with the one I have above described, but the eggs unfortunately had not been laid. The little birds, on this occasion, were quite fearless, hopping from stem to stem of the dense undergrowth which throughout the Bagesur valley fringes both banks of the river, every now and again making a temporary halt for the purpose of picking insects off the leaves, with an occasional 'tchick,' which Hutton resembles to the 'sound emitted by a flint and steel,' but all the time enticing me away from the site of their dwelling-place. In this way they led me a wild-goose chase several times up and down the river-bank before I was able to discover the whereabouts of their nest."

Captain Hutton sent me three eggs of this species. The eggs are otherwise unknown to me, and I describe them only on Captain Hutton's authority. The eggs are rather broad ovals, very smooth and compact in texture, but with little or no gloss. They are pure white, very thinly speckled with reddish and yellowish brown, the markings being most numerous towards the large end, and even there somewhat sparse and very minute. They measure respectively 0·65 by 0·52, 0·65 by 0·51, and 0·62 by 0·51.

367. Acrocephalus agricola (Jerd.). *The Paddy-field Reed-Warbler*.

Acrocephalus agricolus (*Jerd.*), *Jerd. B. Ind.* ii, p. 156.
Calamodyta agricola (*Jerd.*), *Hume, Rough Draft N. & E.* no. 517.

The Paddy-field Reed-Warbler nests apparently occasionally in May and June in the valleys of the Himalayas, the great majority probably going further north-west to breed.

Very little is known about the matter. I have shot the birds in the interior of the hills in May, but I have never seen a nest.

Mr. Brooks, however, says:—"Near Shupyion (Cashmere) I found a finished empty nest of this truly aquatic warbler in a rose-bush which was intergrown with rank nettles. This was in the roadside where there was a shallow stream of beautifully clear water. On either side of the road were vast tracts of paddy swamp, in which the natives were busily engaged planting the young rice-plants. The nest strongly resembled that of *Curruca garrula*. The male with his throat puffed out was singing on the bush a loud vigorous pretty song like a Lesser Whitethroat's, but more varied. I shot the strange songster, on which the female flew from the nest. This was the only pair of these interesting birds that I met with. I think, therefore, that their breeding in Cashmere is not a common occurrence."

This nest, now in my collection, was found on the 13th June, at an elevation of about 5500 feet, in the Valley of Cashmere. It is a deep, almost purse-like cup, very loosely and carelessly put together, of moderately fine grass, in amongst which a quantity of wool has been intermingled.

371. Tribura thoracica (Blyth). *The Spotted Bush-Warbler*.

Dumeticola affinis (*Hodgs.*), *Jerd. B. Ind.* ii, p. 158.
Dumeticola brunneipectus, *Bl., Hume, Rough Draft N. & E.* no. 519 bis.

Mr. Hodgson gives a very careful figure of a female bird of this species, together with its nest and egg, but he labels it underneath *affinis*. As we know, he described *affinis* as having spots on the breast; but he further notes that at the same place at which he obtained the female, nest, and eggs, he also got a male bird with spots on the breast; in fact, in other words, he seems to have come to the conclusion that *Dumeticola affinis* was the male and that *Dumeticola brunneipectus,* which he did not separately name, though he has beautifully figured it, was the female. I have specimens of both, but the sexes were not ascertained; still I doubt whether the two birds can possibly be merely different sexes of the same species. Anyhow, the female bird which he figures (No. 826) is really *brunneipectus*, and under that name I notice the nest and eggs on which the female figured was captured. Mr. Hodgson notes:—" *Gosainthan*. In the snows; female and nest.

" *August 2nd.*—Nest in a bunch of reeds placed slantingly: ovate in shape; aperture at one side; placed about half a foot above the ground, made of grasses and moss, 4 or 5 inches in diameter exteriorly, interiorly between 2 and 3 inches." The eggs are figured as moderately broad ovals, measuring 0·65 by 0·48, of a uniform deep cinnabar-red, reminding one of the eggs of *Prinia socialis*, but much deeper in colour *.

Mr. Mandelli sends me three nests of this species, all found near Yendong, in Native Sikhim, at an elevation of about 9000 feet, on the 15th, 17th, and 21st July. The nests contained two, two, and three fresh eggs respectively, and were placed, two of them in small brushwood, and one in a clump of rush or grass, from 9 to 18 inches above the ground. They seem to have all been rather massive little cups, composed exteriorly of broad grass-blades rather clumsily wound together, and lined with rather finer, but by no means fine grass. In two of them some dead leaves have been incorporated in the basal portion.

They are rather dirty, shabby-looking nests, obviously made of dead materials, old withered and partially-decayed grass, and not with fresh grass; they seem to have measured 3 inches in diameter, and 2·5 in height externally; the cavity was perhaps 1·5 to 1·75 in diameter, and 1 inch more or less in depth.

From Sikhim Mr. Gammie writes:—" Nest among scrub in small bush, 2 feet from ground, at 5000 feet above the sea. Found on the 3rd June, when it contained two eggs; taken on the 5th, with four eggs. I dissected the bird killed off the nest, and found it to be a female; in her stomach were the remains of a few insects. The nest is cup-shaped, loosely made of dry leaves and grass, lined with, for the size of the bird, coarse grass-stalks. Externally it measures 3·5 inches in breadth by 2·5 deep; internally 2 broad by 1·5 deep."

This nest taken by Mr. Gammie near Rungbee on the 5th June, 1875, at an elevation of about 5000 feet, contained four eggs. It was a massive little cup about 3 inches in diameter externally, and with an internal cavity about 2 inches in diameter and 1¾ inch deep; was rather loosely put together, externally composed of dead leaves and broad flags of grass, internally lined with grass-stems.

The eggs of this species are very regular broad ovals, the shells

---

* There can be no doubt, I think, that *T. affinis* and *T. brunneipectus* are the same species as *T. thoracica*. I reproduce Mr. Hodgson's note on the nesting of this species together with Mr. Hume's remarks, but I feel sure that the nest described by Mr. Hodgson and the egg figured by him cannot belong to the present species.—ED.

fine but glossless, the ground-colour a dead white, thickly speckled and spotted about the large end, thinly elsewhere, with somewhat brownish and again purplish red. The markings are all very fine and small, but where they are closely set at the large end there a few little pale purplish-grey specks and spots are intermingled.

The eggs measure 0·68 by 0·55.

The eggs of this species obtained by Mr. Mandelli in the neighbourhood of Darjeeling in July are so similar to those obtained by Mr. Gammie, and of which he sent me the parent bird, that no second description is necessary. They are a shade smaller, but the difference is not more than is always observable in even the same species. They measure 0·67 in length, and 0·53 to 0·55 in breadth.

372. Tribura luteiventris, Hodgs. *The Brown Bush-Warbler.*

Tribura luteoventris, *Hodgs., Jerd. B. Ind.* ii, p. 161; *Hume, Rough Draft N. & E.* no. 522.

A bird unquestionably belonging to this species*, the Brown Bush-Warbler, was sent me along with a single egg from Native Sikhim. The bird was said to have been killed off the nest (which was not preserved), which was found, at an elevation of about 12,000 feet, in low brushwood about 3 feet from the ground.

The egg is a very regular, rather broad oval, has only a faint gloss, and is of a very rich deep maroon-red, slightly darker at the large end.

The egg measures 0·62 by 0·49.

374. Orthotomus sutorius (Forst.). *The Indian Tailor-bird.*

Orthotomus longicauda (*Gm.*), *Jerd. B. Ind.* ii, p. 165; *Hume, Rough Draft N. & E.* no. 530.

The Indian Tailor-bird* breeds throughout India and Burma, alike in the plains and in the hills (*e. g.*, the Himalayas and Nilgiris), up to an elevation of from 3000 to 4000 feet.

The breeding-season lasts from May to August, both months included; but in the plains more nests are to be found in July, and in the hills more, I think, in June, than during the other months.

The nest has been often described and figured, and, as is well known, is a deep soft cup enclosed in leaves, which the bird sews together to form a receptacle for it.

---

* I do not place much confidence in the authenticity of the egg of this bird sent to Mr. Hume. Being a Warbler with twelve tail-feathers, it is unlikely to lay a red egg, and besides this the eggs of the allied species, *T. thoracica*, as found by trustworthy observers like Messrs. Gammie and Mandelli, are known to be white speckled with red, in spite of Mr. Hodgson's figure representing them to be deep cinnabar-red.—Ed.

† The notes on this bird's breeding are so very numerous that I am compelled to omit several of them.—Ed.

It is placed at all elevations, and I have as often found it high upon a mango-tree as low down amongst the leaves of the edible egg-plant (*Solanum esculentum*).

The nests vary much in appearance, according to the number and description of leaves which the bird employs and the manner in which it employs them; but the nest itself is usually chiefly composed of fine cotton-wool, with a few horsehairs and, at times, a few very fine grass-stems as a lining, apparently to keep the wool in its place and enable the cavity to retain permanently its shape.

I have found the nests with three leaves fastened, at equal distances from each other, into the sides of the nest, and not joined to each other at all.

I have found them between two leaves, the one forming a high back and turned up at the end to support the bottom of the nest, the other hiding the nest in front and hanging down well below it, the tip only of the first leaf being sewn to the middle of the second. I have found them with four leaves sewn together to form a canopy and sides, from which the bottom of the nest depended bare; and I have found them between two long leaves, whose sides from the very tips to near the peduncles were closely and neatly sewn together. For sewing they generally use cobweb; but silk from cocoons, thread, wool, and vegetable fibres are also used.

The eggs vary from three to four in number; but I find that out of twenty-seven nests containing more or less incubated eggs, of which I have notes, exactly two thirds contained only three, and one third four eggs.

About the colour of the eggs there has been some dispute, but this is owing to the birds laying two distinct types of eggs, which will be described below. Hutton's and Jerdon's descriptions of the eggs, *white* spotted with rufous or reddish brown, are quite correct, but so are those of other writers, who call them *bluish green*, similarly marked. Tickell, who gives them as " pale greenish blue, with irregular patches, especially towards the larger end, resembling dried stains of blood, and irregular and *broken lines scratched round*, forming a zone near the larger end," had of course got hold of the eggs of a *Franklinia*. I have taken hundreds of both types, and I note that, as in the case of *Dicrurus ater*, eggs of the two types are never found in the same nest. All the eggs in each nest always belong to one or the other type.

The parent birds that lay these very different looking eggs certainly do not differ; that I have positively satisfied *myself*.

I quote an exact description of a nest which I took at Bareilly, and which was recorded on the spot :—

" Three of the long ovato-lanceolate leaves of the mango, whose peduncles sprang from the same point, had been neatly drawn together with gossamer threads run through the sides of the leaves and knotted outside, so as to form a cavity like the end of a netted purse, with a wide slit on the side nearest the trunk beginning near the bottom and widening upwards. Inside this, the real nest, nearly 3 inches deep and about 2 inches in diameter, was neatly constructed of wool and fine vegetable fibres, the bottom being thinly lined with horsehair. In this lay three tiny delicate bluish-white eggs, with a few pale reddish-brown blotches at the large ends, and just a very few spots and specks of the same colour elsewhere."

Dr. Jerdon says :—" The Tailor-bird makes its nest with cotton, wool, and various other soft materials, sometimes also lined with hair, and draws together one leaf or more, generally two leaves, on each side of the nest, and stitches them together with cotton, either woven by itself, or cotton-thread picked up, and after passing the thread through the leaf, it makes a knot at the end to fix it. I have seen a Tailor-bird at Saugor watch till the native tailor had left the verandah where he had been working, fly in, seize some pieces of the thread that were lying about, and go off in triumph with them ; this was repeated in my presence several days running. I have known many different trees selected to build in ; in gardens very often a guava-tree. The nest is generally built at from 2 to 4 feet above the ground. The eggs are two, three, or four in number, and in every case which I have seen were white spotted with reddish brown chiefly at the large end....Layard describes one nest made of cocoanut-fibre entirely, with a dozen leaves of oleander drawn and stitched together. I cannot call to recollection ever having seen a nest made with more than two leaves....Pennant gives the earliest, though somewhat erroneous, account of the nest. He says : ' The bird picks up a dead leaf and, surprising to relate, sews it to the side of a living one.'"

I have often seen nests made between many leaves, and I have seen plenty with a dead leaf stitched to a yet living one ; but in these points my experience entirely coincides with that of the late Mr. A. Anderson, whose note I proceed to quote :—

" The dry leaves that are sometimes met with attached to the nest of this species, and which gave rise to the erroneous idea that the ' bird picks up a dead leaf and, surprising to relate, sews it to the side of a living one,' are easily accounted for.

" I took a nest of the Tailor-bird a short time ago (11th July, 1871) from a brinjal plant (*Solanum esculentum*), which had all the appearance of having had dry leaves attached to it. The nest originally consisted of *three* leaves, but two of them had been pierced (in the act of passing the thread through them) to excess, and had in consequence not only decayed, *but actually separated from the stem of the plant.* These decayed leaves were hanging from the side of the nest by a mere thread, and could have been removed with perfect safety. Perhaps instinct teaches the birds to injure certain leaves in order that they may decay ?

" Jerdon says that he does not remember ever having seen a nest made with more than two leaves. I have found the nest of this species vary considerably in appearance, size, and in the number of leaves employed, and, I would also add, in the site selected, as well as in the markings of the eggs, which latter never exceed four in number.

"The nest already described was built hardly 2 *feet off the ground*, was rather clumsy (if I might use such an expression), and was composed of *three* leaves. The eggs were white, covered with brownish-pink blotches almost coalescing at the large end. Another nest, taken in my presence (July, again, which is the general time) from the *very top of a high tree*, was enclosed inside of *one* leaf, the sides being neatly sewn together, and the cavity at the bottom lined with wool, down, and horsehair. These eggs (four) are covered, chiefly at the larger ends, with minute red spots.

"A third nest seen by me was composed of *seven* or *eight leaves*."

Captain Hutton tells us that he has seen many nests. All were " composed of cotton, wool, vegetable fibre, and horsehair, formed in the shape of a deep cup or purse, enclosed between two long leaves, the edges of which were sewed to the sides of the nest, in a manner to support it, by threads spun by the bird."

He adds that the birds, though common at their bases, do not ascend the hills ; but this is a mistake, for I have repeatedly taken nests at elevations of over 3000 feet; and Mr. Gammie, writing from Sikhim, says :—" We often find nests of this species near my house at Mongphoo (which is at an elevation of about 3500 feet). I took one there on the 16th May, which contained four hard-set eggs. It was in a calicarpa tree and between two of its long ovate leaves, the terminal halves of which were sewn together by the edges, so as to form a purse in which the real nest was placed. Yellow silk of some wild silkworm was the sewing material used."

Again, writing from the Nilgiris, Miss Cockburn remarks :— " The Tailor-bird is seldom met with on the highest ranges, but appears to prefer the warmer climates enjoyed at the elevation of about 3500 or 4000 feet. They often build in the coffee-trees ; a nest now before me was built on a coffee-tree, two of the leaves of which were bent down and sewn together. The threads are of cobweb, and the cavity is lined with the down of seed-pods and fine grass. At the back of the nest the leaves are made to meet, but are a little apart in front, so as to form an opening for the birds to hop in and out. The depth of the nest inside is 2½ inches. It was found in the month of June, and contained four eggs, which were white spotted with light red."

Of its breeding in Nepal, Dr. Scully tells us :—" It breeds freely in the valley at an elevation of 4500 feet. I took many of its nests in the Residency grounds, Rani Jangal, &c., in May, June, and July."

Major C. T. Bingham writes :—" The Indian Tailor-bird breeds in April, May, and June, both at Allahabad and at Delhi. The nest formed of one, two, and occasionally three, leaves neatly sewn so as to form a cone, and lined with the down of the madar, is well known."

Colonel Butler has furnished me with the following note :—

" The Tailor-bird breeds, I fancy, at least twice in the year, as I have seen young birds early in the hot weather both at Mount Aboo and in Deesa, and I have also taken nests in the rains. The nest is usually constructed with much skill and ingenuity. One nest which I took on the 3rd September at Mount Aboo consisted of three leaves cleverly sewn together with raw cotton, leaving a moderate-sized entrance on one side near the top, the inside being lined exclusively with horsehair and fine dry fibres.

" I captured the hen bird with a horsehair noose fixed to the end of a long thin rod as she left the nest. Another nest which I took in Deesa on the 3rd September, 1876, was composed almost entirely of raw cotton with a scanty lining of horsehairs and dry grass-stems. It was fixed to the outside twigs of a lime-tree, two of the leaves of which were sewn to it; two dead leaves were also attached to the nest, one being sewn on each side as a support to the cotton. It was cup-shaped and open at the top, much like a Chaffinch's nest."

Mr. Oates remarks :—" This is a common bird in Burma in the plains, and possibly also on the hills, though I did not observe it on the latter. I found the nest of this species containing young birds in the Thayetmyo cantonment on the 12th August. In the Pegu plains it appears to nest from the middle of May to the end of August."

The eggs are typically long ovals, often tapering much towards the small end. The shells are very thin, delicate, and semitransparent, and have but little gloss.

The ground-colour is either reddish white or pale bluish green. Of the two types, the reddish white is the more common in the proportion of two to one. The markings consist of bold blotchings or sometimes ill-defined clouds (in this respect recalling the eggs of *Prinia inornata*), chiefly confined to the large end; and specks, spots, and splashes, extending more or less over the whole surface, typically of a bright brownish red, varying, however, in different examples both in shade and intensity. The markings have a strong tendency to form a bold, irregular zone or cap at the large end, and in some specimens the markings are entirely confined to this portion of the egg's surface.

The eggs, which have a reddish-white ground, though smaller and of a much more elongated shape, closely resemble those of *Suya fuliginosa*.

In length the eggs vary from 0·6 to 0·7, and in breadth from 0·45 to 0·5; but the average of fifty eggs measured is 0·64 by 0·46.

375. Orthotomus atrigularis, Temm. *The Black-necked Tailor-bird.*

Orthotomus atrigularis, *Temm., Hume, Cat.* no. 530 bis.

Mr. Mandelli sends me a nest which he assures me belongs to this species, and the bird he sent me for identification certainly did so belong. The nest was found near the great Ranjit River on the 18th July, and then contained three fresh eggs. The nest, which is a regular Tailor-bird's, composed entirely of the finest imaginable panicle-stems of flowering grass, is a deep cup placed in between two living leaves, which have been sewn together at the tips and along the margins from the tip for about half their length, so as to provide a perfect pocket in which the nest rests. The leaves of which the pocket is composed were the terminal ones of the twigs of a sapling, and only about 3 feet from the ground. The leaves are large oval ones, each about 7 inches in length; they have been sewn together with wild silk carefully knotted, exactly as is the practice of the common Tailor-bird.

The eggs of this species are not separable from others of *O. sutorius*, and though they may possibly average somewhat larger, I have not seen enough of them to be able to make sure of this; and as regards shape, colours, and markings the description given of the eggs of *O. sutorius* applies equally to eggs of this species.

380. Cisticola volitans, Swinh. *The Golden-headed Fantail-Warbler.*

This species was not known to Jerdon, nor was it known to occur in Burma at the time that I issued my Catalogue. Mr. Oates, writing of the breeding of this bird in Southern Pegu, where it is common, says:—" Breeding-operations commence in the middle of May; on the 28th of this month I found two nests, one containing four eggs slightly incubated, and the other two, quite fresh.

"The nest is a small bag about 4 inches in height and 2 or 3 in diameter, with an opening about an inch in diameter near the top. The general shape of the nest is oval. It is composed entirely of the white feathery flowers of the thatch-grass. The walls of the nest are very thin but strong. The nest is placed about one foot from the ground in a bunch of grass, and, in the two instances where I found it, against a weed, with one or two leaves of which the materials of the nest were slightly bound.

"The eggs are very glossy pale blue, spotted all over with large and small blotches of rusty brown. I have no eggs of *C. cursitans* which match them, in that species the spots being always minute and thickly scattered over the shell, whereas in *C. volitans* the marks are large and fewer in number. Six eggs measured in length from ·54 to ·57, and in breadth from ·42 to ·43."

381. Cisticola cursitans (Frankl.). *The Rufous Fantail-Warbler.*

Cisticola schœnicola, *Bp., Jerd. B. Ind.* ii, p. 174; *Hume, Rough Draft N. & E.* no. 539.

The Rufous Fantail-Warbler breeds pretty well all over India and Ceylon, confining itself, as far as my experience goes, to the low country, and never ascending the mountains to any great elevation.

The breeding-season lasts, according to locality, from April to October, but it never breeds with us in dry weather, always laying during rainy months. Very likely at the Nicobars, where it rains pretty well all the year round, March being the only fairly dry month, it may breed at all seasons.

I have myself taken several, and have had a great many nests sent to me. With rare exceptions all belonged to one type. The bird selects a patch of dense fine-stemmed grass, from 18 inches to 2 feet in height, and, as a rule, standing in a moist place; in this, at the height of from 6 to 8 inches from the ground, the nest is constructed; the sides are formed by the blades and stems of the grass, *in situ*, closely tacked and caught together with cobwebs and very fine silky vegetable fibre. This is done for a length of from 2 to nearly 3 inches, and, as it were, a narrow tube, from 1 to 1·5 in diameter, formed in the grass. To this a bottom, from 4 to 6 inches above the surface of the ground, is added, a few of the blades of the grass being bent across, tacked and woven together with cobwebs and fine vegetable fibre. The whole interior is then closely felted with silky down, in Upper India usually that of the mudar (*Calotropis hamiltoni*). The nest thus constructed forms a deep and narrow purse, about 3 inches in depth, an inch in diameter at top, and 1·5 at the broadest part below. The tacking together of the stems of the grass is commonly continued a good deal higher up on one side than on the other, and it is through or between the untacked stems opposite to this that the tiny entrance exists. Of course above the nest the stems and blades of the grass, meeting together, completely hide it. The dimensions above given are those of the interior of the nest; its exterior dimensions cannot be given. The bird tacks together not merely the few stems absolutely necessary to form a side to the nest, but most of the stems all round, decreasing the extent of attachment as they recede from the nest-cavity. It does this, too, very irregularly; on one side of the nest perhaps no stem more than an inch distant from the interior surface of the nest will be found in any way bound up in the fabric, while on the opposite side perhaps stems fully 3 inches distant, together with all the intermediate ones, will be found more or less webbed together. Occasionally, but rarely, I have found a nest of a different type. Of these one was built amongst the stems of a common prickly labiate marsh-plant which has white and mauve flowers. There was a straggling framework of fine grass, firmly netted together with cobwebs, and a very scanty lining of down. The nest was egg-shaped, and the aperture on one side near the top. Mr. Brooks, I believe, once obtained a similar one; but the vast majority of the others that any of us have ever got have been of the type first described, which corresponds closely with Pässler's account.

Five is the usual complement of eggs; at any rate I have notes of more than a dozen nests that contained this number, and in more than half the cases the eggs were partly incubated. I have no record of more than five, and though I have any number of notes of nests containing one, two, three, and four eggs, yet these latter in almost all these cases were fresh.

Mr. Blyth says that this species is " remarkable for the beautiful construction of its nest, *sewing* together a number of growing stems and leaves of grass, with a delicate pappus which forms also the lining, and laying four or five translucent white eggs, with reddish-brown spots, more numerous and forming a ring at the large end, very like those of *Orthotomus sutorius*. It abounds in suitable localities throughout the country."

I must here note that Mr. Blyth never paid special attention to eggs, or he would have hardly said this, because the character of the markings are essentially different. Those of the Tailor-bird are typically *blotchy*, of the present species *speckly*.

Colonel W. Vincent Legge writes to me from Ceylon that " in the Western Province it breeds from May until September, and constructs its nest either in paddy-fields or in guinea-grass plots attached to bungalows.

" The nest is so beautiful and so neatly constructed that perhaps a short description of it will not be out of place. A framework of cotton or other fibrous material is formed round two or three upright stalks, about 2 feet from the ground, the material being sewn into the grass and passed from one stalk to the other until a complete net is made. This takes the bird from one to two days to construct *. Several blades, belonging to the stalks round which the cotton is passed, are then bent down and interlaced across to form a bottom on which, and inside the cotton network, a neat little nest of fine strips of grass torn off from the blade is built; this is most beautifully lined with cotton or other downy substance, which appears to be plastered with the saliva of the bird, until it takes the appearance and texture of soft felt.

" The average dimensions of the interior or cup are 2 inches in depth by 1¼ in breadth. The whole structure is generally completed in about five days, and the first egg laid on the fifth or sixth day from the commencement. The number of eggs varies from two to four, most nests containing three. The time of incubation is, as a rule, from nine to eleven days.

" I have found but little variation in the eggs of this species either as regards size or colour. They are white or pale greenish white, spotted and blotched in a zone round the larger end with red and reddish grey, a few spots extending towards the point: axis 0·63 inch ; diameter 0·51 inch.

" From close observation I can certify that this and many other small birds do not here sit during the daytime. I scarcely ever found a *Cisticola* on the nest between sunrise and sunset."

---

* Numbers of these birds used to build in a guinea-grass field attached to my bungalow at Colombo, and I had full opportunity of watching the construction of the nest on many occasions.—W. V. L.

Colonel E. A. Butler writing from Deesa says :—" The Rufous Fantail-Warbler breeds in the plains during the monsoon, making a long bottle-shaped nest of silky-white vegetable down, with an entrance at the top, in a tuft of coarse grass a few inches from the ground. I have taken nests on the following dates :—

" July 29, 1875. A nest containing 4 fresh eggs.
" Aug. 1, 1876. „ „ 5 fresh eggs.
" Aug. 5, 1876. „ „ 4 fresh eggs.
" Aug. 5, 1876. „ „ 3 fresh eggs.
" Aug. 5, 1876. „ „ 4 fresh eggs.
" Aug. 5, 1876. „ „ 5 fresh eggs.
" Aug. 7, 1876. „ „ 5 fresh eggs.
" Aug. 8, 1876. „ „ 4 fresh eggs."

And he adds the following note :—" Belgaum, 22nd July, 1879. Four fresh eggs. Same locality, numerous other nests in August and September."

Major C. T. Bingham notes :—" I have not yet observed this bird at Delhi. At Allahabad I procured one nest in the beginning of March, shooting the birds. The nest was made of very fine dry grass, and contained four small white eggs, speckled thickly with minute points of brick-red. The average of the four eggs is 0·60 by 0·41 inch."

Mr. Cripps informs us that in Eastern Bengal this bird is very common and a permanent resident. Eggs are found from the beginning of May to the end of June, in grass-jungle almost on the ground. The nest is a deep cup, externally of fine grasses, internally of the downy tops of the sun-grass.

In the Deccan, Messrs. Davidson and Wenden state that it is " common in all grass-lands. It breeds in the rainy season."

Mr. Oates, writing on the breeding of this bird in Pegu, says :— " The majority of birds begin laying at the commencement of June, and probably nests may be found throughout the rains. I procured a nest on the 2nd of November, a very late date I imagine. It contained four eggs."

I have taken the eggs of this bird myself on many occasions. I have had them sent me with the nest and bird by Mr. Brooks from Etawah, and Mr. F. R. Blewitt from Jhansi. From first to last I have seen fully fifty authentic eggs of this species. All were of one and the same type, and that type widely different from any one of those that Dr. Bree, following European ornithologists, figures. Dr. Bree's three figures all represent a perfectly spotless egg—one pink, the other bluish white, and the third a pretty dark bluish green. Our eggs, on the contrary, are *spotted*; the ground is white with, when fresh and unblown, a delicate pink hue, due not to the shell itself, but to its contents, which partially show through it. Occasionally the white ground has a *faint* greenish tinge.

Every egg is spotted, and most densely so towards the large end, with, as a rule, excessively minute red, reddish-purple, and pale purple specks, thus resembling, though smaller, more glossy, and

far less densely speckled, the eggs of *Franklinia buchanani*. These
are beyond all question the eggs of our Indian species, and the only
type of them that I have yet observed; but the question remains
—Is our Indian *Prinia cursitans*, Franklin, really identical with
the European *C. schœnicola*, Bonaparte? *—and this can only be
settled by careful comparison of an enormous series of good specimens of each bird. For my part I personally have little doubts as
to the identity of the two. At the same time differences in the
eggs may indicate difference of species. Thus of the closely allied
C. volitans, Swinhoe, the latter gentleman informs us that "the
eggs of our bird vary from three to five, are thin and fragile, and of
a pale clear greenish blue" †. He called it *C. schœnicola* when he
wrote, but he really referred to the Formosan bird, which he has
since separated.

The eggs of course vary somewhat. Of one nest I wrote at the
time I found it—"The eggs are a rather short oval, slightly pointed
at one end, with a white ground, thickly sprinkled with numerous
specks and tiny spots of pale brownish red. They measured ·58
by ·46." Of another I say—"The ground had a faint pearly tinge,
and there was a well-marked, though irregular and ill-defined, zone
towards the large end, formed by the agglomeration there of
multitudinous specks, which in places were almost confluent." Of
another set—"The eggs were much glossier and had a china-
white ground; but instead of a multitude of small specks over the
whole surface, they had nearly the whole colouring-matter gathered
together at the large end in a cap of bold, almost maroon-red spots,
only a very few spots of the same colour being scattered over the
rest of the egg.".

The eggs measure from ·53 to ·62 in length, and from ·43 to ·48
in breadth; but the average dimensions of a large number measured
were ·59 by ·46.

382. Franklinia gracilis (Frankl.). *Franklin's Wren-Warbler.*

Prinia gracilis, *Frankl., Jerd. B. Ind.* ii, p. 172; *Hume, Rough
Draft N. & E.* no. 536.
Prinia hodgsoni, *Bl., Jerd. t. c.* p. 173; *Hume, t. c.* no. 538.

I have never myself succeeded in finding a nest of Franklin's
Wren-Warbler, but my friend Mr. F. R. Blewitt has sent me no
less than forty nests and eggs, with the parents; so that, although
the eggs belong to two, I might even say three, very different types,
I entertain no doubt that he is correct in assigning them to the
same species, the more so as, although the eggs vary, the nests

---

* The Indian and European birds are now generally allowed to be perfectly
identical, notwithstanding the alleged difference in the colour of the eggs; and
Mr. Hume is now, I think, of this opinion.—ED.
† But *C. volitans*, or the closely allied race which occurs in Pegu, assuredly
lays spotted eggs. I found two nests of this bird, both with spotted eggs (*vide*
p. 236).—ED.

are identical. He has sent me several notes in regard to this species. He says:—"On the 1st July, three miles south of the village of Doongurgurh in the Raipoor District, I found a nest of Franklin's Wren-Warbler, containing three fresh eggs. It was on rocky ground between a footpath and a water-course, about 2 feet from the ground, and firmly sewn to a single leaf of a murori plant. The nest was constructed exclusively of very fine grass, with spiders' web affixed in places to the exterior. It was somewhat cup-shaped, 3·3 inches in depth and 2·4 in breadth externally. The egg-cavity was about 1·4 in diameter, and about the same depth. The eggs were a delicate pale unspotted blue.

"About 100 yards from the first, a second precisely similar, and similarly situated, nest of this same species was found, which contained three hard-set eggs, exactly similar in shape, texture, and ground-colour to those in the first nest, but everywhere excessively finely and thickly speckled with red, the specks exhibiting a strong tendency to coalesce in a zone round the large end.

"On the 12th and 13th July we obtained ten nests of Franklin's Wren-Warbler, all in the neighbourhood of Doongurgurh. From what I have seen, I gather that this species breeds from the middle of June to the middle of August in this part of the country. They appear to resort to tracts at some little elevation, where the murori and kydia bushes are abundant, and where grass grows rapidly in the early part of the rains. The nests, very ingeniously made, are invariably sewn to one or two leaves in the centre of one of the above-named bushes, the entrance above, just as in the nest of an *Orthotomus*. They are placed at heights of from a foot to 3 feet from the ground. Fine grass, vegetable fibres, and other soft materials are chiefly used in their construction, a little cobweb being often added. The eggs are laid daily, and four is the normal number, though three hard-set ones are sometimes found. The nest is prepared annually. As far as I know they have only one brood. Both parents unite in building the nest and in hatching and feeding the young.

"Of the ten nests now taken four contained speckled and six unspeckled eggs. The two types are never found in the same nest. I send all the nests, eggs, and birds."

Dr. Jerdon says:—"I found the nest of this species at Saugor, very like that of the Tailor-bird but smaller, made of cotton, wool, and various soft vegetable fibres, and occasionally bits of cloth, and I invariably found it sewn to one leaf of the kydia, so common in the jungles there. The eggs were pale blue, with some brown or reddish spots often rarely visible."

Colonel E. A. Butler writes from Deesa :—

"July 26, 1876. A nest containing 3 fresh eggs.
"Aug. 1, 1876.   ,,   ,,   4 fresh eggs.
"Aug. 15, 1876.   ,,   ,,   2 fresh eggs.
"Sept. 3, 1876.   ,,   ,,   4 incubated eggs,

"All of the above nests were exactly alike, being composed of fine dry grass without any lining, felted here and there exteriorly

with small lumps of woolly vegetable down, and built between two
leaves carefully sewn to the nest in the same way as the nests of
Orthotomus sutorius.   The eggs, three or four in number, are white,
sparingly speckled with light reddish chestnut, with a cap more or
less dense of the same markings at the large end.   All of the eggs
in the above-mentioned nests were of this type.   I found the nests
in a grass Beerh near Deesa, studded over with low ber bushes
(*Zizyphus jujuba*), generally about 2 or 3 feet from the ground, and
in similar situations to those selected by *Prinia socialis,* often
amongst dry nullahs overgrown with low bushes and long grass."

Mr. Vidal notes in his list of the Birds of the South Konkan :—
" Common in mangrove-swamps, reeds, hedgerows, thickets, and
bush-jungle throughout the district.   Breeds during the rainy
months."

Mr. Oates writes from Pegu :—" Nest with three fresh eggs on
the 19th August ; no details appear necessary except the colour
of the eggs, since this bird appears to lay two kinds of eggs. ' My
eggs are very glossy, of a light blue speckled with minute dots of
reddish brown, more thickly so at the large end than elsewhere."

The nests sent by Mr. Blewitt are regular Tailor-birds' nests,
composed chiefly of very fine grass, about the thickness of fine
human hair, with no special lining, carefully sewn with cobwebs,
silk from cocoons, or wool, into one or two leaves, which often
completely envelop it, so as to leave no portion of the true nest
visible.

The eggs belong to at least two very distinct types.   Both are
typically rather slender ovals, a good deal compressed towards one
end ; but in both somewhat broader and more or less pyriform
varieties occur.   In both the shell is exquisitely fine and glossy ;
in some specimens it is excessively glossy.   In both the ground-
colour is a very delicate pale greenish blue, *occasionally* so pale that
the ground is all but white—in one type entirely unspeckled and
unspotted, in the other finely and thickly speckled everywhere, and
towards the large end more or less spotted, with brownish or
purplish red.   The markings are densest towards the large end,
where they either actually form, or exhibit a strong tendency to
form, a more or less conspicuous speckled, semi-confluent zone.

Out of fifty-six eggs, twenty-one belong to the latter type.   As
in *Dicrurus ater*, the two types never appear to be found in the
same nest ; but the nests in which the two types are found are
precisely similar, and the parent birds are identical.

In length the eggs vary from 0·53 to 0·62, and in width from
0·4 to 0·45 ; but the average of fifty-six eggs is 0·58 by 0·42.
There is no difference whatever in the size of the two types.

383. **Franklinia rufescens** (Blyth).   *Beavan's Wren-Warbler.*

Prinia beavani, *Wald., Hume, Cat.* no. 538 bis.

Mr. Oates, who found the nest of this Warbler in Pegu, says :—

" June 29th. Found a nest sewn into a broad soft leaf of a weed in forest about 2 feet from the ground. The edges of the leaf are drawn together and fastened by white vegetable fibres. The nest is composed entirely of fine grass, no other material entering into its composition. For further security the nest is stitched to the leaves in a few places; the depth of the nest is about 3 inches, and internal diameter all the way down about 1½. Eggs three, very glossy, pale blue, with specks and dashes of pale reddish brown, chiefly at the larger end, where they form a cap. Size ·58, ·62, ·61, by ·47."

Mr. Mandelli sends me a regular Tailor-bird's nest as that of this species. It was found below Yendong in Native Sikhim on the 1st May, and contained three fresh eggs. The nest itself is a beautiful little cup, composed of silky vegetable down and excessively fine grass-stems, and a very little black hair firmly felted together, and is placed between two living leaves of a sapling neatly sewn together at the margins with bright yellow silk.

The eggs are rather elongated, very regular ovals. The shell stout for the size of the egg, but very fine and compact, and with a moderate gloss. The ground-colour is a very delicate pale greenish blue. At or round the larger end there is very generally a mottled cap or zone (more commonly the latter) of duller or brighter brownish red, while irregular blotches, streaks, spots, and specks of the same colour, but usually a slightly paler shade, are more or less sparsely scattered over the rest of the surface of the egg, sometimes they are almost wholly wanting. Occasionally the zone is at the small end.

The eggs measure from 0·60 to 0·62 in length, by 0·43 to 0·48 in breadth ; but the average of six eggs is 0·61 by 0·45.

384. **Franklinia buchanani** (Blyth).   *The Rufous-fronted Wren-Warbler.*

Franklinia buchanani (*Blyth*), *Jerd. B. Ind.* ii, p. 186 ; *Hume, Rough Draft N. & E.* no. 551.

The Rufous-fronted Wren-Warbler breeds throughout Central India, the Central Provinces, the North-western Provinces, the Punjab, and Rajpootana. It affects chiefly the drier and warmer tracts, and, though said to have been obtained in the Nepal Terai, has never been met with by *me* either there or in any very moist, swampy locality. The breeding-season extends from the end of May until the beginning of September.

The nests, according to my experience, are always placed at heights of from a foot to 4 feet from the ground, in low scrub-jungle or bushes. They vary greatly in size and shape, according to position. Some are oblate spheroids with the aperture near the top, some are purse-like and suspended, and some are regular cups. One of the former description measured externally 5 inches in diameter one way by 3¼ inches the other. One of the suspended

16*

nests was 7 inches long by 3 wide, and one of the cup-shaped nests was nearly 4 inches in diameter and stood, perhaps, at most $2\frac{1}{2}$ inches high. The egg-cavity in the different nests varies from $1\frac{3}{4}$ to $2\frac{1}{4}$ inches in diameter, and from less than 2 to fully 3 inches in depth. Externally the nest is very loosely and, generally, raggedly constructed of very fine grass-stems and tow-like vegetable fibre used in different proportions in different nests; those in which grass is chiefly used being most ragged and straggling, and those in which most vegetable fibre has been made use of being neatest and most compact. In all the nests that I have seen the egg-cavity has been lined with something very soft. In many of the nests the lining is composed of small felt-like pieces of some dull salmon-coloured fungus, with which the whole interior is closely plastered; in others there is a dense lining of soft silky vegetable down; and in others the down and fungus are mingled. They lay from four to five eggs, never more than this latter number according to my experience.

"At the end of June 1867," writes Mr. Brooks, "I took two nests of this bird at Chunar in low ber bushes about 2 feet from the ground. They were little spheres of fine grass with a hole at the side. One contained four eggs; these were of a greyish-white ground or nearly pure white, finely speckled over with reddish brown, some of the eggs exhibiting a tendency to form a zone round the large end, and others with a complete zone."

"At Sambhur," Mr. Adam says, "this Wren-Warbler is always found wherever there are low bushes. It breeds just before the rains, but I have not recorded the date. I had a nest with the bird and five eggs sent to me. The eggs are pale bluish white, with reddish-brown spots and freckles all over them."

"During July, August, and the early part of September," remarks Mr. W. Blewitt, "I found a great number of the nests and eggs of this bird in the jungle-preserves of Hansie and its neighbourhood. The nests, of which I have already sent you several, were mostly in ber (*Zizyphus jujuba*) and hinse (*Capparis aphylla*) bushes, at heights of from 3 to 4 feet from the ground. Five was the largest number of eggs that I found in any one nest."

Major C. T. Bingham remarks:—"I found several nests of this bird in the beginning of October at Delhi in the jherberry bushes so plentiful on the Ridge. Both nests and eggs are very like those of *Cisticola cursitans* before described; the only difference I could find was that the entrance in the nest of *C. cursitans* that I found was at the top, and in all the nests of *F. buchanani* at the side rather low down; the nests of the latter are also firmer and more globular in shape. The eggs are, to my eye, identical in colour and form."

Mr. G. Reid informs us that at Lucknow it is fairly common and a permanent resident. It makes an oblong, loosely constructed nest with the aperture near the top, and lays three or four white eggs minutely spotted with dingy red.

Mr. J. Davidson writes that in Western Khandeish this Warbler

is the commonest bird, breeding about Dhulia in July, August, and
September.

Colonel E. A. Butler writes :—" I found a nest of the Rufous-
fronted Wren-Warbler at Deesa on the 27th July, 1875. It was
in a grass beerh, and placed in a heap of dead thorns overgrown
with grass and about a foot from the ground. It was composed
externally of dry grass-stems, with lumps of silky white vegetable
down (*Calotropis*) scattered sparingly over the whole nest. The
lining consisted of very fine dry grass neatly put together and
felted with silky down, and a considerable amount of the dull
salmon-coloured fungus or lichen referred to in the ' Rough Draft
of Nests and Eggs,' p. 359. In shape the nest is nearly spherical,
being slightly oval however, with a small aperture near the top.
The entrance was 1½ inches in diameter, and the nest itself roughly
measured from the outside 4½ inches in length and 4 in width.
The eggs, usually four in number, are white, closely speckled over
with pale rusty red, intermingled with a few pale washed-out inky
markings, in some cases at the large end, which is surrounded by a
zone clear and well-marked in some instances, less distinct in
others. I found other nests in the same neighbourhood as
below :—

" Aug. 24, 1875. A nest containing 4 fresh eggs.
" July 20, 1876.   „   „   4   „   „
" July 28, „   „   „   4 young birds.
" Aug. 4,   „   „   „   4 fresh eggs.
" Aug. 5,   „   „   „   4   „   „
" Aug. 5,   „   „   „   4   „   „
" Aug. 5,   „   „   „   5   „   „
" Aug. 8,   „   „   „   5   „   „
" Aug. 14, „   „   „   5   „   „

" In every one of the above instances the nest was exactly similar
to the one I have described, and built in the same kind of situation,
i. e. in heaps of dead thorns overgrown with long grass. The
eggs are all much the same, the spots being larger in some than in
others and more numerous in some cases than in others. In one
set I have the ground is very pale bluish white (skimmed milk)
instead of being pure white. As a rule the eggs are almost
exactly like the eggs of *C. cursitans*, and if mixed I doubt very much
if any person could separate them. On examining the salmon-
coloured fungus-lining it appears to me to be nothing more nor
less than small pieces of dried ber leaves, and I have never
examined a nest without finding some of this material at the
bottom of it."

" The Rufous-fronted Wren-Warbler," writes Lieut. Barnes,
" breeds in Rajpootana during July, August, and the early part of
September. The nest, composed of grass, is loosely constructed,
and placed in low bushes or scrub."

The eggs vary somewhat in size and shape ; a moderately broad
oval, slightly compressed towards the larger end, being, however,

the commonest type. Examining a large series, it appears that variations from this type are more commonly of an elongated than a spherical form. The eggs are of the same character as those of *Cisticola cursitans* (p. 236), but yet differ somewhat. The eggs are many of them fairly glossy, the shells very delicate and fragile; the ground-colour white, usually slightly greyish, but in some specimens faintly tinged with very pale green or pink. Typically they are very thickly and very finely speckled all over with somewhat dingy red or purplish red. In three out of four eggs the markings are densest and largest towards the large end; and, to judge from the large series before me, at least one in four exhibits a more or less well-defined mottled zone or cap at this end, formed by the partial confluence of multitudinous specks.

In some specimens the markings are pale inky purple, and in some slightly purplish brown, but these are abnormal varieties. In one or two eggs fairly-sized spots and blotches are intermingled with the minute specklings, but this also is rare. Of course in different specimens the density of the speckling varies greatly: in some eggs not a fifth of the surface is covered with the markings, while in some it appears as if there were more of these than of the ground-colour.

In length the eggs vary from 0·55 to 0·66, and in breadth from 0·43 to 0·52; but the average of eighty-seven eggs is 0·62 by 0·48.

385. Franklinia cinereicapilla (Hodgs.).   *Hodgson's Wren-Warbler.*

Prinia cinereocapilla, *Hodgs., Jerd. B. Ind.* ii, p. 172; *Hume, Rough Draft N. & E.* no. 537.

Captain Hutton says * :—" In this species the structure of the nest is somewhat coarser than in *P. stewarti*, and it is more loosely put together, but like that species it is also a true Tailor-bird.

" In the specimen before me two large leaves are stitched together at the edges, and between these rests the cup-shaped nest composed of grass-stalks and fine roots, as in *P. stewarti*, and without any lining, while, being more completely surrounded by or enfolded in the leaves, the cottony seed-down which binds together the fibres in the others is here dispensed with.

" The eggs were three in number, of a pale bluish hue, irrorated with specks of rufous-brown, and chiefly so at the larger end, where they form an ill-defined ring.

---

* I reproduce this note as it appeared in the 'Rough Draft,' but I have no faith in the identification of this rare bird by Capt Hutton. Mr. Hume is apparently of the same opinion, as he does not quote the Dhoon as one of the localities in which this species occurs (S. F. ix, p. 286). It may be well, however, to point out that Mr. Brooks procured this species at Dhunda, in the Bhagirati valley, so that it is not unlikely to occur in the Dhoon.—ED.

"The eggs measured 0·62 by 0·44.

"The nest was found hanging on a large-leafed annual shrub growing in the Dhoon, and was placed about 2 feet from the ground. It was taken on 22nd July."

386. **Laticilla burnesi** (Bl.). *The Long-tailed Grass-Warbler.*

Eurycercus burnesii, *Bl., Jerd. B. Ind.* ii, p. 74.

Mr. S. B. Doig appears to be the only ornithologist who has found the nest of the Long-tailed Grass-Warbler. Writing of the Eastern Narra District, in Sind, he says :—

"This bird is in certain localities very numerous, but invariably confines itself to dense thickets of reed and tamarisk jungle. The discovery of my first nest was as follows:

"On the 13th March, while closely searching some thick grass along the banks of a small canal, I heard a peculiar twittering which I did not recognize. After standing perfectly still for a short while, I at length caught sight of the bird, which I at once identified as *L. burnesi*. Leaving the bed of the canal in which I was walking and making a slight detour, I came suddenly over the spoil-bank of the canal on to the place where the bird had been calling. My sudden appearance caused the bird to get very excited, and it kept on twittering, approaching me at one time until quite close and then going away again a short distance; I at once began searching for its nest, and out of the first tussock of grass I touched, close to where I was standing, flew the female, who joined her mate, after which both birds kept up a continuous and angry twittering. On opening out the grass, I found the nest with three fresh eggs in it, placed right in the centre of the tuft and close to the ground. The eggs were of a pale green ground-colour, covered with large irregular blotches of purplish brown, and not very unlike some of the eggs of *Passer flavicollis*. After this I found several nests, but they were all building, and were one and all deserted, though in many instances I never touched the nest, often never saw it, as on seeing the birds flying in and out of the grass with building material in their bills I left the place and returned in ten days' time, but only to find the nest deserted. In one case where a single egg had been laid, I found that the bird before deserting the nest had broken the egg. In July I again got a nest and shot the parent birds; the eggs in this nest were quite of a different type, being of a very pale cream ground-colour, with large rusty blotches, principally confined to the larger end. The nests of this bird are composed of coarse grass, the inside being composed of the finer parts ; they are 4 to 5 inches external diameter and 2½ inches internal diameter, the cavity being about 1½ inches deep. The months in which they breed are, as far as I at present know, March, June, and September. The eggs vary in size from ·65 to ·80 in length and from ·50 to ·55 in breadth. The average of seven eggs is ·72 in length and ·54 in breadth."

The eggs of this species vary somewhat in size and shape, but they are typically regular rather elongated ovals, rather obtuse at both ends, and often slightly compressed towards the small end. The shell is fine and compact and has a slight gloss; the ground-colour is sometimes greenish white, sometimes faintly creamy. The eggs are generally pretty thickly and finely speckled and scratched all over, and besides the fine markings there are a greater or smaller number of more or less large irregular blotches and splashes, chiefly confined to the large end. These markings, large and small, are brown, very variable in shade, in some eggs reddish, in some chocolate, in some raw sienna, &c. Besides these primary markings most eggs exhibit a number of paler subsurface secondary markings, varying in colour from sepia to lavender or pale purple; these are mostly confined to the large end (though tiny spots of the same tint occur occasionally on all parts of the egg), where with the large blotches they often form a more or less conspicuous and more or less confluent but always ill-defined zone or even cap. Here and there an egg absolutely wants the larger blotches, but even in such cases the specklings are more crowded about the large end, and these with the lilac clouds still combine to indicate a sort of zone.

The eggs I possess of this species, sent me by Mr. Doig, vary from 0·71 to 0·81 in length by 0·52 to 0·59 in breadth; but the average of seven eggs is 0·72 by 0·55.

388. Graminicola bengalensis, Jerd. *The Large Grass-Warbler*.

Graminicola bengalensis, *Jerd. B. Ind.* ii, p. 177.
Drymoica bengalensis (*Jerd.*), *Hume, Rough Draft N. & E.* no. 542.

Long ago the late Colonel Tytler gave me the following note on this species:—"I shot these birds at Dacca in 1852, and sent a description and a drawing of them to Mr. Blyth. They were named after my esteemed friend Jules Verreaux, of Paris. They are not uncommon at Dacca in grass-jungle. I think the bird Dr. Jerdon gives in his 'Birds of India' as *Graminicola bengalensis*, Jerdon, No. 542, p. 177, vol. ii., is meant for this species. The genus *Graminicola*, under which he places this bird, appears to be a genus of Dr. Jerdon's own, for it is not in Gray's 'Genera and Subgenera of Birds in the British Museum,' printed in 1855. If it is the same bird as Dr. Jerdon's, then my name, which I communicated in 1851–52 not only to Mr. Blyth but also to Prince Bonaparte and M. Jules Verreaux, and which was published in my Fauna of Dacca, has, it seems to me, the priority."

The birds *are* identical. Jerdon gave me one of his Cachar specimens, and I compared it with Tytler's types, and certainly Tytler's name was published ten years before Jerdon's (*vide* Ann. & Mag. Nat. Hist., Sept. 1854, p. 176); but no description was published, and I fear therefore that the name given by Colonel Tytler cannot be maintained, unless indeed, which I have been unable to ascertain, either Bonaparte or Verreaux figured or described the specimens Tytler sent them in some French work.

I have only one supposed nest of this species, brought me from Dacca by a native collector who worked there for me under Mr. F. B. Simson. He did not take it himself; it was brought to him with one of the parent birds by a shikaree. The evidence is, therefore, very bad, but I give the facts for what they are worth.

The nest is a rather massive and deep cup, the lower portion prolonged downwards so as to form a short truncated cone. It is fixed between three reeds, is constructed of sedge and vegetable fibre firmly wound together and round the reeds, and is lined with fine grass-roots. It measures externally 5 inches in height and nearly 4 inches in diameter, measuring outside the reeds which are incorporated in the outer surface of the nest. The cavity is about 2½ inches in diameter and nearly 2 inches deep. It contained four eggs, hard-set; only one could be preserved, and that was broken in bringing up-country; so I could not measure it, but the shell was a sort of pale greenish grey or dull greenish white, rather thickly but very faintly speckled and spotted with very dull purplish and reddish brown, with some grey spots intermingled. The nest was obtained (no date noted) between the middle of July and the middle of August. I note that the eggs were on the point of hatching, so that the fresh egg would probably be somewhat brighter coloured.

389. Megalurus palustris, Horsf. *The Striated Marsh-Warbler.*

Megalurus palustris, *Horsf., Jerd. B. Ind.* ii, p. 70; *Hume, Rough Draft N. & E.* no. 440.

Nothing has hitherto been recorded of the nidification of the Striated Marsh-Warbler, although it has a very wide distribution and is very common in suitable localities.

The Striated Marsh-Babbler, as Jerdon calls it, has nothing of the Babbler in it. It rises perpendicularly out of the reeds, sings rather screechingly while in the air, and descends suddenly. It has much more of a song than any of the Babblers, a much stronger flight, and its sudden, upward, towering flight and equally sudden descent are unlike anything seen amongst the Babblers.

Mr. E. C. Nunn procured the nest and an egg of this species (which along with the parent birds he kindly forwarded to me) at Hoshungabad on the 4th May, 1868. The nest was round, composed of dry grass, and situated in a cluster of reeds between two rocks in the bed of the Nerbudda. It contained a single fresh egg.

Writing from Wau, in the Pegu District, Mr. Oates remarks :—
" I found a nest on the 19th May containing four eggs recently laid. The female flew off only at the last moment, when my pony was about to tread on the tuft of grass she had selected for her home.

" The nest was placed in a small but very dense grass-tuft about a foot above the ground. It was made entirely of coarse grasses, and assimilated well with the dry and entangled stems among which it lay. The nest was very deep and purse-shaped. It was about 8 inches in total height at the back, and some 2 inches lower in front, the upper part of the purse being as it were cut off slantingly, and thus leaving an entrance which was more or less circular. The width is 6½ inches, and the breadth from front to back 4 inches. The interior is smooth, lined with somewhat finer grass, and measures 4 inches in depth by 3 inches from side to side, and by 2 inches from front to back.

"*Megalurus palustris* is very common throughout the large plains lying between the Pegu and Sittang Rivers. At the end of May they were all breeding. The nest is, however, difficult to find, owing to the vast extent of favourable ground suited to its habits. Every yard of the land produces a clump of grass likely enough to hold a nest, and as the female sits still till the nest is actually touched, it becomes a difficult and laborious task to find the nest."

He subsequently remarks:—"May seems to be the month in which these birds lay here. The nest is very often placed on the ground under the shelter of some grass-tuft."

Mr. Cockburn writes to me:—"I found a nest of this bird on the north bank of the Bramaputra, near Sadija. One of the birds darted off the nest a foot or two from me in an excited way, which led me to search. The nest was almost a perfect oval, with a slice taken off at the top on one side, built in a clump of grass, and only 9 or 10 inches from the ground. It was made of sarpat-grass, and lined internally with finer grasses. The grass had a bleached and washed-out appearance, while the clump was quite green. This was on the 29th May. I noticed at the same time that the nest was not interwoven with the living grass. I removed it easily with the hand."

Mr. Cripps says:—"They breed in April and May in the Dibrugarh district, placing their deep cup-shaped nests in tussocks of grass wherever it is swampy, in some instances the bottoms of the nests being wet. Four seems to be the greatest number of eggs in a nest."

The eggs are much the same shape and size as those of *Acrocephalus stentoreus*. They have a dead-white ground, thickly speckled and spotted with blackish and purplish brown, and have but a slight gloss; the speckling, everywhere thick, is generally densest at the large end, and there chiefly do spots, as big as an ordinary pin's head, occur. At the large end, besides these specklings, there is a cloudy, dull, irregular cap, or else isolated patches, of very pale inky purple, which more or less obscure the ground-colour. In the peculiar speckly character of the markings these eggs recall doubtless some specimens of the eggs of the different Bulbuls, but their natural affinities seem to be with those of the *Acrocephalinæ*.

The eggs vary from 0·8 to 0·97 in length, and from 0·61 to 0·69 in breadth ; but the average of twelve eggs is 0·85 by 0·64.

390. Schœnicola platyura (Jerd.).   *The Broad-tailed Grass-Warbler.*

Schœnicola platyura (*Jerd.*), *Jerd. B. Ind.* ii, p. 73.

Colonel E. A. Butler discovered the nest of the Broad-tailed Grass-Warbler at Belgaum. He writes:—

" On the 1st September, 1880, I shot a pair of these birds as they rose out of some long grass by the side of a rice-field; and, thinking there might be a nest, I commenced a diligent search, which resulted in my finding one. It consisted of a good-sized ball of coarse blades of dry grass, with an entrance on one side, and was built in long grass about a foot from the ground. Though it was apparently finished, there were unfortunately no eggs, but dissection of the hen proved that she would have laid in a day or two. On the 10th instant I found another nest exactly similar, built in a tussock of coarse grass, near the same place; but this was subsequently deserted without the bird laying. On the 19th September I went in the early morning to the same patch of grass and watched another pair, soon seeing the hen disappear amongst some thick tussocks. On my approaching the spot she flew off the nest, which contained four eggs much incubated. The nest was precisely similar to the others, but with the entrance-hole perhaps rather nearer the top, though still on one side. The situation in the grass was the same—in fact it was very similar in every respect to the nest of *Drymœca insignis.* The eggs are very like those of *Molpastes hæmorrhous*, but smaller, having a purplish-white ground, sprinkled all over with numerous small specks and spots of purple and purplish brown, with a cap of the same at the large end, underlaid with inky lilac.

" These birds closely resemble *Chætornis striatus* in their actions and habits, and in the breeding-season rise constantly into the air, chirruping like that species, and descending afterwards in the same way on to some low bush or tussock of grass, sometimes even on to the telegraph-wires. They are fearful little skulks, however, if you attempt to pursue them, and the moment you approach disappear into the grass like a shot, from whence it is almost impossible to flush them again unless you all but tread on them. It is perfectly marvellous the way they will hide themselves in a patch of grass when they have once taken refuge in it; and although you may know within a yard or two of where the bird is, you may search for half an hour without finding it. If you shoot at them and miss, they drop to the shot into the grass as if killed, and nothing will dissuade you from the belief that they are so until, after a long search, the little beast gets up exactly where you have been hunting all along, from almost under your feet, and darts off to disappear, after another short flight of fifteen or twenty yards, in another patch of grass, from whence you may again try in vain to dislodge it."

The eggs of this species, though much smaller, are precisely of the same type as those of *Megalurus palustris* and *Chætornis*

striatus; moderately broad ovals with a very fine compact shell,
with but little gloss, though perhaps rather more of this than in
either of the species above referred to.   The ground-colour is
white, with perhaps a faint pinkish shade, and it is profusely
speckled and spotted with brownish red, almost black in some spots,
more chestnut in others.   Here and there a few larger spots or
small irregular blotches occur.   Besides these markings, clouds,
streaks, and tiny spots of grey or lavender-grey occur, chiefly about
the large end, where, with the markings (often more numerous
there than elsewhere), they form at times a more or less confluent but irregular and ill-defined cap.

One egg measured 0·73 by 0·6.

391. **Acanthoptila nepalensis** (Hodgs.).   *The Spiny Warbler.*

Acanthoptila nipalensis (*Hodgs.*), *Jerd. B. Ind.* ii, p 57.
Acanthoptila pellotis, *Hodgs., Hume, Rough Draft N. & E.* no. 431 bis.

According to Mr. Hodgson's notes and figures, this species
builds, in a fork of a tree, a very loose, shallow grass nest.   One
is recorded to have measured 4·87 in diameter and 1·75 in height
externally, and internally 3·37 in diameter and an inch in depth.
The eggs are verditer-blue, and are figured as 1·1 by 0·65.

I may here note that *Acanthoptila pellotis* and *A. leucotis* are
totally distinct, as Mr. Hodgson's figures clearly show.   Hodgson
published *A. leucotis* apparently under the name of *A. nipalensis*, so
that the two will stand as *A. pellotis* and *A. nipalensis* *.

392. **Chætornis locustelloides** (Bl.).   *The Bristled Grass-Warbler.*

Chætornis striatus (*Jerd.*), *Jerd. B. Ind.* ii, p. 72; *Hume, Rough
   Draft N. & E.* no. 441.

Dr. Jerdon remarks that Mr. Blyth mentions that the nest of
the Grass-Babbler, as he calls it, nearly accords with that of *Malacocercus*, and that the eggs are blue.

I cannot find the passage in which Blyth states this, and I
cannot help doubting its correctness.   This bird, like the preceding,
is not a bit of a Babbler.   I have often watched them in Lower
Bengal amongst comparatively low grass and rush along the
margins of ponds and jheels, not, as a rule, affecting high reed or
seeking to conceal themselves, but showing themselves freely
enough, and with a song and flight wholly unlike that of any
Babbler.

They are very restless, soaring about and singing a monotonous
song of two notes, somewhat resembling that of a Pipit, but clear
and loud.   They do not soar in one spot like a Sky-Lark, as
Jerdon says, but rise to the height of from 30 to 50 yards, fly

---

* I do not agree with Mr. Hume on this point.   It seems to me that this
bird has both a summer and a winter plumage, and Hodgson's two names refer
to one and the same bird.—ED.

rapidly right and left, over perhaps one fourth of a mile, and then suddenly drop on to the top of some little bush or other convenient post, and there continue their song.

Mr. Brooks remarks:—" On the 28th August, 1869, I observed at the side of the railway, at Jheenjuck Jheel, on the borders of the Etawah and Cawnpoor Districts, several pairs of *Chætornis*. A good part of the jheel was covered with grass about 18 inches high, and to this they appeared partial, though occasionally I found them among the long reeds. The part of the jheel where they were found was drier than the rest, there being only about an inch of water in places, while other portions were quite dry.

" I noticed the bird singing while seated on a bush or large clump of grass, and sometimes it perched on the telegraph-wires alongside of the line of railway, continuing its song while perched.

" By habits and song it seems more nearly allied to the Pipits than the Babblers. Males shot early in September were obviously breeding, and a female shot on the 13th of that month contained a nearly full-sized egg."

It does not do to be too positive, but I should be inclined to believe that the eggs are not uniform coloured, blue and glossy like a Babbler's, but dull, dead, or greenish white, with numerous small specks and spots *.

Colonel E. A. Butler, who was the first to discover the eggs of the Bristled Grass-Warbler, writes:—

" The Grass-Babbler is not uncommon about Deesa in the rains, at which season it breeds. I found a nest containing four eggs on the 18th August, 1876. It consisted of a round ball of dry grass with a circular entrance on one side, near the top, was placed on the ground in the centre of a low scrubby bush in a grass Bheerh, and when the hen-bird flew off, which was not until I almost put my foot on the nest, I mistook her for *Argya caudata*. On looking, however, into the bush, I saw at once by the eggs that it was a species new to me. I left the spot and returned again in about an hour's time, when, to my disappointment, I found that three of the eggs had hatched. The fourth egg being stale, I took it and added it to my collection. The eggs are about the size of the eggs of *A. caudata*, but in colour very like those of *Franklinia buchanani*, namely, white, speckled all over with reddish brown and pale lavender, most densely at the large end. This bird has a peculiar habit in the breeding-season of rising suddenly into the air and soaring about, often for a considerable distance, uttering a loud note resembling the words ' chirrup, chirrup-chirrup,' repeated all the time the bird is in the air, and then suddenly descending slowly into the grass with outspread wings, much in the style of *Mirafra erythroptera*. This bird is so similar in appearance, when flying and hopping about in the long grass, to *A. caudata*, that I have no

* The discovery of this bird's eggs has proved Mr. Hume to be right in his conjecture.—ED.

doubt it is often mistaken for that species. I have invariably found it during the rains in grass Bheerhs overgrown with low thorny bushes (*Zizyphus jujuba*, &c.). Whether it remains the whole year round I cannot say; at all events, if it does, its close resemblance to *A. caudata* enables it to escape notice at other seasons."

Mr. Cripps, writing from Fureedpore, says :—" Very common in long grass fields. Permanent resident. It utters its soft notes while on the wing, not only in the cold season but the year through; it is very noisy during the breeding-time. Breeds in clumps of grass a few inches above as well as on the ground. I found five nests in the month of May from 23rd to 28th : one was on the ground in a field of indigo ; the rest were in clumps of 'sone' grass and from the same field composed of this grass. One nest contained three half-fledged young, and the rest had four eggs slightly incubated in each. Although they nest in 'sone' grass which is rarely over three feet in height, it is very difficult to find the nest, as the grass generally overhangs and hides it. Only when the bird rises almost from your feet are you able to discover the whereabouts. On several occasions I have noticed this species perching on bushes."

The eggs, which, to judge from a large series sent me by Mr. Cripps, do not appear to vary much in shape, are moderately broad ovals, more or less pointed towards one end. The shell is fine and fragile but entirely devoid of gloss; the ground-colour is white with a very faint pinky or lilac tinge, and they are thickly speckled all over with minute markings of two different shades—the one a sort of purplish brown (they are so small that it is difficult to make certain of the exact colour), and the other inky purple or grey. In most eggs the markings are most dense at or about the large end, and occasionally a spot may be met with larger than the rest, as big as a pin's head say, and some of these seem to have a reddish tinge, while some are more of a sepia.

The eggs vary from 0·75 to 0·86 in length and from 0·59 to 0·62 in breadth, but the average of twelve eggs is almost exactly 0·8 by 0·6.

394. Hypolais rama (Sykes). *Sykes's Tree-Warbler.*

Phyllopneuste rama (*Sykes*), *Jerd. B. Ind.* ii, p. 189.
Iduna caligata, *Licht.*, *Hume, Rough Draft N. & E.* no. 553.

I have never myself obtained the nest and eggs of Sykes's Tree-Warbler, *P. rama, apud Jerd.*[*] On the 1st April, at Etawah, my friend Mr. Brooks shot a male of this species off a nest ; and I saw the bird, nest, and eggs within an hour, and visited the spot later. The nest was placed in a low thorny bush, about a foot from the ground, on the side of a sloping bank in one of the large dry ravines that in the Etawah District fringe the River Jumna for a breadth

---

[*] I reproduce the note on this bird as it appeared in the 'Rough Draft,' but I think some mistake has been made, as Mr. Hume himself suggests. Full reliance, however, may be placed on Mr. Doig's note, which is a most interesting contribution.—ED.

of from a mile to four miles. The nest was nearly egg-shaped, with a circular entrance near the top. It was loosely woven with coarse and fine grass, and a little of the fibre of the "sun" (*Crotalaria juncea*), and very neatly felted on the whole interior surface of the lower two thirds with a compact coating of the down of flowering-grasses and little bits of spider's web. It was about 5 inches in i's longest and 3½ inches in its shortest diameter. It contained three fresh eggs, which were white, very thickly speckled with brownish pink, in places confluent and having a decided tendency to form a zone near the large end. Three or four days later we shot the female at the same spot.

A similar nest and two eggs, taken in Jhansi on the 12th August, were sent me with one of the parent birds by Mr. F. R. Blewitt, and, again, another nest with four eggs was sent me from Hoshungabad.

There ought to be no doubt about these nests and eggs, the more so that I have several specimens of the bird from various parts of the North-Western Provinces and Central Provinces killed in August and September, but somehow I do not feel quite certain that we have not made some mistake. Beyond doubt the great mass of this species migrate and breed further north. I have never obtained specimens in June or July; and if these nests really, as the evidence seems to show, belonged to the birds that were shot on or near them, these latter must have bred in India before or after their migration, as well as in Northern Asia.

Though one may make minute differences, I do not think either of the three nests or sets of eggs could be certainly separated from those of *Franklinia buchanani*, which might well have eggs about both in April and August; and I am not prepared to say that in each of these three cases *Hypolais rama*, which frequents precisely the same kind of bushes that *F. buchanani* breeds in, may not accidentally have been shot in the immediate proximity to a nest of the latter, the owner of which had crept noiselessly away, as these birds so often do.

Dr. Jerdon says :—" I have obtained the nest and eggs of this species on one occasion only at Jaulnah in the Dekhan ; the nest was cup-shaped, made of roots and grass, and contained four pure white eggs."

I do not attach undue weight to this, for Dr. Jerdon did not care about eggs, and was rather careless about them ; but still his statement has to be noted, and the whole matter requires careful investigation.

Mr. Doig found this species breeding on the Eastern Narra in Sind. He writes :—" I first obtained eggs of this bird in March 1879. The first nest was found by one of my men, who afterwards showed me a bird close to the place he got the eggs, which he said was either the bird to which the nest and eggs belonged or one of the same kind. This I shot and sent to Mr. Hume with one of the eggs to identify. Some time after I again came across a lot of these birds breeding, and this time lay in wait myself for the bird to come to the nest and eggs, and when it did I shot it. This I also sent to Mr. Hume to identify. Some time after I heard from Mr. Hume, who said that there must be some mistake, as the birds sent belonged to two different species, viz. *Sylvia affinis* and *Hypolais rama*, and were both, he believed, only cold-weather visitants. This year I again 'went for' these birds and again sent specimens of birds and eggs to Mr. Hume, who informed me that the birds now sent were *H. rama*, and that the eggs must belong to this species. Soon after this Mr. Brooks saw the eggs with Mr. Hume and identified them as being those of *H. rama* and identical with eggs he saw at home collected by, I think, Mr. Seebohm of this species in Siberia. Only fancy a bird breeding on the Narra of all places, especially in May, June, and July, in preference to Siberia! Locally they are very numerous, as I collected upwards of 90 to 100 eggs in one field about eight acres in size. They build in stunted tamarisk bushes, or rather in bushes of this kind which originally were cut down to admit of cultivation being carried on, and which afterwards had again sprouted. These bushes are very dense, and in their centre is situated the nest, composed of sedge, with a lining of fine grass, mixed sometimes with a little soft grass-reed. The eggs are, as a rule, four in number, of a dull white ground-colour with brown spots, the large end having as a rule a ring round it of most delicate, fine, hair-like brown lines, something similar to the tracing to be seen on the eggs of *Drymœca inornata*. The egg in size is also similar to those of that species."

The eggs of this species vary from broad to moderately elongated ovals, but they are almost always somewhat pointed towards the small end; the shell is fine but as a rule glossless; here and there, however, an egg exhibits a faint gloss. The ground-colour is whitish, never pure white, with an excessively faint greenish, greyish, creamy, or pinky tinge. The markings are very variable in amount and extent, but they are always black or nearly so and pale inky grey; perhaps typically the markings consist of a zone of black hair-lines twisted and entangled together, in which irregular shaped spots and small blotches of the same colour appear to have been caught, which zone is underlaid and more or less surrounded by clouds, streaks, and spots of pale inky grey. This zone is typically about the large end, but in one or two eggs is near the middle of the egg and in one or two is about the small end. Outside this zone a few small specks and spots, and rarely one or two tiny blotches, of both black and grey are thinly scattered; occasionally, however, the hair-lines so characteristic of this egg are almost entirely wanting, there is no apparent zone, and the markings, spots, and specks are thinly and irregularly distributed about the entire surface; here and there the whole of the dark markings on the egg are entirely confined to the zone, elsewhere only pale lilac specks are visible. Occasionally together with a well-defined zone numerous specks, spots, and a few hair-line scratches of black are intermingled with faint purplish-grey spots, and pretty thinly scattered everywhere.

The eggs vary from 0·53 to 0·68 in length and from 0·46 to 0·51 in breadth; but the average of a very large number is 0·61 by 0·49.

402. **Sylvia affinis** (Blyth). *The Indian Lesser White-throated Warbler.*

Sylvia curruca (*Gm.*), *apud Jerd. B. I.* ii, p. 209.
Sterparola curruca (*Lath.*), *Hume, Rough Draft N. & E.* no. 583.

Of the nidification of the Lesser Whitethroat within our limits, I only know that it was found in May, breeding abundantly in Cashmere in the lower hills, by Mr. Brooks. He did not notice it comparatively high up; for instance at Goolmerg, which, though not above 9000 feet high, is at the base of a snowy range, he did not see it at all.

It builds a loose, rather shallow, cup-shaped nest, composed chiefly of grass, coarser on the exterior and finer interiorly, which it places in low bushes and thickets at no great elevation from the ground. The nest is more or less lined with fine grass and roots.

It lays four or sometimes five eggs.

Mr. Brooks writes:—" I found this Whitethroat tolerably numerous in Cashmere, where it appears generally distributed, occurring at from 5500 to 6500 feet elevation or thereabouts. It frequents places where there is abundance of brushwood or underwood, especially along the banks of rivers or near them.

" I found several nests, and they were all placed in small bushes, and from 4 to 6 feet above the ground. One was in a bush on a small island in the Kangan River, which runs into the Sind River; and this nest I well remember was just so high that I could not look into it as I stood. The nests precisely resembled in size and structure those of *C. garrula* which I have seen at home, being formed of grasses, roots, and fine fibres, and I think scantily lined with a few black horsehairs; but I forget this now. They were slight, thinly formed nests, very neat but strong, and had bits of spider's web stuck about the outside here and there. This appears to be the decoration this bird and *C. garrula* are partial to. They were not added, I think, for the purpose of rendering the nest inconspicuous, for there were just enough to give the nest a spotted appearance.

" The song of this species strongly resembles that of its congener, and is full, loud, and sweet. I found the nests by the song of the male, for he generally sings near the nest. The eggs don't differ from those of *C. garrula* in my collection."

Major Wardlaw Ramsay says, writing of Afghanistan:—" This Warbler was very common and was breeding by the 27th May. All the nests found were shallow cups, composed entirely of dried grass, and situated in small bushes, frequently juniper, about 2½ feet from the ground. The eggs vary much both in size and colour—some being long ovals, nearly pure white, spotted with pale brown towards the larger end, and others of a much rounder form and a pale greenish white, thickly spotted in a broad zone near the thicker end and smeared with very pale brown, or else spotted and smeared with olive-brown over the whole of the thicker end."

The eggs are somewhat broad ovals, typically a good deal pointed towards the lesser end. They vary, however, much both in size and shape: some are short and broad, decidedly pointed at the small end; others are more elongated, and some are almost regular ellipsoids. The eggs have little or no gloss; the ground-colour is white, with a more or less perceptible though very faint greenish tinge. Typically they are very Shrike-like in their markings, the majority of these being gathered together in a more or less dense zone near the large end. The markings consist of small spots, blotches, and specks of pale yellowish brown, more or less intermingled with spots and specks of dull inky purple or grey; in many eggs there are very few markings, and these are mere spots except in the zone, while in others full-sized markings are scattered, though thinly, more or less over the whole surface of the egg. In some the zone is confluent and blurred; in others composed of small sharply defined specks and spots. Here and there a pretty large yellowish-brown cloud may be met with partially or entirely bounded by a narrow hair-like black line. Tiny black specks now and then occur, and little zigzag lines that might have been borrowed from a Bunting's egg; but these are not met with in probably more than one out of ten eggs.

In length the eggs vary from 0·6 to 0·75, and in breadth from 0·48 to 0·55; but the average of sixteen eggs is 0·66 by 0·5.

406. **Phylloscopus tytleri**, Brooks. *Tytler's Willow-Warbler.*

Phylloscopus tytleri, *Brooks, Hume, Rough Draft N. & E.* no. 560 bis.

Tytler's Willow-Warbler, as yet a rare bird in collections, and which appears only to straggle down to the plains of Upper India during the cold season, was found by Captain Cock breeding at Sonamerg (9400 feet elevation) in the Sindh Valley, Cashmere, in June.

Mr. Brooks, who discriminated the bird, said of it and its nidification :—" In plumage resembling *P. viridanus*, but of a richer and deeper olive; it is entirely without the 'whitish wing-bar,' which is always present in *viridanus*, unless in very abraded plumage. The wing is shorter, so is the tail; but the great difference is in the bill, which is much longer, darker, and of a more pointed and slender form in *P. tytleri*. The song and notes are utterly different, so are the localities frequented. *P. viridanus* is an inhabitant of brushwood ravines, at 9000 and 10,000 feet elevation; while *P. tytleri* is exclusively a pine-forest *Phylloscopus*. In the places frequented by *P. viridanus*, it must build on the ground, or very near it; but our new species builds, 40 feet up a pine-tree, a compact half-domed nest on the side of a branch.

"Captain Cock shot one of this species off the nest at Sonamerg with four eggs. The bird he sent to me, and gave me two of the eggs. Regarding the nest he says : 'I took a nest, containing four eggs, about 40 feet up a pine, on the outer end of a bough, by means of ropes and sticks, and I shot the female bird. I do not know what the bird is. I thought it was *P. viridanus*, but I send it to you. The nest was very deep, solidly built, and cup-shaped. Eggs, plain white.' In conversation with Captain Cock he afterwards told me that he had watched the bird building its nest. It was rather on the side of the branch, and its solid formation reminded him of a Goldfinch's nest. It was composed of grass, fibres, moss, and lichens externally and thickly lined with hair and feathers. The eggs were pure unspotted white, rather smaller than those of *Reguloides occipitalis*. Two of them measured ·58 by ·48 and ·57 by ·45. They were taken on the 4th June."

Captain Cock himself writes to me :—" Of all the birds' nests that I know of, this is one of the most difficult to find. One day in the forest at Sonamerg, Cashmere, I noticed a Warbler fly into a high pine with a feather in its bill. I watched with the glasses and saw that it was constructing a nest, so allowing a reasonable time to elapse (nine days or so) I went and took the nest. It was placed on the outer end of a bough, about 40 feet up a high pine, and I had to take the nest by means of a spar lashed at right angles to the tree, the outer extremity of which was supported by a rope fastened to the top of the pine. The nest was a very solid, deep cup, of grass, fibres, and lichens externally, and lined with hair and feathers. It contained four white eggs, measuring 0·58 by 0·48.

" I shot the female, which I sent to Mr. Brooks for identification.

" I forgot to add that this nest, the only one I ever found, was taken early in June."

The egg of this species closely resembles that of some of the species of *Abrornis*—a moderately broad oval, slightly pointed at the small end, pure white, and almost glossless. The only specimen I have seen measures 0·58 by 0·45.

410. Phylloscopus fuscatus (Blyth). *The Dusky Willow-Warbler.*

Phylloscopus fuscatus (*Blyth*), *Jerd. B. I.* ii, p. 191.
Horornis fulviventer, *Hodgs., Hume, Rough Draft N. & E.* no. 523.

Mr. Blyth long ago stated in ' The Ibis ' that *Horornis fulviventris* was identical with *P. fuscatus* *.
Subsequently I procured several specimens which were quite

---

* It is with considerable hesitation that I reproduce this note. *Horornis fulviventris* with which Jerdon identified the bird, the nest of which he describes, is certainly *P. fuscatus.* The only doubt I have is whether Jerdon, who apparently had not seen a specimen of *H. fulviventris*, rightly identified his bird with it. With this explanation the note is republished as it appeared in the ' Rough Draft.'—ED.

distinct from *P. fuscatus*, structurally as well as in plumage answering perfectly to Hodgson's description.

I wrote to Dr. Jerdon mentioning this fact, and he replied:—" I also am not satisfied of the identity of this species (*H. fulviventris*) with *Phylloscopus fuscatus*. I have recently got at Darjeeling what I take to be *Horornis fulviventris*, and it is somewhat smaller in all its dimensions, but I had not a typical *P. fuscatus* with which to compare it. Specimens measured $4\frac{3}{4}$ to $4\frac{7}{8}$ inches ; expanse $6\frac{1}{2}$ inches ; wing 2 to $2\frac{1}{16}$ inches. I procured the nest and eggs in July ; the nest, cup-shaped, on a bank, composed of grass chiefly, with a few fibres ; and the eggs, three in number, pinky white, with a few reddish spots."

It is certainly not *P. fuscatus* (though possibly some specimens of *P. fuscatus* in the British Museum may bear a label formerly attached to a bird of this species), nor any other *Horornis* or *Horeites* included in Dr. Jerdon's work, all of which I have. Mr. Blyth possibly went by Mr. Hodgson's specimens in the British Museum, but some confusion has, it is known, somehow crept in amongst these ; and I have no doubt myself that *Horornis fulviventris* is a good species, and that it was the nest and eggs of this species which Dr. Jerdon found *.

415. **Phylloscopus proregulus** (Pall.). *Pallas's Willow-Warbler*.

Reguloides chloronotus (*Hodgs.*), *Jerd. B. I.* ii, p. 197.
Reguloides proregulus (*Pall.*), *Hume, Rough Draft N. & E.* no. 566.

Captain Cock has the honour of being the first to take, and, I believe, up to date the *only* oologist who has ever taken, the nest and eggs of Pallas's Willow-Warbler. Mr. Brooks tried hard for the prize, but he searched on the ground and so missed the nest. He wrote to me from Cashmere, just about the time (June 1871) that Captain Cock found the nest he obtained:—" I have been utterly unable to do anything with *P. proregulus*. I shot a female, with an egg nearly ready to lay, when I first went to Goolmerg, but though I often heard the males singing, I never could find any indication of the nesting female. The feeble song, like that of *P. sibilatrix*, alluded to by Blyth as being that of *P. superciliosus*, is not that of this latter bird, but of *P. proregulus*."

Later, in the Journal of the Asiatic Society, he noted that " Captain Cock took the nest and eggs at Sonamerg. It builds, like the Golden-crested Regulus, up a fir-tree, at from 6 to 40 feet elevation, on the outer ends of the branches. The nest is of moss, wool and fibres, and profusely lined with feathers. Eggs, four or

---

* I omit the article on *Abrornis chloronotus*, Hodgs., which appeared in the 'Rough Draft' under number 574 bis. There is no manner of doubt that Hodgson got the wrong nest, a nest of a Sunbird, and figured it as that of this bird.—ED.

five, pure white, profusely spotted with red and a few spots of purple grey. Size, 0·53 by 0·43."

Later still he added in 'The Ibis :'—" Captain Cock writes from Sonamerg: 'The second day I found my first nest with eggs. It was the nest of *P. proregulus*. I shot the old bird. Three eggs. These nests are often placed on a bough high up in a pine-tree, and are domed or roofed, made of moss and lined with feathers. I took another one to day with five eggs, and shot the bird just as it was entering its nest. This was on a bough of a pine, but low down. I know of two more nests of *P. proregulus*, all on pine-trees, from which I hope to take eggs.'

" After describing the nest of *P. humii*, and saying that it was lined with the hair of the musk-deer, he adds : 'In this the nest differs from that of *P. proregulus*, which lines its nest with feathers and bits of thin birch-bark ; and the nest of *P. proregulus* is only partly domed.'

"I measured four eggs of *P. proregulus* which Captain Cock kindly gave me, and the dimensions are as follows : ·55 by ·44, ·53 by ·43, ·53 by ·43, and ·54 by ·43. They are pure white, richly marked with dark brownish red, particularly at the larger end, forming there a fine zone on most of the eggs. Intermingled with these spots, and especially on the zone, are some spots and blotches of deep purple-grey. The egg is very handsome, and reminds one strongly of those of *Parus cristatus* on a smaller scale. The dates when the eggs were taken are 30th May and 2nd June, and the place Sonamerg, which is four marches up the valley of the Sindh River."

Captain Cock himself tells me that he " took several nests of this bird at Sonamerg in Cashmere in pine-forests. It breeds in May and June, making a partially domed nest, which is sometimes placed low down on the bough of a pine-tree, sometimes on a small sapling pine where the junction of the bough with the stem takes place, and at other times high up on the outer end of a bough. It lays five eggs, like those of *P. humii* only smaller. The nests I found were all lined with feathers and thin birch-bark strips. I never found a hair-lining in any of this bird's nests. The outer portions of the nest consisted of moss and lichen, arranged so as to harmonize with the bough on which it was placed. The nests are compact little structures."

Mr. Brooks, writing of the valley of the Bhagirati river, says :— " Common in the alpine parts of the valley. It breeds about Derali, Bairamghati, and Gangaotri, in the large moss-grown deodars."

The eggs of this species closely resemble those of *P. humii*, but are smaller, and, to judge from a few specimens taken by Captain Cock that I have seen, they are somewhat shorter and broader.

Texture smooth, without any perceptible gloss. Ground-colour pure white, spotted freely and principally towards the larger end with red : brick-dust red would perhaps scarcely be a correct term. The colour would be obtained by mixing a little brown and a good deal of purple with vermilion, or by mixing Indian red with a little Venetian red. At the larger end they have an irregular zone of small, more or less confluent, spots and specks of this red, mingled with reddish or brownish purple, and a few specks and spots of the red scattered over the rest of the surface of the egg.

This egg may also be well described, as regards colour and mode of marking, by saying that it resembles the illustration in Hewitson's work of the eggs of *Parus cristatus*, except that the egg of *P. proregulus* has a distinct zone of nearly confluent spots, and their colour is more of a brownish red than those shown in the plate above referred to, which by-the-by do not correctly represent the colour of the spots upon the eggs of *P. cristatus* which I have seen. These spots are coloured with too much of a tendency towards crimson instead of brownish red.

Three of the eggs taken by Captain Cock varied from 0·53 to 0·55 in length, and from 0·43 to 0·44 in breadth.

416. Phylloscopus subviridis (Brooks). *Brooks's Willow-Warbler.*

Reguloides subviridis, *Brooks, Hume, Cat.* no. 566 bis.

Colonel Biddulph remarks that this species is common in Gilgit at 5000 feet in March, April, May, and beginning of June, and that it breeds in the Nulter valley in July at 10,000 feet. Young birds were shot in August fully fledged.

Major Wardlaw Ramsay observes on the label of a specimen procured by him at Bian Kheyl in Afghanistan in April, "evidently breeding"; and on that of another specimen shot in May at the same place, "contained eggs nearly ready to lay."

418. Phylloscopus humii (Brooks). *Hume's Willow-Warbler.*

Reguloides humii, *Brooks, Hume, Cat.* no. 565 bis.
Reguloides superciliosus (*Gm.*), *Hume, Rough Draft N. & E.* no. 565.

Mr. Brooks and Captain Cock are the only persons I know of who have taken the eggs and nests of this species. The nest and eggs sent to and described by me in 'The Ibis' as belonging to this bird cannot really have pertained to it.

Mr. Brooks tells us that *P. humii* "is very abundant in Cashmere, and I believe in all hills immediately below the snows. It would be vain to look for this bird at elevations below 8000 feet, or at any distance from the snows. It was common even in the birch woods above the upper line of pines. I found many nests. It builds a globular nest of coarse grass on a bank side, always on the ground, and never up a tree. The nest is lined with hair in greater or lesser quantities. The eggs, four or five in number, average ·56 by ·44, are pure white, profusely spotted with red, and sometimes have also a few spots of purplish grey. On the 15th June I found a nest with four young ones on the south side of the Pir-Pinjal Pass. This bird has no song, only a double chirp in addition to its call-note. The double chirp, which is very loud, is intended for a song, for the male bird incessantly repeats it as he feeds from tree to tree near where the female is sitting upon her nest."

Nests of this species obtained in Cashmere towards the end of May and during June near Goolmerg, and brought me by Mr. Brooks, were certainly by no means worthy of this pretty little Warbler. They are very loosely made, more or less straggling cups of somewhat coarse grass, only slightly lined interiorly with fine moss-roots. The egg-cavity is very small compared with the size of the nest, some of which look like balls of grass with a small hole in the centre. They average from 4 to 5 inches in external diameter, and from 2 to 3 inches in height. The egg cavity does not exceed 2 inches in diameter, and seems often to be less, and is from an inch to half an inch in depth.

From Cashmere, when in the thick of the nests of this species, Mr. Brooks wrote to me as follows:—

"From Goolmerg, which is at the foot of a snowy range, I went up to the foot of the snows through pine-forests. The pines ceased near the snow and were replaced by birch wood on tremendously rocky ground, which bothered me greatly to get over. I had missed *P. humii* after leaving the foot of the hill, where water was plentiful, but here again the bird became abundant. I could not, however, find a nest here, though I watched several pairs. I think in the cooler country they breed later. Flowers which had gone out of bloom below I again met with up here in full flower.

"Blyth says: '*R. superciliosus* has not any song, unless a sort of double call, consisting of two notes, can be called a song.' This the males vigorously uttered all day long, but I did not notice this much; but as soon as the female sharply and rapidly uttered the well-known bell-like call, I knew she was disturbed from her nest, or had left it of her own accord. Whichever of us heard this rushed quickly to the spot, and the female once sighted was kept in view as she flitted from tree to tree, apparently carelessly feeding all the while; soon she came lower down to the bushes below, and now her note quickened and betokened anxiety; generally before half an hour would elapse she would make a dash at a particular spot, and wish to go in but checked herself. This would be repeated two or three times, and now the nest was within the compass of 2 or 3 yards. At last down she went and her note ceased. When all had been quiet for a minute or two, the male meanwhile continuing his double note in the trees above, I cautiously approached the place. Sometimes the nest was very artfully concealed, but other times there it was—the round green ball with the opening at one side. I often saw the female put her head out and then partially draw it in again. Her well-defined supercilium was very distinct. I thought I could catch her on the nest once, and went round above her, but out came her head a little further, and she bolted as I brought down my pocket handkerchief on the nest. I shot one or two from the nest, but this I found unnecessary. In every case the female shouted vigorously on leaving the nest or immediately after, and by her very peculiar note fully authenticated the eggs."

Elsewhere Mr. Brooks has remarked:—" Goolmerg is one o those mountain downs, or extensive pasture lands, which are nume rous on the top of the range of hills immediately below the Pir-Pinja Range, which is the first snowy range. It is a beautiful mountaii common, about 3000 feet above the level of Sirinugger, whicl latter place has an elevation of 5235 feet. This common is abou 3 miles long and about a couple of miles wide, but of very irregula shape. On all sides the undulating grass-land is surrounded b' pine-clad hills, and on one side the pine-slopes are surmounted b' snowy mountains. On the side near the snow the supply of water ii the woods is ample. The whole hill-side is intersected by smal ravines, and each ravine has its stream of pure cold water—wate: so different from the tepid fluid we drink in the plains. In sucl places where there were water and old pines P. humii was ver] abundant: every few yards was the domain of a pair. The male: were very noisy, and continually uttered their song. This song i: not that described by Mr. Blyth as being similar to the notes of the English Wood-Wren (P. sibilatrix) but fainter—it is a loud double chirp or call, hardly worthy of being dignified with the name o song at all. While the female was sitting, the male continuec vigorously to utter his double note as he fed from tree to tree. T(this note I and my native assistants paid but little attention; bu' when the female, being off the nest, uttered her well-known 'tiss-yip,' as Mr. Blyth expresses the call of a Willow-Wren, we repairec rapidly to the spot and kept her iu view. In every instance, befor(an hour had passed, she went into her nest, first making a few im patient dashes at the place where it was, as much as to say—' Ther(it is, but I don't want you to see me go in.'

"The nest of P. humii is always, so far as my observation goes placed on the ground on some sloping bank or ravine-side. Th(situation preferred is the lower slope near the edge of the wood and at the root of some very small bush or tree; often, however, or quite open ground, where the newly growing herbage was so shor' that it only partially concealed it. In form it is a true Willow Wren's nest—a rather large globular structure with the entranc(at one side. Regarding the first nest taken, I have noted that i was placed on a sloping bank on the ground, among some low fern: and other plants, and close to the root of a small broken fir tree which, being somewhat inclined over the nest, protected it fron being trodden upon. It was composed of coarse dry grass and moss and lined with finer grass and a few black hairs. The cavity was about 2 inches, and the entrance about 1½ inch in diameter. Abou: 20 yards from the nest was a large, old, hollow fir tree, and ir this I sat till the female returned to her nest. My attendant ther quietly approached the spot, when she flew out of the nest and sat on a low bank 2 or 3 yards from it: then she uttered her 'tiss-yip,' which I know so well, and darted away among the pines My man retired, upon which she soon returned, and having callec for a few minutes in the vicinity of the nest, she ceased her note and quickly entered. Again she was quietly disturbed, and sat or a twig not far from the nest. I heard her call once more, and then shot her. There were five eggs, which were slightly incubated.

*   *   *   *

" My second nest was placed on the side of a steep bank on the ground. The third was similarly placed, and composed of coarse grass and moss, and lined with black horsehair. In each of these nests the number of eggs was five.

" Another nest, taken on the 1st June, with four eggs, was placed on the ground on a sloping bank, at the foot of a small thin bush. It was composed as usual of coarse dry grass and moss, and lined with finer grasses and a few hairs. The eggs were five or six days incubated.

" Another nest, with four eggs, was placed on the ground, under the inclined trunk of a small fir. The same materials were used.

" Another nest, containing four eggs, was placed on a sloping bank and quite exposed, there being little or no herbage to conceal it. It was composed as before, with the addition of a few feathers in the outer portion of the nest.

" Another nest was at the roots of a fern growing on a very steep bank. The new shoots of the fern grew up above the nest, and last year's dead leaves overhung it and entirely concealed it.

" Another was placed on a sloping bank, immediately under the trunk of a fallen and decayed pine. On account of the irregularities in the ground, the trunk did not touch the ground where the nest was by about 2 feet. This was again an instance of contrivance for the nest's protection. It was composed of the same materials as usual.

" Another was among the branches of a shrub, right in the centre of the bush and on the ground, which was sloping as usual.

" Another nest, with four eggs, taken on 3rd June, was placed in the steep bank of a small stream only 3 feet 6 inches above the water.

" The above examples will give a very fair idea of the situation of the nest; and it now remains only to describe the eggs, which average ·56 long by ·44 broad. The largest egg which was measured was ·62 long and ·45 broad, and the smallest measured ·52 long and ·43 broad. The ground-colour is always pure white, more or less spotted with brownish red, the spots being much more numerous and frequently in the form of a rich zone or cap at the larger end. Intermixed with the red spots are sometimes a few purplish-grey ones. Other eggs are marked with deep purple-brown spots, like those of the Chiffchaff, and the spots are also intermingled with purplish grey. Some eggs are boldly and richly marked, while others are minutely spotted. The egg also varies in shape ; but, as a general rule, they are rather short and round, resembling in shape those of *P. trochilus*. In returning from Cashmere, on the south face of the Pir-Pinjal Mountain and close to the footpath, I found on the 15th June a nest of this bird with four young ones. This nest was placed in an unusually steep bank. Half an hour after finding the nest, and perhaps 1000 feet lower down the hill,

I stood upon a mass of snow which had accumulated in the bed of a mountain-stream."

Captain Charles R. Cock writes to me that he " took numbers of nests at Sonamerg, in the Sindh Valley in Cashmere, during a nesting trip that I took in 1871 with my valued and esteemed friend W. E. Brooks, Esq. Although at the time of our finding the nest of this Warbler we were about 80 miles apart, yet we both found our first nest on the same day—the 31st May. I believe he was by a couple of hours or so the winner, as I do not think the egg had ever been taken before.

" Breeds in May or June on the ground in banks ; makes a globular nest of moss, well lined with fine grass, musk-deer hair, or horse-hair. It lays five eggs, white spotted with rusty red, inclining to a zone at the larger end."

Typically the eggs of this species are broad ovals, slightly compressed towards one end ; the ground pure white and almost perfectly devoid of gloss, speckled and spotted with red or purplish red, the markings, most dense about the large end, often forming an irregular mottled cap or zone. These are the general characters, but the eggs vary very much in shape, size, colour, and density of markings. Some eggs are almost spherical ; others are somewhat elongated ; others slightly pyriform. As a body, alike in shape and coloration, they remind one of the eggs of many species of Indian Tit, especially those of *Lophophanes melanolophus*. In some eggs the markings are a slightly brownish brickdust-red, moderate sized spots and specks scattered pretty thickly over the whole surface, but gathered into a dense, more or less confluent, zone or cap towards the large end. Intermingled with these primary markings a few pale purple spots are scattered towards the large end of the eggs. In other eggs the markings are mostly mere specks, and in this type of egg the specks are mostly brownish purple, in some almost black. Occasionally an egg is almost entirely spotless, having only towards the large end a clouded dingy reddish-purple zone. In some eggs again the colour of the markings is pale and washed out. As a rule, the eggs in which the markings are of the brickdust-red type have these larger, bolder, and more numerous ; while those in which the markings are purple have them of a more minute character.

The shape of the eggs, as already noticed, varies much, being sometimes longer than those of *P. trochilus*, and at other times very much of the same rounded shape. Frequently they are more pointed at the smaller end than those of *P. trochilus* usually are. The texture of the egg is similar to that of *P. trochilus*, with scarcely any gloss. The ground-colour is always pure white, and the markings, which are always more or less plentiful, are either reddish brown or purple-brown, intermingled sparingly with lighter or darker purple-grey.

Some eggs contain hardly a speck of the purple-grey, while others have considerable blotches of that colour scattered amongst the red spots.

Some eggs are scantily marked, and have the spots very small ;

while others are densely spotted and blotched, the spots often being more or less confluent at the larger end. Frequently they accumulate round the larger end in the form of a confluent zone. The variety with deep purple-brown spots, which is the rarest, resembles those of *P. rufa* in miniature; but, as a rule, the egg bears a much stronger resemblance to that of *P. trochilus*, though it is of course much smaller. *As far as the colour goes*, the representations in Hewitson's work of the eggs of *Parus cristatus, Parus cœruleus*, and *Phylloscopus trochilus* will give a very correct idea of the different varieties of the egg of the present bird.

The greatest number of eggs found in any nest by Captain Cock and Mr. Brooks was five; frequently, however, four was the number upon which the bird was sitting; eggs partially incubated. On the Pir-Pinjal Mountain, just below the snows, a nest with four young ones was found on the 15th June, so that, though five seems to be the usual number, the bird frequently lays only four.

In length the eggs vary from 0·52 to 0·62, and in breadth from 0·43 to 0·47; but the average of fifty eggs carefully measured was 0·56 full by 0·44.

428. Acanthopneuste occipitalis, Jerd. *The Large Crowned Willow-Warbler.*

Reguloides occipitalis (*Jerd.*), *Jerd. B. I.* ii, p. 196; *Hume, Rough Draft N. & E.* no. 563.

The Large Crowned Willow-Warbler breeds in Cashmere and the North-west Himalayas generally, during the latter half of May, June, and the first half of July, apparently at any elevation from 4000 to 8000 feet.

Mr. Brooks says:—"This is perhaps the commonest bird in Cashmere, even more so than *Passer indicus*. It is found at almost all elevations above the valley where good woods occur.

"I only took three nests, as the little bird is very cunning, and, unlike the simple *P. humii*, is very careful indeed how it approaches its nest when an enemy is near.

"The nest is placed in a hole under the roots of a large tree on some steep bank-side. I found one in a decayed stump of a large fir-tree, inside the rotten wood. It was placed on a level with the ground, and could not be seen till I had broken away part of the outside of the stump. It was composed of green moss and small dead leaves, a scanty and loosely formed nest, and not domed. It was lined with fine grass and a little wool, and also a very few hairs. There were five eggs.

"Another nest was also placed in a rotten stump, but under the roots. A third nest was placed in a hole under the roots of a large living pine, and in front of the hole grew a small rose-bush quite against the tree-trunk. This nest was most carefully concealed, for the hole behind the roots of the rose-bush was most difficult to find.

"The eggs, four or five in number, are of a rather longer form than those of *P. humii*, and are pure white without any spots. They average ·65 by ·5."

He added *in epist.* :—" This is a much shier bird than *P. humii.*
I watched many a one without effect.   The nest is a loose structure
of moss lined with a little wool, and would not retain its shape after
coming out of the hole.   It is a most amusing bird, very noisy,
with a short poor song, and utters a variety of notes when you are
near the nest."

Certainly the nests he brought me are nothing but little pads of
moss, 3 to 4 inches in diameter and perhaps an inch in thickness.
There is no pretence for a lining, but a certain amount of wool and
excessively fine moss-roots are incorporated in the body of the nest.
In situ they would appear to be sometimes more or less domed.

Captain Cock writes to me :—" I have taken numbers of nests of
this bird in Cashmere and in and about the hill-station of Murree.
They commence breeding in May and have finished by July.   The
nests are placed under roots of trees, in crevices of trees, between
large stems, and a favourite locality is, where the road has a stone
embankment to support it, between the stones.   The nest is globular, made of moss, and the number of eggs is four.   I have often
caught the old bird on the nest.   The nests are easy to find, as the
birds are very noisy and demonstrative when any one is near their
nests."

Colonel C. H. T. Marshall also very kindly gives me the following
most interesting note on the nidification of this species in the
vicinity of Murree.   He says :—

" This little Willow-Warbler, so far as my own experience goes,
always prefers a pretty high elevation for breeding.   Out of the
dozen nests found by Captain Cock and myself in the neighbourhood of Murree, none were at an elevation of less than 6500 feet
above the sea ; and my shikaree, who was always on the look out for
me in the lower ranges, never came across the nest of this species.

" The nest is generally placed in holes at the foot of the large
spruce firs.   It is a difficult nest to find, as the bird selects holes
into which the hand will not go, and outside there are no signs of
there being any nest within.

" The cock bird spends most of his time at the tops of trees,
coming down at intervals.   The only chance of success in taking
the eggs is to watch carefully any that may be flying low in the
bushes, until they disappear cautiously into the holes where they
are breeding.   I should mention that we have also found some nests
in the rough stone walls on the hill road-sides.

" The nest is as neatly and carefully built as if it had to be exposed
on the branch of a tree.   It is globular in shape, made of moss,
and lined with feathers.   The eggs are pure white.   They apparently
rear two broods in the year.   In the first nest, which we found
under the root of an old spruce-fir on the 17th May, the eggs were
quite hard-set; and I may remark that immediately over this nest,
about 8 feet up the tree in a crack in the wood, a little *Muscicapula
superciliaris* was sitting on five eggs.   Later at the end of June we
found *fresh* eggs in several nests.   The eggs in our collection were
all taken between the 17th May and the 10th July."

They do not always, however, select such situations as those referred to in the above accounts. Sir E. C. Buck, C.S., says :—" I found a nest on 11th June in the roof of Major Batchelor's bungalow at Nachar, in the Sutlej Valley; it contained young birds. I was not allowed to disturb the nest, which was composed externally of moss. I noticed a second half-made nest near the other."

The eggs of this species are, as might be expected, somewhat larger than those of *P. humii*, and they are of a different character, being spotless, white, and slightly glossy. In shape the eggs vary from a nearly perfect, moderately elongated oval to a slightly pyriform shape, broad at the large end, and a good deal compressed and somewhat pointed towards the small end (*vide* the representation of the eggs of *Ruticilla tithys* in Hewitson's work).

In length they vary from 0·63 to 0·68, and in breadth from 0·48 to 0·53; but the average of fifteen eggs measured is 0·65 by 0·5.

430. Acanthopneuste davisoni, Oates. *The Tenasserim White-tailed Willow-Warbler.*

Reguloides viridipennis (*Blyth*), *apud Hume, Cat.* no. 567 *.

It was on the 2nd of February, just at the foot of the final cone of Mooleyit, at an elevation of over 6000 feet, that Mr. Davison came upon the nest of this species. He says:—

"In a deep ravine close below the summit of Mooleyit I found a nest of this Willow-Warbler. It was placed in a mass of creepers growing over the face of a rock about 7 feet from the ground. It was only partially screened, and I easily detected it on the bird leaving it. I was very much astonished at finding a nest of a Willow-Warbler in Burmah, so I determined to make positively certain of the owner. I marked the place, and after a short time returned very quietly. I got within a couple of feet of the nest; the bird sat still, and I watched her for some time; the markings on the top of the head were very conspicuous. On my attempting to go closer the bird flew off, and settled on a small branch a few feet off. I moved back a short distance and shot her, using a very small charge.

"The nest was a globular structure, with the roof slightly projecting over the entrance. It was composed externally chiefly of moss, intermingled with dried leaves and fibres; the egg-cavity was warmly and thickly lined with a felt of pappus.

"The external diameter of the nest was about 4 inches; the egg-cavity 1 inch at the entrance, and 2 inches deep.

"The nest contained three small pure white eggs."

The three eggs here mentioned measured 0·59 and 0·6 in length, by 0·49 in breadth.

---

* Mr. Hume is of opinion that this bird is the true *P. viridipennis* of Blyth. I have elsewhere stated my reasons for disagreeing with him.—ED.

Abrornis albosuperciliaris, *Blyth, Jerd. B. Ind.* ii, p. 202; *Hume, Rough Draft N. & E.* no. 573.

Throughout the Himalayas south of the first snowy ranges, and in all wooded valleys in rear of these, from Darjeeling to Murree, this Warbler appears to be a permanent resident.

I have received its nests and eggs from several sources, and have taken them in the Sutlej and Beas Valleys myself. They lay in the last week of March, and throughout April and May, constructing a large globular nest of moss, more or less mingled exteriorly with dry grass and lined thinly with goat's hair, and then inside this thickly with the softest wool or, in one nest that I found, with the inner downy fur of hares. The entrance to the nest is sometimes on one side, sometimes almost at the top, and is rather large for the size of the bird. The nest is almost without exception placed on a grassy bank, at the foot of some small bush, and usually contains four eggs.

Talking of this species, and writing from Almorah on the 17th May, Mr. Brooks said :—" I have just taken a nest. It was placed on a sloping bank-side near the foot of a small bush. The bank was overgrown with grass. The nest, which was on the ground, was a large ball-shaped one, composed of very coarse grass, moss-roots, and wool, and lined with hair and wool. It contained four pure white glossy eggs, which were much pointed at the small end. I shot the bird off the nest. I had already frequently met with fully-grown young birds of this species."

Writing from Dhurmsala, Captain Cock remarked :—" On the 8th April I found a nest of this species containing four white eggs ; it was placed on the ground, under a bush, on a steep bank. The nest was globular, with rather a large entrance-hole, and was made of moss, with dry grass outside, then black hair of goats, and thickly lined with the softest of wool : *no feathers* in the nest. I caught the bird on the nest ; it is common here."

Colonel G. F. L. Marshall tells us :—" A nest found on the 22nd May at Naini Tal, about 7000 feet above the sea, contained three hard-set eggs. The eggs were pure white. The nest was a most beautiful little structure of moss, lined with wool ; it was globular, with the entrance at one side, and placed on a bank among some ground-ivy, the outer part of the nest having a few broad grass-blades interwoven so as to assimilate the appearance of the nest to that of the bank against which it lay. It was at the side of a narrow glen with a northern aspect, and about four feet above the pathway, close to the spring from which my *bhisti* daily draws water, the bird sitting fearlessly while passed and repassed by people going down the glen within a foot or two of the nest."

The eggs are pure white, and generally fairly glossy. In texture the shells are very fine and compact. The eggs are moderately broad ovals, much pointed towards the small end, and vary from 0·6 to 0·65 in length, and from 0·48 to 0·52 in breadth; but the average of twenty eggs measured is 0·63 by 0·5 nearly.

435. Cryptolopha jerdoni (Brooks). *Brooks's Grey-headed Flycatcher-Warbler.*

Abrornis xanthoschistos (*Hodgs.*), *Jerd. B. Ind.* ii, p. 202; *Hume, Rough Draft N. & E.* no. 572.

This Warbler breeds, according to Mr. Hodgson's notes [*], both in Nepal and Sikhim up to an elevation of 6000 or 7000 feet. They lay in May three or four pure white eggs. They make their nest on the ground in thick bushes, or in holes in banks, or under roots of trees. The nest is a large mass of moss and dry leaves, somewhat egg-shaped, with the entrance at one end, some 6 inches in length, 4 inches in breadth, and 3·5 in height externally, and with an oval entrance about 1·5 high and 2·25 wide. Inside it is carefully lined with moss-roots. Both sexes assist in hatching and rearing the young, which are ready to fly in July.

From Sikhim Mr. Gammie says:—"I found one nest of this species at Rishap, at an elevation of 5000 feet, on the 20th May. The nest was in thin forest, near its outer edge, and placed on the ground beside a small stem. It was domed, and composed entirely of moss, with the exception of a few fibres in the hood or dome portion, and was lined with thistle-down. The exterior diameter was 3·3, the height 3·2: the cavity was 1·6 in diameter, and only an inch in depth below the lower margin of the entrance, which was the rim of the true cup, over which the hood was drawn. The nest contained four fresh eggs."

Several nests of this species that have been sent me from Sikhim were all of the same type—beautiful little cups, some placed on the ground, some amongst the twigs of brushwood a little above the ground, composed entirely of fine moss and a little fern-root, and with the interior of the cavity not indeed regularly lined but dotted about with tufts of silky seed-down.

The eggs are very similar to but smaller than those of the preceding species—very broad ovals, a good deal pointed towards one end, pure white, and faintly glossy. In length they vary from 0·53 to 0·58, and in breadth from 0·45 to 0·49.

---

[*] Mr. Hodgson's specimens in the British Museum are *C. xanthoschista*; but *C. jerdoni* also occurs in Nepal, and Mr. Hodgson *may* have found the nests of both. I leave the note as it appeared in the 'Rough Draft,' as the two species are not likely to differ in their habits, and it matters little to which species Mr. Hodgson's note refers, provided the above remarks are borne in mind.—ED.

436. Cryptolopha poliogenys (Blyth). *The Grey-cheeked Flycatcher-Warbler.*

Abrornis poliogenys (*Blyth*), *Jerd. B. Ind.* ii, p. 203.

From Sikhim Mr. Gammie writes :—"A nest of the Grey-cheeked Flycatcher-Warbler, taken on the 8th May in large forest at 6000 feet, contained three hard-set eggs. It was suspended to a snag among the moss growing on the stem of a small tree at five feet up. The moss supported it more than did the snag. It is a solid cup-shaped structure, made of green moss and lined with very fine roots. Externally it measures 3½ inches across and 2¼ deep ; internally 2 inches wide and 1¾ deep."

The eggs of this species, like those of *C. xanthoschista* and *C. jerdoni*, are pure white. They are not, I think, separable from the eggs of these two species. Those sent me by Mr. Gammie measure 0·66 and 0·67 in length by 0·5 in breadth.

437. Cryptolopha castaneiceps (Hodgs.). *The Chestnut-headed Flycatcher-Warbler.*

Abrornis castaneiceps, *Hodgs., Jerd. B. Ind.* ii, p. 205 ; *Hume, Rough Draft N. & E.* no. 578.

According to Mr. Hodgson's notes and figures, the Chestnut-headed Flycatcher-Warbler breeds in the central hill-region of Nepal from April to June, laying three or four eggs, which are neither figured nor described. The nest itself is a beautiful structure of mosses, lichens, moss- and fern-roots, and fine stems worked into the shape of a large egg, measuring 6 and 4 inches along the longer and shorter diameters ; it is placed on the ground in the midst of a clump of ferns or thick grass, with the longer diameter perpendicular to the ground. The aperture, which is about halfway between the middle and the top of the nest, and on one side, is oval, about 2 inches in width and 1·75 in height. Both sexes are said to assist in hatching and rearing the young.

438. Cryptolopha cantator (Tick.). *Tickell's Flycatcher-Warbler.*

Culicipeta cantator (*Tick.*), *Jerd. B. Ind.* ii, p. 200.
Abrornis cantator (*Tick.*), *Hume, Rough Draft N. & E.* no. 570.

A nest containing a single egg has been sent me as that of Tickell's Flycatcher-Warbler. It was found in May in Native Sikhim, at an elevation, it is said, of 12,000 feet. It was suspended to the tip of a branch of a tree at a height of about 8 feet from the ground. The nest is a most lovely one ; but I confess that I have doubts as to its really belonging to this species.

The nest is, for the size of the bird, a large watch-pocket, some 6 inches in total length and 3·5 in breadth, composed entirely of white, satiny seed-down, densely felted together to the thickness of half an inch. The lower part, sides, and back very thinly, and the upper portion and the margin of the mouth of the pocket thickly, coated with excessively fine green moss and very fine soft vegetable fibre.

My sole reason for doubting the authenticity of the nest is that another *precisely* similar one was sent me by another collector, a European, as belonging to an *Æthopyga*, together with the female which he shot off the nest.

The present nest contained a pure white egg; the other spotted eggs. Both collectors I have no doubt were fully assured of the correctness of their identification, and it may be that both species of birds construct similar nests; but I entertain considerable doubts on this subject, and think it right to note the fact.

The egg is a very broad oval, pure white, and very glossy, and measures 0·6 by 0·49.

Mr. Mandelli sends me a lovely nest, which he says belongs to this species. It was found in May in Native Sikhim, at about 12,000 feet elevation. It was suspended from the tiny branch of a tree at a height of about 8 feet from the ground. The nest is a perfect watch-pocket, composed entirely of white silky down belonging to one of the bombaxes, thinly coated here and there with strings of moss to keep it together, and more thickly so with this and vegetable fibre at and about the point of suspension and round the rim of the mouth of the pocket. The nest is altogether about 6 inches long and about 3 inches in diameter at its broadest; the lower edge of the aperture into the pocket is 2 inches from the bottom of the nest, and the aperture is about 2 inches wide. It is altogether one of the loveliest nests I have ever seen: but I cannot feel certain that the nest really belongs to this species; for I have had a precisely similar nest, also found in Sikhim, on the 20th May, similarly suspended at a height of about 5 feet from the ground, sent me as belonging to another species of *Abrornis*; and though Mr. Mandelli is usually right, I think the matter requires further confirmation.

440. Abrornis superciliaris, Tick. *The Yellow-bellied Flycatcher-Warbler.*

Abrornis flaviventris, *Jerd. B. Ind.* ii, p. 203.

Writing from Tenasserim, Major T. C. Bingham says :—

" I have shot this bird on the Zammee choung, where I got a nest with eggs; and I have more than once seen it in the Thoungyeen forests.

" The following is an account of the nest I found, recorded in my note-book :—

" Khasat village—Khasat choung, Zammee river, 9th March, 1878.—My camp to-day was pitched in the midst of a dense bamboo-break, close to a path leading to the village.

" About ten feet from my tent on this path, passers-by had cut one of the bamboos in a clump and left it leaning up against the clump ; between two knots of this a rough hack had broken an irregular hole into a joint.

" Sitting outside my tent and looking carelessly about, my attention was attracted by what I took to be a leaf flutter down close to the above-mentioned bamboo, and to my surprise disappear before it reached the ground. Wondering at this, I got up and approached the place, when from the aforementioned hole in the bamboo out darted a little bird ; and looking in I saw a neat little nest of fibres placed on the lower knot with three eggs, white densely speckled, chiefly in a ring at the larger end, with pinkish claret spots.

" I went back to my tent, watched the bird return, and shot her as on being frightened off she flew out a second time. It proved to be the above species.

" I took the nest and eggs. The latter, I regret to say, were lost subsequently through the carelessness of a servant, but I had luckily measured and taken a description of them.

" Their dimensions were respectively 0·57 × 0·42, 0·59 × 0·42, and 0·59 × 0·44."

From Sikhim Mr. Gammie writes :—" I took a nest of this Warbler on the 15th June at 1800 feet elevation. It was inside a bamboo-stem near the banks of the Ryeng stream. Just under a node some one had cut out a notch, which the birds made their entrance. The nest rested on the node below and fitted the hollow of the bamboo. It was made of dry bamboo-leaves, and lined with soft, fibrous material. It measured 5 inches deep and 3 inches wide, with an egg cavity of 2 inches in depth, by 1¾ inch in width. The eggs, which were hard-set, were but three in number."

The eggs are rather long ovals, the shell fine but with very little gloss ; the ground-colour is a dull white or pinky white, and it is thickly freckled and mottled about the large end and thinly elsewhere with red, in some cases slightly browner, in others purple. The markings have a tendency to form a cap or zone about the large end, and here, where the markings are densest, some little lilac or purplish-grey spots and clouds are intermingled.

An egg measures 0·61 by 0·43.

441. Abrornis schisticeps (Hodgs.).   *The Black-faced Flycatcher-Warbler.*

Abrornis schisticeps, *Hodgs., Jerd. B. Ind.* ii, p. 201; *Hume, Rough Draft N. & E.* no. 571.

Captain Hutton tells us that the Black-faced Flycatcher-Warbler is " a common species in the neighbourhood of Mussoorie, at 5000 feet, and commences building in March. A pair of these birds selected a thick China rose-bush trained against the side of the house, and had completed the nest and laid one egg when a rat destroyed it. I subsequently took two other nests in May, both placed on the ground in holes in the side of a bank by the road-side. In form the nest is a ball, with a round lateral entrance, and is composed externally of dried grasses and green moss, lined with bits of wool, cotton, feathers, thread, and hair. The eggs are three in number."

Two eggs of this species, sent to me by Captain Hutton, are very perfect ovals, pure white *, and rather glossy.

They both measure 0·62 by 0·48.

From Sikhim Mr. Gammie writes:—"The only nest I ever found of this Warbler was in a natural hole in a small tree in an open part of a large forest, at 5500 feet above the sea. In a cleft, five feet from ground, where a limb had been lopped off, there was a small hole, barely large enough at entrance to admit the bird, but gradually widening out for the seven or eight inches of its depth. In the bottom of this cavity was a loose lining of dry bamboo-leaves, on which lay five eggs. They do not agree with those taken by Captain Hutton, which were 'pure white,' but I am absolutely certain of the authenticity of the eggs taken by me. They were well-set, so five is probably the full complement. They were taken on the 26th May."

The eggs sent by Mr. Gammie, for the authenticity of which he vouches, are moderately broad ovals, somewhat compressed and pyriform towards the small end. They have but little gloss, and are of the same type as *A. superciliaris* and *A. albigularis*. The ground is a dull pinkish white, and they are profusely mottled and streaked with red, which in some eggs is brownish, in some purplish. The markings are densest at the large end, where they have a tendency to form an irregular zone, which in some speci-mens is very conspicuous.

These eggs vary from 0·56 to 0·57 in length, and from 0·41 to 0·42 in breadth.

442. Abrornis albigularis, Hodgs. *The White-throated Flycatcher-Warbler.*

Abrornis albigularis, *Hodgs., Jerd. B. Ind.* ii, p. 204.

A nest of this species found in Native Sikhim, below Namtchu, on the 28th July, is a regular Tailor-bird's nest, absolutely undis-tinguishable from the one also sent me by Mr. Mandelli as belonging to *Orthotomus atrigularis*, so that for the moment I have some doubts as to the authenticity of this nest. Two leaves, precisely of the same species as those made use of by the Tailor-bird in question, have been sewn together with the same bright yellow silk, and the little deep cup-shaped nest within is composed exactly of the same excessively fine grass. Another nest, also said to

---

* There can be little doubt that Capt. Hutton's eggs were wrongly identi-fied.—Ed.

belong to this species, but of a very different character, has been sent me by Mr. Mandelli. This was found at Yendong, in Native Sikhim, on the 6th July, and contained four fresh eggs precisely of the type of those of *A. schisticeps*. The nest was placed in the cavity of a truncated bamboo about 4 feet from the ground, and was a loose cup, the basal portion composed of dry bamboo-leaves, and the rest of the nest being made of excessively fine grass, flower-stems, similar to those used in the Tailor-bird-like nest above described, but with a quantity of feathers mingled with this in the lining of the nest.

The eggs of this species are of precisely the same type as those of *A. schisticeps* and *A. superciliaris*, but they are the smallest of all. They are little regular oval eggs, with a white, greyish, or pinky white ground, with deep red freckled and mottled markings, which are densely set about the large end, where they generally form a cap or zone, and usually much less dense elsewhere.

The eggs sent me measured 0·55 and 0·57 by 0·43.

445. Scotocerca inquieta (Cretzschm.). *The Streaked Scrub-Warbler.*

Scotocerca inquieta (*Rüpp.*), *Hume, Rough Draft N. & E.* no. 550 bis.

The Streaked Scrub-Warbler is a permanent resident of the bare stony hills which, under many names and broken into multitudinous ranges, run down from the Khyber Pass to the sea, dividing the Punjab and Sind from Afghanistan and Khelat.

An account of its nidification is contained in the following note furnished me by the late Captain Cock :—

" I first discovered this bird breeding in February in the Khuttuck Hills. It is common throughout the range of stony hills between Peshawur and Attock, and I have seen it on the hills between Jhelum and Pindi, but never took their nest in this latter locality. At Nowshera it is very common, and towards the end of February a collector could take four or five nests in a day. It builds in a low thorny shrub, about 1½ feet from the ground, makes a largish globular nest of thin dry grass-stems, with an opening in the side, thickly lined with seed-down, and containing four or five eggs. Their nesting-operations are over by the end of March."

Lieut. H. E. Barnes, who observed the bird at Chaman in Afghanistan, says :—" These birds are quite common about here on the plains, but I have not observed them on the hills. They commence breeding towards the end of March ; the nest is globular in shape, not unlike that of *Franklinia buchanani*, but somewhat larger, built invariably in stunted bushes about two feet from the ground. It is well lined with feathers and fine grass, the outer portion being composed of fibres and coarse grass. The normal number of eggs is six. I have found less, but never more, and whenever a lesser number has been taken they have always proved to be fresh laid.

"The eggs are oval in shape, white, with a pinkish tinge when fresh, very minutely spotted and speckled with light red, most densely at the larger end. The average of twelve eggs is 0·62 by 0·43."

The eggs are moderately broad and regular ovals, usually somewhat compressed towards one end, but occasionally exhibiting no trace of this. The shell is very fine and delicate, but, as a rule, entirely devoid of gloss. The ground-colour varies from pure to pinky white. The markings are always minute, but in some they are comparatively much bolder and larger than in others, and they vary in colour from reddish pink to a comparatively bright red. In many eggs the markings are much more dense towards the large end, where they form, or exhibit a strong tendency to form, an irregular, more or less confluent zone; and wherever the markings are dense there a certain number of tiny pale purple or lilac spots or clouds will be found intermingled with and underlying the red markings. Some eggs show none of these spots and exhibit no tendency to form a zone, being pretty uniformly speckled and spotted all over. Some are not very unlike eggs of the Grasshopper and Dartford Warblers; others, again, are almost counterparts of the eggs of *Franklinia buchanani.*

In length the eggs vary from 0·6 to 0·68, and in breadth from 0·46 to 0·51.

446. Neornis flavolivaceus, Hodgs.* *The Aberrant Warbler.*

Neornis flavolivacea, *Hodgs., Jerd. B. Ind.* ii, p. 188.

Mr. W. Theobald makes the following remarks on the breeding of this bird at Darjeeling :—"Lays in the second week in July. Eggs three in number, blunt, ovato-pyriform. Size 0·69 by 0·55. Colour deep dull claret-red, with a darker band at broad end. Nest, a deep cup, outside of bamboo-leaves, inside fine vegetable fibres, lined with feathers."

From Sikhim Mr. Gammie writes :—"I have found this Tree-Warbler (though why it should be called a Tree-Warbler I cannot imagine, for it sticks closely to grass and low scrub, and never by any chance perches on a tree) breeding from May to July at elevations from 3500 up to 6000 feet. All the nests I have seen were of a globular shape with entrance near the top. Both in shape and position the nest much resembles that of *Suya atrigularis*, and is, I have no doubt, the one brought to Jerdon as belonging to that bird. It is placed in grassy bushes, in open country, within a foot or so of the ground, and is made of bamboo-leaves and, for the size of the bird, coarse grass-stems, with an inner layer of fine grass-panicles, from which the seeds have dropped, and

---

* I have transferred Hodgson's notes under this title in the 'Rough Draft' to *Horornis fortipes,* to which bird Hodgson's account of the nidification undoubtedly relates, his type-birds No. 900 being *Neornis assimilis.*—En.

lined with feathers. Externally it measures about 6 inches in depth by 4 in width. The egg-cavity, from lower edge of entrance, is 2¼ inches deep by 1¾ wide. The entrance is 2 inches across. The usual number of eggs is three."

The eggs sent by Mr. Gammie are very regular, rather broad, oval eggs, with a decided but not very strong gloss. In colour they are a uniform deep chocolate-purple. In length they vary from 0·63 to 0·69, and in breadth from 0·49 to 0·52.*

---

* I cannot identify the following bird, which appears in the 'Rough Draft' under the number 552 bis. I reproduce the note together with some additional matter furnished later on by Mr. Gammie. *Neornis assimilis* is nothing but *Horornis fortipes*; but I cannot reconcile Mr. Gammie's account of the nest with that of *H. fortipes*, inasmuch as nothing is said about a lining of feathers, which appears to be an unfailing characteristic of the nest of *H. fortipes*.—ED.

No. 552 bis.—NEORNIS ASSIMILIS, *Hodgs.*

Mr. Gammie sent me a bird unmistakably of this species—Blyth's Aberrant Tree-Warbler—together with the lining of a nest and three eggs.

He says :—"The nest, eggs, and bird were brought to me on the 18th May by a native, who said the nest was placed in a shrub, about 6 feet from the ground, in a place filled with scrub near Rishap, at about 3500 feet above the sea. I noted at the time the man's account, but as I did not take the nest myself, I kept no account of it. All I know about it is written on the ticket attached to the nest sent to you. The bird was snared on the nest. Though I did not take it myself, I have little doubt that it is quite correct."

The lining of the nest is a little, soft, shallow saucer 2½ inches in diameter, composed of the finest and softest brown roots.

The eggs are somewhat of the same type as those of *N. flavolivaceus*, but in colour more resembling those of some of the ten-tail-feathered *Prinias*. They are very short broad ovals, pulled out and pointed towards one end, *approximating* to the peg-top type. They are very glossy and of a uniform Indian red; duller coloured rather than those of the *Prinias*; not so deep or purple as those of *N. flavolivaceus*.

They measured 0·65 by 0·52.

From Sikhim Mr. Gammie writes further :—"This bird, I find, does not build in bushes, but on the ground, or rather on low leaf or weed heaps. It not unfrequently takes advantage of the small weed heaps collected round the edges of native cultivations. On the tops of these heaps it collects a lot of dry leaves, and places its nest among them. It sits exceedingly close, only rising when almost stepped on.

"The nest is a rather deep cup, neatly made of dry grass and a few leaves, and lined with fine roots, and the bare twigs of fine grass-panicles. It measures externally about 3·2 inches in diameter by 2·8 in depth; internally 2 inches by 1·75.

"The eggs are three or four in number, and are laid in May from low elevations up to about 3500 feet."

The eggs of this species, of which Mr. Gammie has now sent me two nests, are of the regular *Prinia* type—typically broad ovals, approximating to the peg-top type, but sometimes more elongated and pointed towards the small end. They are very glossy and of a uniform dull Indian red, deeper coloured than any *Prinia's* that I have seen.

They vary from 0·65 to 0·69 in length, and from 0·48 to 0·52 in breadth.

448. Horornis fortipes, Hodgs. *The Strong-footed Bush-Warbler*.

Horornis fortipes, *Hodgs.*, *Jerd. B. Ind.* ii, p. 162.
Dumeticola fortipes, *Hodgs.*, *Hume, Rough Draft N. & E.* no. 526.

According to Mr. Hodgson *, this Tree-Warbler breeds from May to July in the central region of Nepal. They build a tolerably compact and rather shallow cup-shaped nest of grass and dry bamboo-leaves, mingled with grass-roots and vegetable fibre and lined with feathers.

A nest taken on the 29th May measured externally 3·5 in diameter and 2 inches in height, and internally 2 inches in diameter by 1·37 in depth. It contained four eggs, which are figured as deep dull purple-red. Dr. Jerdon gave me two eggs, as I now feel certain, belonging to this species; there is no mistaking them, as they are the most wonderful coloured eggs I ever saw; but as he was not certain to what species they belonged, I unfortunately threw them away. Mr. Hodgson figures the egg as a moderately broad oval, a good deal pointed towards one end, slightly glossy, and measuring 0·65 by 0·47.

Two nests and eggs, together with one of the parent birds, of the Strong-footed Bush-Warbler were sent me from Sikhim. Both nests were found in thick brushwood or low jungle, at elevations of 5000 to 5500 feet—the one at Lebong on the 12th June, the other on another spur of the same hill in July.

The nests were very similar—small massive cups, composed exteriorly of dry blades of grass and leaves, and lined internally with fine grass and a few feathers. Both nests exhibit this lining of feathers, so that it is no accident but a characteristic of the bird's architecture. In one nest a good deal more of the fine flower-panicle stems of grasses are intermingled than in the other. Externally the nests are about 4·5 in diameter and 2·5 in height; the cavity 2 inches in diameter and about 1·25 in depth.

Five more nests of this species have been taken by Mr. Mandelli in the neighbourhood of Lebong, between the 18th May and 15th July; with one exception, where there were only three slightly set eggs, all the nests contained four more or less incubated ones. All the nests were placed in amongst the twigs of low brushwood at heights of from 1 to 3 feet from the ground, and all present the invariable characteristic feature of this species, namely, a greater or less admixture of feathers in the lining of the cavity. Examining the nests carefully, it will be seen that they are composed of three layers—exteriorly everywhere coarse blades of grass and straw loosely put together, inside this a mass of extremely fine

---

* This note of Mr. Hodgson's refers to his plate No. 900. The birds in his collection bearing this number are *Neornis assimilis*, and are the same as *Horornis fortipes.*—ED.

panicle-stems of flowering grass, and then inside this the lining
of moderately fine grass mingled with feathers. The nests vary
a good deal in size, according to the thickness of the coarse outer
layer and the extent to which this straggles; but they seem
to be generally from 4 to 5 inches in diameter, and 2·5 in height,
whilst the cavity is about 2 inches in diameter, and 1, or a little
more than 1, in depth.

The eggs (each nest contained four) are *sui generis*, moderately
broad regular ovals, with a decided but not brilliant gloss, and of a
nearly uniform chocolate-purple. The eggs of one nest are of a
a slightly deeper shade than those of another, probably in consequence of one set being more incubated than the other. They
vary in length from 0·66 to 0·69, and from 0·49 to 0·52 in
breadth.

I do not entertain the slightest doubt of these nests and eggs.

Mr. Mandelli has sent me many more eggs of this species, mostly
deep chocolate-purple, but here and there an egg somewhat paler,
what might be called a pinkish chocolate. They vary from 0·61 to
0·70 in length, and from 0·48 to 0·53 in breadth; but the average
of fifteen eggs is 0·67 by 0·51 nearly.

450. Hororis pallidus (Brooks).   *The Pale Bush-Warbler.*

Horeites pallidus, *Brooks, Hume, Rough Draft N. & E.* no. 527 bis.

The Pale Bush-Warbler breeds in Cashmere, according to
Mr. Brooks, during May. I know nothing either of the bird or
its nidification myself. I have never even closely examined a
specimen, and merely accept the species on Mr. Brooks's authority.

He tells me that he found a nest on the 25th May at Kangan in
Cashmere.

Mr. Brooks writes:—" The nest of *Hororis pallidus*, which I
found near Kangan in Cashmere, up the Sind Valley, was placed in
tangled brushwood, and about five feet above the ground. It was
on a slightly sloping bank, and close to the edge of a patch of jungle,
not far from the right bank of the river.

"It was composed of coarse dry grass externally, with fine
roots and fibres towards the inside of the nest, and was profusely
lined with feathers. It was large for the bird, being 7 or 8 inches
in external diameter, of a globular form, with the entrance at the
side. I don't remember the size of the cavity of the nest, but its
walls were very thick.

"In external appearance it was rough and clumsy, and looked
more like a Sparrow's nest than that of a small Sylvine bird. The
entrance was about 1¾ inch in diameter, and was with the interior
of the nest neat and strong. *Hororis pallidus* occurs at from
5600 feet elevation up to 7000 and even 8000 feet. It was
abundant at Suki up the Bhagirutti Valley, and I heard of one even
at Gangootree."

The shape of the egg is peculiar, being rather flattened in outline at the sides and then suddenly rounded at the smaller end. There is a considerable amount of gloss on the surface, which is of a dull purple-brown, rather darker in tint at the large end. There are a very few indistinct cloudy markings of brown scattered here and there over the egg. In general appearance the egg puts one in mind of a *Prinia's*.

The egg measured 0·64 by 0·49.

451. Horornis pallidipes (Blanf.). *Blanford's Bush-Warbler.*

Horeites pallidipes (*Blanf.*), Hume, *Cat.* no. 527 quat.

Mr. Mandelli sent me two nests of this species. The one was found on the 24th May at Ging, near the Rungnoo River, Sikhim, and contained four fresh eggs; it was placed on the ground amongst coarse grass. The other, which was similarly placed, was found on the 29th June below Lebong at an elevation of about 4000 feet, and contained three fresh eggs. Both nests are rather coarse untidy little cups, some 3 inches in diameter, and 1·75 in height exteriorly, lined and mainly composed of very fine grass, but coated exteriorly everywhere with dry flags, bits of bamboo spathes, and with one or two dead leaves incorporated at the bottom of the structure.

452. Horornis major (Hodgs.). *The Large Bush-Warbler.*

Horeites major, *Hodgs.*, Hume, *Rough Draft N. & E.* no. 529 (err. 629).

A nest said to belong to the Large Bush-Warbler was sent in with one of the parent birds in July from near Lachong in Native Sikhim, where it was found at an elevation of about 14,000 feet. It was placed at a height of about a foot from the ground in a stunted thorny shrub common at these high elevations. It was a very warm little cup, about 3 inches in diameter, composed of the finest fern and moss-roots, tiny fern-leaves, wool, and numbers of the coarse white crinkly hairs of the burhel. It contained three fresh eggs, regular, slightly elongated ovals, a little pointed towards the small end; the shell fine and compact, but with scarcely any gloss.

The ground-colour is white with a faint greenish-blue tinge, and on the larger half of the egg excessively minute specks of brownish red are thinly sprinkled, except just at the crown of the egg, where the specks are denser and exhibit a tendency to form a tiny cap. On the smaller half of the egg very few, if any, specklings are to be traced.

In length the eggs measure 0·7 and 0·71, and in breadth 0·53 to 0·55.

454. Phyllergates coronatus (Jerd. & Bl.). *The Golden-headed Warbler.*

Orthotomus coronatus, *Jerd. & Bl.*, *Jerd. B. Ind.* ii, p. 168; *Hume, Rough Draft N. & E.* no. 531.

Dr. Jerdon says :—" A nest and eggs were brought to me, said to be those of this bird. The nest was similar to that of the last [*O. sutorius*], but not so carefully made ; the leaves were loosely attached, and with fewer stitches. The eggs were two in number, white, with rusty spots."

455. Horeites brunneifrons, Hodgs. *The Rufous-capped Bush-Warbler.*

Horeites brunneifrons, *Hodgs.*, *Jerd. B. Ind.* ii, p. 163.

The egg is a rather broad oval, a good deal pointed towards the small end ; the shell is pretty stout for the size of the egg, and is entirely devoid of gloss. The ground-colour is a pale drabby stone-colour, and all about the large end is a broad dense zone of dull brownish purple. The zone consists of a nearly confluent mass of extremely minute ill-defined speckles, and outside the zone similar speckles and tiny spots occur, though nowhere very noticeable unless closely examined.

Two eggs of this species were brought from Native Sikhim, together with one of the parent birds ; they are regular ovals, slightly pointed towards the small end.

The ground-colour is dull, glossless, pinky white ; the markings consist chiefly of a broad ill-defined zone of dull dark purple ; the other parts of the egg are sparingly, but pretty evenly speckled and spotted with pale purple.

The eggs measure 0·66 by 0·49 and 0·64 by 0·48*.

458. Suya crinigera, Hodgs. *The Brown Hill-Warbler.*

Suya criniger, *Hodgs.*, *Jerd. B. Ind.* ii, p. 183 ; *Hume, Rough Draft N. & E.* no. 547.

The Brown Hill-Warbler breeds throughout the Himalayas, at elevations of from 2000 to 6000 feet, at any rate from Sikhim, where it is comparatively rare, to the borders of Afghanistan.

The breeding-season lasts from the beginning of May until the middle of July, but the majority of the birds lay during May.

A nest which I took at Dilloo, in the Kangra Valley, on the 26th May, was situated near the base of a low bush on the side of a steep hill ; it was placed in the fork of several twigs near the centre of the bush, about 2 feet from the ground. It was an

---

* I cannot find any note about the nest of this species amongst Mr. Hume's papers. There is nothing beyond the above two notes on the eggs.—ED.

excessively flimsy deep cup, about 3 inches in diameter, and $2\frac{1}{2}$ inches in depth internally. It was composed of downy seeds of grass held together externally by a few very fine ,blades of grass, and irregularly and loosely lined with excessively fine grass-stems.

Many other nests subsequently obtained were similar in their materials, the great body of the nest consisting of grass-down, slightly felted together and wound round with slender blades of grass. The nest, however, is by no means always cup-shaped; it is often covered in above, an aperture being left on one side near the top.

A nest which I found near Kotegurh is composed of fine grass *very* loosely and slightly put together, all the interspaces being carefully filled in with grass-down firmly felted together. The nest is nearly the shape of an egg, the entrance being on one side, and extending from about the middle to close to the top. The exterior dimensions of the nest are about $5\frac{1}{4}$ inches for the major axis, and 3 inches for the minor. The entrance-aperture is circular, and about 2 inches in diameter. The thickness of the nest is a little over three eighths of an inch; but the lower portion, which is lined with *very* fine grass-stems, is somewhat thicker. The nest was in a thorny bush, partly suspended from just above the entrance-aperture and partly resting against, though not attached to, some neighbouring twigs. It contained seven eggs, and was taken at Kirlee (Kotegurh) on the 30th May. Of course, the position of the nest was that of an egg standing on end and not lying on its side.

They lay from five to seven eggs, and have, *I think*, two broods.

Dr. Jerdon states that "it makes a large, loosely constructed nest of fine grass, the opening near the top a little at one side, and lays three or four eggs of a fleshy white, with numerous small rusty-red spots tending to form a ring at the large end."

Writing about a collection of eggs made at Murree, Messrs. Cock and Marshall tell us :—" Nest built in high jungle-grass, loosely but neatly made of very fine grass and cobwebs, opening at one side near the top. Breeds late in June at about 4000 feet elevation."

From Almorah Mr. Brooks writes that this species was " common on hill-sides where low bushes were numerous. One nest found was suspended in a low bush, and was a very neat purse-shaped one, with an opening near the top and rather on one side. It was composed of fine soft grass of a kind which had dried green, and was intermixed with the down of plants and lined with finer grass. The eggs were four in number; the ground-colour white, speckled sparingly with light red, but having also a broad zone or ring of deeper reddish brown very near the large end—on the top of the larger end, in fact.

" Laying in Kumaon in May."

From Mussoorie Captain Hutton remarks :—" This little bird appears on the hill, at about 5000 feet, in May. A nest taken

much lower down in June was composed of grasses neatly interwoven in the shape of an ovate ball, the smaller end uppermost
and forming the mouth or entrance; it was lined first with cottony
seed-down, and then with fine grass-stalks; it was suspended
among high grass, and contained five beautiful little eggs of a
carneous white colour, thickly freckled with deep rufous, and with
a darkish confluent ring of the same at the larger end. I have
seen this species as high as 7000 feet in October. It delights to
sit on the summit of tall grass, or even of an oak, from whence it
pours forth a loud and long-continued grating note like the filing
of a saw."

Writing of Nepal, Dr. Scully says:—" A nest taken on the 29th
June contained only two fresh eggs. The nest was of the shape of a
mangoe, the small end being uppermost, and the entrance on one
side, near the top; its measurements externally were, in height
5·2, in breadth 3·6 in one direction and 2·65 in the other; the
opening was nearly circular, 1·8 in diameter. It was rather flimsy
in structure, composed of grass-down, more or less felted together,
and bound round externally with dry green grass-blades; internally
it was scantily lined with fine grass-stems, which were used to
strengthen the lower lip of the entrance-hole. The eggs were
fairly glossy, moderate or longish oval in shape, and measured 0·65
by 0·5 and 0·7 by 0·49; the ground-colour was pinkish white, the
small end nearly free from markings, the middle portion with
faint streaks and tiny indistinct spots of brownish red, and the
large end with a zone of bright brownish red or a confluent cap
of the same colour."

From Sikhim Mr. Gammie writes:—"This Suya breeds from
May to June in the warmest valleys up to 3500 feet. It affects
open grassy tracts, and builds its nest in a bunch of grass, within
a foot or two of the ground. The nest is an extremely neat egg-
shaped structure, with entrance at side, made of fine grass-stems
thickly felted over with the white seeds of a tall flowering grass,
which gives it a very pretty appearance. Externally it measures
5 inches in height by 3 in diameter; the cavity is 2·25 wide and 2
deep, from lower edge of entrance. The entrance is about 2·25
across.

"The usual number of eggs is four. I have never found more,
but on several occasions as few as two and three well-incubated
eggs."

A nest of this species taken by Mr. Gammie near Mongphoo, on
the 18th April, at an elevation of about 3000 feet, contained three
fresh eggs. It closely resembles nests that I have taken of S. crinigera in shape, somewhat like an egg, with the entrance on one
side, near the top, exteriorly about 5 inches in length, and 2¾
inches in diameter, with an aperture a little less than 2 inches
across. It was built amongst grass, of which a few fine stalks
constitute the outer framework, and the whole body of the nest
inside this framework consists solely of the flower-down of grass

firmly felted together. It is lined pretty thickly everywhere with the excessively fine stalks which bear this down.

Taking a large series, I should describe the eggs as typically regular but somewhat elongated ovals, often fairly glossy, at times almost glossless. The ground varies from pale pinky white to pale salmon-colour. A dense, more or less mottled, zone or cap at the large end, varying in different specimens from reddish pink to almost brick-red, and more or less of speckling, mottling, or freckling of a somewhat lighter shade than the zone spreads in some thinly, in some densely over the rest of the egg.

In length they vary from 0·63 to 0·75, and in breadth from 0·46 to 0·55 ; but the average of sixty-five eggs is 0·69 by 0·52.

459. Suya atrigularis, Moore *. *The Black-throated Hill-Warbler.*

Suya atrogularis, *Moore, Jerd. B. Ind.* ii, p. 184; *Hume, Rough Draft N. & E.* no. 549.

The Black-throated Hill-Warbler breeds in Kumaon and the Himalayas eastwards from thence, at elevations of 4000 to 6000 feet.

The breeding-season lasts from April to July, but the birds mostly lay in May and June. Open grassy hillsides dotted about with scrub, thin forests, or gardens are the localities it affects. The nest is placed at times in some low bush surrounded with and grown through by grass, more commonly in clumps of grass, and never at any great height from the ground. It is more or less egg-shaped, and placed with the longer diameter vertical, the entrance being on one side above the middle. It is composed exteriorly sometimes of fine grass-roots, sometimes of the finest possible grass, loosely but sufficiently firmly interwoven, a little moss being often incorporated in the upper portion, and internally always, I think, exclusively of fine grass.

Four is perhaps the usual number of the eggs, but I have found five.

Mr. Gammie, writing from Sikhim, says :—" I have found four nests of this species this year in the Chinchona reserves, at elevations of from 4500 to 5500 feet, during the months of May and June. The nests were all in open grassy country, in grass by the sides of low banks, and not above a foot off the ground. They are globular, with a lateral entrance, composed of grass, and with a little moss about the dome. One I measured was 5·5 high, and

* I reproduce this article nearly as it appears in the 'Rough Draft;' but I have great doubts as to the occurrence of this bird in Kumaon, and I further doubt the identification of Hodgson's notes with this species. It is quite clear, from his specimens in the British Museum, that Hodgson confounded *S. atrigularis* in winter plumage with *S. crinigera*, and his plate of the former in summer plumage contains no note on nidification.—ED.

4·5 in diameter externally; internally the nest was 2·4 in diameter, and the cavity had a total height of 3·9, of which 2 inches was below the lower edge of the entrance. According to my experience four is the regular complement of eggs. I have repeatedly (three times this year) shot the female off the nest, and beyond question Jerdon is wrong about this bird's laying Indian-red eggs."

According to Mr. Hodgson's notes, this species breeds in groves and open forest in Sikhim and the central region of Nepal from April to June, building a large globular nest in clumps of grass, of dry grass, roots, and moss, lined with fine grass and moss-roots. The entrance, which is circular, is at one side; the nest is egg-shaped, the longer diameter being perpendicular, and is placed at a height of about 6 inches from the ground. A nest taken on the 30th May measured 6·12 in height and 3·5 in diameter externally, and the circular aperture, which was just above the middle, was 1·75 in diameter. It contained four eggs, which are represented as ovals, a good deal pointed towards one end, measuring 0·69 by 0·55. The ground-colour is a pale green, and they are speckled and spotted with bright red, the markings being most numerous towards the large end, where they have a tendency to form a zone or cap.

Dr. Jerdon says that "it makes its nest of fine grass and withered stalks, large, very loosely put together, globular, with a hole near the top, and lays three or four eggs of an entirely dull Indian-red colour." This undoubtedly is a mistake; the eggs he refers to are, I think, those of *Neornis flavolivaceus*. He gave them to me, but was not certain of the species they belonged to.

The eggs of the present species are of much the same shape as those of the preceding, and there is a certain similarity in the colour of both; but in these eggs the ground-colour instead of being pink or pinky white, is a pale, delicate, sometimes greyish, green. Then though there is the same kind of zone round the large end, it is a purple or purplish, instead of a brick-red, and it is manifestly made up of innumerable minute specks, and has not the cloudy confluent character of the zone in *S. crinigera*. Outside the zone minute specks of the same purplish red are scattered, in some pretty thickly, in others sparsely, over the whole of the rest of the surface. As a body the eggs have a faint gloss, decidedly less, however, than those of *S. crinigera*, but some few are absolutely glossless.

In length the eggs vary from 0·63 to 0·79, and in breadth from 0·46 to 0·43; but the average of forty-five eggs is 0·68 by 0·5.

460. Suya khasiana, Godw.-Aust. *Austen's Hill-Warbler.*

Suya khasiana, *Godw.-Aust., Hume, Cat.* no. 549 bis.

I found this bird high up in the eastern hills of Manipur, frequenting dense herbaceous undergrowth of balsams and the like in forest. On the 11th of May I caught a female on her nest, containing four well-incubated eggs. The nest was placed in a wild ginger-plant, about two feet from the ground, in forest at the very summit of the Makhi hill.

462. Prinia lepida, Blyth. *The Streaked Wren-Warbler*

Burnesia lepida (*Blyth*), Jerd. B. Ind. ii, p. 185.
Burnesia gracilis, *Rüpp.*, Hume, *Rough Draft N. & E.* no. 550.

I have never happened to meet with the nest of the Streaked Wren-Warbler, and all the information I possess in regard to its nidification I owe to others.

The late Mr. Anderson remarked :—" Although this species was far from uncommon, I found it very local and confined entirely to the tamarisk-covered islands and ' churs ' along the Ganges.

" The first nest was taken on the 13th March last, and contained three well-incubated eggs ; of these I saved only one specimen, which is now in the collection of Mr. Brooks. The second was found on the following day, and contained two callow young and one perfectly fresh egg.

" The nest is domed over, having an entrance at the side ; and the cavity is comfortably lined, or rather felted, with the down of the madar plant. It is fixed, somewhat after the fashion of that of the Reed-Warbler, in the centre of a dense clump of surpat grass, about 2 feet above the ground. On the whole the structure is rather large for so small a bird, and measures 6 inches in height by 4 inches in breadth.

" But while the *nest* corresponds exactly with Canon Tristram's description * of those taken by him in Palestine, there are differences, oologically speaking, which induce me to hope that our Indian bird may yet be restored to specific distinction †. In the first place, my single eggs from each nest have a *green* ground-colour, and are covered all over with reddish-brown spots. Now Mr. Tristram describes his Palestine specimens as ' richly coloured *pink* eggs, with a zone of darker red near the larger end, and in shape and colour resembling some of the *Prinia* group.' Is it possible for the same birds to lay such widely different eggs ? If I had taken only one specimen, it might have been looked upon as a mere variety. Again, our Indian bird lays three eggs, and I have never seen the parent birds feeding more than this number of young ones, occasionally only two. Mr. Tristram, *per contra*, mentions having met with as many as five and six. The egg is certainly the prettiest, and one of the smallest, I have ever seen ; indeed, I found it too small to risk measurement."

He adds :—" Since writing the above, which appeared in ' The Ibis,' I have discovered that this species breeds in September and

---

* Tristram on the Ornithology of Palestine, P. Z. S. 1864, p. 437 ; Ibis, 865, pp. 82, 83.
† The two birds are now considered distinct by all ornithologists.—ED.

October, as well as in February and March, so some of them probably have two broods in the year. I took a nest on the 9th
October at Futtegurh, which contained two callow young and one
(*fresh*) egg, which I send you, and which is exactly similar to all
the others I have taken from time to time."

The egg sent me by Mr. Anderson is a very broad oval in shape,
a good deal compressed however, and pointed towards the small
end. The shell is very fine and has a decided gloss. In colouring
the egg is exactly like those of some of the Blackbirds—a pale
green ground, profusely freckled and streaked with a bright, only
slightly brownish, red; the markings are densest round the large
end, where they form a broad, nearly confluent, well-marked, but
imperfect and irregular, zone. It measures 0·55 by 0·41.

Colonel C. H. T. Marshall says:—"The Streaked Wren-Warbler
breeds in great numbers near Delhi in March; Mr. C. T. Bingham
has found several of them in the clumps of surpat grass that had
been cut within three feet of the ground on the alluvial land of the
Jumna. It was when out with him in the end of March 1876
that I first saw the nest of this species. The locality of the nest
is exactly that described by Mr. Anderson; it is oval in shape,
with a large side entrance near the top; it is built of fine grass
and seed-down, no cobweb being employed in the structure; it is
loosely made, and there are always a few feathers in the egg-cavity.
The whereabouts is generally pointed out by the cock bird, who,
seated on the top of the highest blade of grass he can find near
where his hen is sitting, pours out with untiring energy his feeble
monotonous song, little knowing that by so doing he has betrayed
the spot where he has fixed his nest to the marauder. The eggs,
of which I have seen about fifteen or twenty, answer the description given in 'Stray Feathers' exactly."

Major C. T. Bingham tells us:—"Between the 12th and 31st
March this year I found ten nests of this bird, which is very
common in the grass-covered land of the Jumna. These nests
were all alike, of fine dry grass mixed with the down of the surpat,
which also thickly lined the inside. In shape the nests are blunt
ovals, with a tiny hole for entrance a little above the centre.
Seven out of the ten nests contained four eggs each, the rest three
each. The eggs in colour are a pale yellowish white with a tinge
of green, thickly speckled with dashes rather than spots of rusty
red, tending in some to form a cap, in others a zone round the
large end. The average of twenty eggs measured is 0·53 by 0·44
inch. The nests were all, with one exception, supported by
stems of the grass being worked into the sides. The one exception
was a nest I found in the fork of a tamarisk bush. It is not a
difficult nest to find, for when you are in the vicinity of one, one
of the birds will flit about the stems of the surrounding clumps of
grass and above you freely, opening its tiny mouth absurdly wide,
but giving forth the feeblest of feeble sounds."

Writing on the Avifauna of Mt. Abu and N. Guzerat, Colonel
E. A. Butler says:—"I found a nest in a tussock of coarse grass

in the sandy bed of a river, amongst a number of tamarisk-bushes, on the 8th July, 1875, in the neighbourhood of Deesa. It was composed of fine dry fibrous roots and grass-stems exteriorly, and lined with silky vegetable down. It was a long bottled-shaped structure with a small entrance on one side. The nest, eggs, situation, locality, &c. all agree so exactly with the descriptions quoted by Dr. Jerdon and with Mr. Anderson's note in 'Nests and Eggs,' *Rough Draft*, that I should have found it difficult to avoid copying these two gentlemen in describing my own nest.

"The nest contained three hard-set eggs and one young one just hatched."

Referring to its occurrence in the Eastern Narra District, Mr. Doig tells us :—" This little Warbler is very common. I took the first nest in March and again in May ; they build in stunted tamarisk-bushes ; the nest is circular dome-shaped, with the entrance on one side the top, the inside being very beautifully and softly lined with the pappus of grass-seeds. Four is the usual number of eggs in one nest."

The Blackbird type of egg above described is by no means the commonest one ; the great mass of the eggs have the ground greyish, greenish, or pinkish white, and they are very thickly and finely freckled and speckled all over, but most densely about the large end, with a slightly brownish, rarely a slightly purplish grey. Occasionally when the markings are very dense in a cap at the large end there is a distinct purplish-grey tinge there, and on the rest of the surface of the egg the markings are somewhat less thickly set, leaving small portions of the ground-colour clearly visible. Typically the eggs are moderately broad ovals, a little compressed towards the small end, and though none are very glossy, the great majority have a fair amount of gloss.

463. Prinia flaviventris (Deless.). *The Yellow-bellied Wren-Warbler.*

Prinia flaviventris *(Deless.), Jerd. B. Ind.* ii, p. 160; *Hume, Rough Draft N. & E.* no. 532.

Of the Yellow-bellied Wren-Warbler's nidification I know personally nothing.

Tickell describes the nest as pensile but quite open, being a hemisphere with one side prolonged, by which it is suspended from a twig. The eggs, he says, are bright brick-red without a spot.

Mr. H. C. Parker tells me that " this bird breeds in the Salt-Water Lake, or rather on the swampy banks of the principal canals that intersect it. The nest is nearly always placed on an ash-leaved shrub-like plant growing on the banks of the canal and overhanging the water. One taken on the 26th July, 1873, containing four nearly fresh eggs, was almost touching the water at high tide. The male has the habit, when the female is sitting, of hopping to the extreme point of a tall species of cane-like grass

which grows abundantly in these swamps, whence he gives forth a
rather pleasing song, erecting his tail at the same time, after which
he drops into the jungle and is seen no more. It is almost
impossible to make him show himself again."

The nest, which I owe to Mr. Parker, and which was found in
the neighbourhood of the Salt-Water Lake, Calcutta, on the 26th
July, is of an oval shape, very obtuse at both ends, measuring
externally 4 inches in length and about 2¾ inches in diameter.
The aperture, which is near the top of the nest, is oval, and measures about 1 inch by 1½ inch. The nest is fixed against the side
of two or three tiny leafy twigs, to which it is bound lightly in one
or two places with grass and vegetable fibre ; and two or three leafy
lateral twiglets are incorporated into the sides of the nest, so that
when fresh it must have been entirely hidden by leaves. The nest
was in an upright position, the major axis perpendicular to the
horizon. It is a very thin, firm, close basket-work of fine grass,
flower-stalks, and vegetable fibre, and has no lining, though the
interior surface of the nest is more closely woven and of still finer
materials than the outside. The cavity is nearly 2½ inches deep,
measuring from the lower edge of the entrance, and is about
2 inches in diameter.

During this present year (1874) Mr. Parker obtained several
more nests of this species, all built in the low jungle that fringes
the mud-banks of the congeries of channels and creeks that are
known in Calcutta by the name of the " Salt Lake."

This jungle consists chiefly of the blue-flowered holly-leaved
Acanthus ilicifolia and of the trailing semi-creeper-like *Derris
scandens*. It is in amongst the drooping twigs of the latter that
the nest is invariably made.

The nests vary a good deal in shape ; some are regular cylinders
rounded off at both ends, with the aperture on one side above the
centre—a small oval entrance neatly worked. Such a nest is
about 4·5 inches in length externally from top to bottom, and 2·75
in diameter ; the aperture 1·3 in height, and barely 1·0 in width.

Others are still more egg-shaped, with a similar aperture near
the top, and others are more purse-like. The material used
appears to be always much the same—fine grass-stems intermingled
with blades of grass, and here and there dry leaves of some rush, a
little seed-down, scraps of herbaceous plants, and the like ; the
interior, always of the finest grass-stems, neatly arranged and
curved to the shape of the cavity. The nests are firmly attached
to the drooping twigs, to and between which they are suspended,
sometimes by fine vegetable fibre, but more commonly by cobwebs
and silk from cocoons, a good deal of both of which are generally
to be seen wound about the surface of the nest near the points of
suspension or attachment.

Four appears to be the full number of the eggs.

Mr. Doig, writing from Sind, says :—" This bird is tolerably
common all along the Narra, but as it keeps in very thick jungle it
is not often seen unless looked for. I took my first nest on the

12th, and my second on the 17th of May. This evidently is the second brood, as I noticed on the same day a lot of young birds which must have been fully six weeks old. One nest was lined with horsehair and fine grasses. Four was the normal number of eggs."

Mr. Oates writes:—"The Yellow-bellied Wren-Warbler is very abundant throughout Lower Pegu in suitable localities. In the plains between the Sittang and Pegu rivers they are constant residents, breeding freely from May to August and September. In Rangoon also, all round the Timber Depôt at Kemandine, and in the low-lying land between the town proper and Monkey Point, they are very numerous."

The eggs are of the well-known *Prinia* type—broad regular ovals, of a nearly uniform mahogany-red, and very glossy. To judge from the few specimens I have seen, they average a good deal smaller, and are somewhat less deeply coloured, than those of *P. socialis.* They vary from 0·52 to 0·6 in length, and from 0·43 to 0·48 in breadth.

464. Prinia socialis, Sykes. *The Ashy Wren-Warbler.*

Prinia socialis, *Sykes, Jerd. B. Ind.* ii, p. 170; *Hume, Rough Draft N. & E.* no. 534.
Prinia stewarti, *Blyth, Jerd. B. Ind.* ii, p. 171; *Hume, Rough Draft N. & E.* no. 535.

Prinia socialis.

The Ashy Wren-Warbler breeds throughout the southern portion of the Peninsula and Ceylon, alike in the low country and in the hills, up to an elevation of nearly 7000 feet.

The breeding-season extends from March to September, but I am uncertain whether they have more than one brood.

Dr. Jerdon says:—"Colonel Sykes remarks that this species has the same ingenious nest as *O. longicauda.* I have found the nest on several occasions, and verified Colonel Sykes's observations; but it is not so neatly sewn together as the nest of the true Tailor-bird, and there is generally more grass and other vegetable fibres used in the construction. The eggs are usually reddish white, with numerous darker red dots at the large end often coalescing, and sometimes the eggs are uniform brick-red throughout."

Now, first, as regards the eggs, it is clearly wrong to say that the eggs are usually reddish white; that such eggs, as exceptions, may have occurred I do not doubt, but I have seen more than fifty eggs of this bird taken by Miss Cockburn, Messrs. Carter, Davison, Wait, Theobald, and others, and all were without exception mahogany- or brick-red, at times mottled, somewhat paler and darker here and there, but making no approach, even the most distant, to what Dr. Jerdon says is the *usual* type. Moreover, I have taken *many hundreds* of the eggs of *P. stewarti* (the northern, rather smaller form), which is not only *most* closely allied but really *very* doubtfully distinct, and yet I never met with one single

19*

egg of this type. At the same time Mr. Swinhoe ('Ibis,' 1860, p. 50) tells us that *P. sonitans* also at times exhibits a reddish-white egg; so I do not for a moment question that Dr. Jerdon had seen such eggs, only it must be understood that, so far from constituting the *usual type*, it is in reality a most abnormal and rare variety. Out of eight correspondents who have collected for me in Southern India, I cannot learn that any one has ever yet even seen an egg of this type.

As regards the nest, this species often constructs a Tailor-bird nest, the true nest being filled in between two or more leaves carefully stitched together to the nest; but it also, like that species, often builds a very different structure.

A nest now before me, sent from Conoor, is a loosely-made cup—a very slight fabric of grass-stems, matted with a quantity of the downy seed of some flowering grass and with a lining of fine grass-roots. It is an irregular cup about 2½ inches in diameter and 2 inches in depth.

Four seems to be the regular number of the eggs.

From Kotagherry Miss Cockburn writes that "the Ashy Wren-Warbler builds a neat little hanging nest very much in the Tailor-bird style, for it draws the leaves of the branch on which the nest is constructed close together, and sews them so tightly as sometimes to make them nearly touch each other, while a small quantity of fine grass, wool, and the down of seed-pods is used as a lining and also placed between the leaves. These nests are built very low, and contain three *beautiful* little bright red eggs, a shade darker at the thick end. They are easily discovered; for the birds get so agitated if any one approaches the bush on which they have built that they invariably attract one to the very spot they most wish to conceal. They build in the months of June and July."

Mr. Davison says:—"This bird breeds on the Nilghiris in March, April, and May, and sometimes as late as the earlier part of June. The nest is generally placed low down near the roots of a bush or tuft of grass. It is made of grass beautifully and closely woven, domed, and with the entrance near the top. The eggs, three or four in number, are of a deep brick-red, darker at the larger end, where there is generally a zone, and are very glossy. I once obtained a nest made of grass and bits of cotton, but instead of being built as above described it was placed between, and sewn to, two leaves of the *Datura stramonium*. It contained three eggs of a deep brick-red; in fact, precisely like those described above."

Mr. Wait tells us that "in September I found two nests, the one deeply cup-shaped, the other domed, both constructed of similar materials. The latter of the two was placed at the bottom of a large bunch of lemon-grass, and was constructed of root-fibre and grass, grass-bents, and down of thistle and hawkweed, all intermixed. Exteriorly it measured between 3 and 4 inches in diameter. The nests contained three and five eggs, all highly glossy and of a deep brownish-red, deeper than brick-red, mottled with a still deeper shade."

Colonel W. V. Legge, writing from Ceylon, tells us that
" *P. socialis* breeds with us in the commencement of the S.W.
monsoon during the months of May, June, and July. It nests in
long grass on the Patnas in the Central Province, in guinea-grass
fields, and in sugarcane-brakes where these exist, as in the Galle
District for instance. I can scarcely imagine that Jerdon is
correct about this Warbler's nesting.

" Nothing can be more un-Tailorbird-like than the nest which
it builds in *this* country, and this led me to think that ours was a
different species until my specimens were identified by Lord
Walden. In May 1870 a pair resorted to a large guinea-grass
field attached to my bungalow at Colombo, for the purpose of
breeding. I soon found the nest, which was the most peculiarly
constructed one I have ever seen. It was, in fact, an almost
shapeless ball of guinea-grass roots, *thrown* as it were between the
upright stalks of the plant at about 2 feet from the ground : I say
' thrown,' because it was scarcely attached to the supporting
stalks at all. It was formed entirely of the roots of the plant,
which, when it is old, crop out of the ground and are easily plucked
up by the bird, the bottom or more solid part being interwoven with
cotton and such-like substances to impart additional strength. The
entrance was at the side in the upper half, and was tolerably neatly
made ; it was about an inch in diameter, the whole structure measuring about 6 inches in depth by 5 inches in breadth. I found
the nest in a partial state of completion on the 10th of May ; by
the 19th it was finished and the first of a clutch of three eggs laid.
The nest and eggs were both taken on the evening of the 24th, and
the following day another was commenced close at hand. This was
somewhat smaller, but constructed in the same peculiar manner as
the first. This was completed, and the first of another clutch
laid. The eggs are somewhat pointed at the smaller end, and
of an almost uniform dull mahogany ground-colour, showing indi-
cations of a paler underground at the point."

Birds like these, that build half-a-dozen different kinds of nests,
ought to be abolished ; they lead to all kinds of mistakes and differences of opinion, and are more trouble than they are worth.

Colonel E. A. Butler writes :—" Found numerous nests of this
species at Belgaum on the following dates :—

" July 13.   A nest containing 4 fresh eggs.

"	22.	"	"	3 "
"	25.	"	"	4 "
"	26.	"	"	3 "
"	26.	"	"	3 "
"	28.	"	"	2 slightly incubated eggs.
Aug.	5.	"	"	4 fresh eggs.
"	6.	"	"	4 "

" All of the above nests were built in sugarcane-fields or in
corn-fields ; and most of them were stitched up in leaves of various
plants, after the fashion of Tailor-birds' nests ; but in some instances
they were of the other type, simply supported by the blades of

sugarcane or corn they were built in. In addition to the above I found numerous other nests all through August, many of which were destroyed by something or other—what, I do not know! In fact, it has always been a puzzle to me what it is that takes the eggs of these small birds: three out of four nests, when visited a second time, are either empty, gone altogether, or pulled down; and how the birds ever manage to hatch off a brood at all with so many enemies I do not know.

"I found a nest of the Ashy Wren-Warbler at Deesa on the 21st July, containing three fresh eggs, of a highly polished deep mahogany-red colour, with an almost invisible cap of the same colour a shade darker at the large end. The nest, which was placed in the centre of a low bush and fixed to a few small twigs, was oval in shape, measuring $3\frac{3}{4}$ inches in length exteriorly and $2\frac{5}{8}$ in width, with a small round entrance near the top about $1\frac{1}{4}$ inch in diameter. It was composed of fine dry fibrous grass, with silky vegetable down (*Calotropis gigantea*) and cobwebs smeared over the exterior. The walls were very thin, but the bottom of the nest somewhat solid. The whole well woven and compactly built. Later on I got nests on the following dates:—

"Aug. 1. A nest containing 3 fresh eggs.
"   1.   „   „   2   „
"   5.   „   „   4   „
"   5.   „   „   4   „
"   8.   „   „   3   „
"   9.   „   „   4   „
"   26.   „   „   3   „

"In addition to the above, I found nests containing young birds on the 15th, 17th, and 23rd August.

"The nests are of two distinct types. One as above described; the other, which is the commoner of the two, a regular Tailor-bird's nest stitched between two leaves but without any lining. The eggs vary a good deal in shade, some being paler than others. Some eggs I have look almost like little balls of red carnelian. Creepers (convolvulus &c.) growing up low thorny bushes in grass-beerhs are a favourite place for the nest."

Lieut. H. E. Barnes informs us that in Rajputana this Warbler breeds from July to September.

Messrs. Davidson and Wenden state that this bird is common in the Deccan and breeds in August.

Mr. Rhodes W. Morgan, writing from South India, says:—"It builds in March, constructing a very neat pendent nest, which is artfully concealed, and supported by sewing one or two leaves round it. This is very neatly done with the fine silk which surrounds the eggs of a small brown spider. The nest is generally built of fine grass, and contains three eggs of a bright brick-colour with a high polish. The entrance to the nest is at the top and a little on one side. An egg measured 0·7 inch in length by 0·48 in breadth."

As for the eggs, it is unnecessary to describe them; they are precisely similar to those of *P. stewarti*, fully described below. All that can be said is that as a body they are slightly larger, and *possibly*, as a *whole*, the least shade less dark. In length they vary from 0·52 to 0·72, and in breadth from 0·45 to 0·52; but the average of twenty-one eggs measured is 0·64 by rather more than 0·47*.

Prinia stewarti.

Stewart's Wren-Warbler is one of those forms in regard to which at present great difference of opinion prevails as to whether or no they merit specific separation. *P. stewarti* from the N.W. Provinces and *P. socialis* from the Nilghiris differ only in size; the latter is somewhat more robust, and probably weighs one fifth more than the former. But then in the Central Provinces you meet with intermediate sizes, and I have plenty of birds which might be assigned indifferently to either race as a rather small example of the one or rather large one of the other. I myself consider all to belong to one species, but as this is not the general view I have kept my notes on their nidification separate.

This species or race breeds almost throughout the plains of Upper India and in the Sub-Himalayan ranges to an elevation of 3000 or 4000 feet. In the plains the breeding-season extends from the first downfall of rain in June (I have never found them earlier) to quite the end of August. In the moist Sub-Himalayan region, the Terais, Doons, Bhaburs, and the low hills, they commence laying nearly a month earlier.

This species often constructs as neatly sewn a nest as does the *Orthotomus*; in fact, many of the nests built by these two species so closely resemble each other that it would be difficult to distinguish them were there not very generally a difference in the lining. With few exceptions all the innumerable nests of *O. sutorius* that I have seen were lined with some soft substance—cotton-wool, the silky down of the cotton-tree (*Bombax heptaphyllum*), grass-down, soft horsehair, or even human hair, while the nests of *P. stewarti* are almost without exception *lined* with fine grass-roots.

Our present bird does not, however, invariably construct a "tailored" nest. When it does, like *O. sutorius*, it sews two, three, four, or five leaves together, as may be most convenient, filling the intervening space with down, fine grass, vegetable fibre, or wool, held firmly into its place by cross-threads, sometimes composed of cobwebs, sometimes made by the bird itself of cotton, and sometimes apparently derived from unravelled rags. It also, however, often makes a nest entirely composed of fine vegetable fibre, cotton, and grass-down, and lined as usual with fine grass-roots. Sometimes these nests are long and purse-like, and sometimes globular, either attached to, or pendent from, two or more twigs. One nest before me, a sort of deep watch-pocket, suspended from

---

* As a matter of convenience I keep the notes on *P. socialis* and *P. stewarti* separate, as is done in the 'Rough Draft'; but there is no doubt whatever now that the two birds are the same species.—ED.

five twigs of the jhao (*Tamarix dioica*), measures externally 2·75 inches in diameter, is a good deal longer at what may be called the back than the front, and at the back fully 5·5 long. Internally the diameter is about 1·5, and the cavity, measuring from the lowest portion of the external rim, is 2·5. This is a *very* large nest. Another, built between three leaves, has an external diameter of about 2½ inches, and is externally not above 3 inches long. It is unnecessary here to describe the beautiful manner in which, when it makes use of leaves, this bird sews them together, as this has already been well described by others where *O. sutorius* is concerned, and *P. stewarti* is, in some cases, when forming a nest with leaves, fully as neat a workman.

The nests vary so much, and I have heard so much discussion about them, that having seen at least a hundred and having taken full notes of some twenty of them, I shall reproduce a few of these notes :—

" *Agra, July 17th.*—Two nests—one nearly globular, composed entirely of fibrous roots, hair, wool, and thread, and lined with fine grass, suspended by a few fibres and hairs between the fork of a branchlet in a little dense bush of Indian box ; the other, suspended from the tendril of an elephant creeper, was principally formed by one of the leaves of this, to which, to form the remaining third of the exterior, a second leaf of the same plant was carefully sewn. Interiorly there was a little wool, and at the bottom fine grass.

" *July 20th.*—On a furash-tree (*Tamarix furas*), beautifully made of fine soft wool, shreds of tow and string, very fine grass and grass-roots, and the bottom neatly lined with very fine grass-roots. In shape the nest is like one half of a long old-fashioned silk purse, round-bottomed and very compact, with a long slit-like opening on one side towards the top. It contained five eggs.

" *July 26th.*—Two nests, one formed almost entirely in a single mango-leaf, the sides of which are curled round so as nearly to meet, and then laced by a succession of cross-threads of cobweb, carefully knotted at each place where the margin of the leaf is pierced. The intervening space is closed by fine tow, wool, and the silky down of the cotton-tree, with just the top of a small mango-leaf caught in from above so as to form an arched roof. The other nest was rounder in form, having less of a leafy structure. It had, however, the leaf of the *Phalsa* forming the back and sides (partly), whilst the whole of the front was composed of soft wool, tow, dry grass-roots, thread, and a few pieces of the soft tree-cotton. It had a neighbouring leaf just caught in on one side. This contained four fresh eggs.

" *July 30th.*—A beautiful nest between three twigs, several of the leaves of each of which had been tacked on to the outside of the nest. The nest itself was firmly put together with fine grass-roots, and was nearly globular in shape, with one side continued upwards into a sort of hood overhanging the greater portion of the aperture. It contained four eggs of the usual deep red colour.

" *August 8th.*—At Bichpoori found a number of nests, and some of them of a strangely different type. One was inside a tiny hut on the line, about 3 feet above the head of the chaprassie's bed. It had no leaves about it, and was composed of thread, wool, and a few very fine grass-stems, and lined thinly with fine grass-stems and a little black horsehair. It was about two thirds of a sphere, the external diameter of which was about $3\frac{1}{4}$ inches, and the internal $2\frac{1}{2}$ inches. The bird was on the nest, so that there could be no mistake, otherwise it would have been impossible to believe that it belonged to *P. stewarti*, of which we have taken so many sewn in leaves. A little further on another nest of the same species, built in the ragged caves of a thatch, externally composed almost entirely of cotton-wool, with a little tow-fibre binding the structure together, internally as usual lined with very fine grass-roots with a few horsehairs. Another nest of the *Prinia* was in one respect even more remarkable. It was built in the usual situation in a low herbaceous plant, sewn to and suspended from two leaves, and two or three others worked into its sides. It was constructed almost entirely of fine grass-roots and fibres, with a few tiny tufts of cotton-wool, and the leaves as usual firmly tacked on with threads and cobweb fibres. It would seem that, after constructing the nest, but before laying, a large female spider took possession of the bottom of the nest, and shut herself in by constructing a diaphragm of web horizontally across the nest, thus occupying the whole of the cavity of the nest. The little bird accepted this change of circumstances, built the nest a little higher at the sides, and over the spider's web placed a false bottom of fine grass-roots, on which she laid her four eggs, and there she was sitting when the nest was taken, the spider, alive and apparently happy in the cell below, plainly visible through the interstices of the grass, with a huge sac of eggs which she was incubating. Her chamber is fully one half of the nest.'

I may add that this latter nest, with the *now* dead spider, *in situ*, is still in our museum.

In number the eggs are sometimes four, sometimes five, and I have *heard* of six being found.

They rear usually two broods; if their eggs are taken they will lay three or four sets; sometimes they use the same nest twice; sometimes, directly the first brood is at all able to shift for themselves, the parents leave them in the old nest, and commence building a new one at no great distance.

The late Mr. A. Anderson remarked:—" Owing to the inclemency of the weather (August) the geranium-pots in the garden were placed in the verandah of the house I am at present living in, and, strange to say, a pair of these Warblers commenced building in the leaves of one of the plants immediately under my window.

" When the nest was about half-finished the birds forsook it without apparently any reason, as they were never molested in any way. On examining the nest, however, the cause was evident, and afforded a remarkable instance of instinct on the part of the little architects. The leaves that had been pierced and sewn together had actually commenced to *wither*, and in the course of a few days later the whole structure came down bodily.

"This is the only *Prinia* to be found at Futtehgurh, and they are one of our most common garden-birds. Their beautiful brick-red eggs and neatly-sewn nests are too well known to require description.

"Four generally, and five frequently, is the number of eggs they lay. I have *one* record of *six* on the 17th August, 1873; in this case one egg was laid daily, the first having been laid on the 12th, and the sixth on the 17th."

Captain Hutton remarks:—"This is a true Tailor-bird in respect to the construction of the nest, which is composed of one leaf as a supporting base stitched to two others meeting it perpendicularly, the apices of all three being neatly sewn together with threads roughly spun from the cottony down of seeds. Between or within these leaves is placed the nest, very slightly and loosely constructed of fine roots, grass-stalks, and seed-down, the latter material being interwoven to hold the coarser fibres of the nest together. There is no finer lining within, and the edges of the exterior leaves are drawn together round the nest and held there partly by roughly-spun threads of down, and partly by the ends of the stiff fibres being thrust through them. The whole forms a very light and graceful fabric. Within this nest were four beautiful and highly polished eggs of a deep brick-red colour, darkest at the larger end, faint specks and blotches of a deeper colour being indistinctly discernible beneath the surface of the shell, which shines as if it had been varnished. The nest is not closed above, but is open and deeply cup-shaped. This was taken in the Dhoon on the 30th May."

Major C. T. Bingham says:—"Breeds at Allahabad in June, July, and August. At Delhi I have not yet found its nest. I once found in July three nests all attached together in a sort of triangle, but whether built by separate pairs of birds I cannot say. Only one nest contained eggs."

Colonel G. F. L. Marshall writes:—"A nest found in July in the Cawnpoor district was built of grass, a deep oblong domed nest with the entrance at the side near the top. It was placed close to the ground in a tuft of surkerry grass sloping rather backwards. The position is, I believe, unusual. The old birds were still putting finishing touches to the building when I found it."

The eggs are ovals, as a rule, neither very broad nor much elongated. Pyriform examples occur, but a somewhat perfect oval is the usual type, and the examination of a large series shows that the tendency is to vary to a globular and not to an elongated shape. The eggs are brilliantly glossy, and, though considerably smaller, strongly resemble, as is well known, those of the little short-tailed Cetti's Warbler. .

In colour they are brick-red, some, however, being paler and yellower, others deeper and more mahogany-coloured. There is a strong tendency to exhibit an ill-defined cloudy cap or zone, of far greater intensity than the colour of the rest of the egg, at or towards the large end.

In length the eggs vary from 0·6 to 0·68, and in breadth from 0·45 to 0·5; but the average of seventy eggs measured is 0·62 by 0·46.

465. Prinia sylvatica, Jerd. *The Jungle Wren-Warbler.*

Drymoipus sylvaticus, *Jerd. B. Ind.* ii, p. 181; *Hume, Rough Draft N. & E.* no. 545.

Drymoipus neglectus, *Jerd. B. Ind.* ii, p. 182; *Hume, Rough Draft N. & E.* no. 546.

Dr. Jerdon says:—" I found the nest in low jungle near Nellore, made chiefly of grass, with a few roots and fibres, globular, large, with a hole at one side near the top, and the eggs white, spotted very thickly with rusty red, especially at the thick end."

Mr. Blewitt appears to have taken many eggs of this species in the Raipoor District, and he has sent me the following notes, together with numerous eggs. He says:—

" The Jungle Wren-Warbler breeds in the Raipoor District from about the middle of June to the middle of August. Low thorn-bushes on rocky ground are chiefly selected for the nest, and both parent birds assist in building it and in hatching and rearing the young. A new nest is made each year, and four is the maximum number of eggs.

" On the 1st July this year I found a nest of this species in the centre of a low thorny bush, growing in rocky ground, about two miles north of Doongurgurh in the Raipoor District.

" The nest was about 4 feet from the ground, firmly attached to and supported by the branches. It was of a deep cup shape, 3·6 in diameter and 4·9 in height, composed of coarser and finer grasses firmly interwoven, and contained four fresh eggs. In the same locality we secured a second similarly situated nest, about 2½ feet from the ground, and it contained a single fresh egg. It was rather more neatly and massively made than the former. It was about 4 inches in diameter and 5 inches in height, and the egg-cavity was nearly 3 inches deep. The lining is of fine grass-stalks well interwoven. The exterior is composed of coarse grass mixed with a little greyish-white fibre.

" Subsequently several other similar and similarly situated nests were found."

Colonel E. A. Butler writes:—" The Jungle Wren-Warbler breeds in the neighbourhood of Deesa in the months of July, August, and September. The following are the dates upon which I found nests this year (1876):—

" July 28.   A nest containing 4 young birds.
  „   29.   „   5 fresh eggs.
  Aug. 1.   „   4   „
  „   5.   „   5   „

" Aug. 13.   A nest containing 5 fresh eggs.
  ,, 16.   ,,   4 young birds fledged.
  ,, 17.   ,,   5   ,,
  ,, ,,   ,,   3   ,,
  ,, 19.   ,,   4   ,,
  ,, ,,   ,,   5   ,,
  ,, 30.   ,,   5   ,,
 Sept. 3.   ,,   5   ,,

"In addition to the above, I found nests in the same neighbourhood in 1875. One on the 14th August containing four young birds almost ready to leave the nest. It was placed in the middle of a tussock of coarse grass on the side of a nullah on a bank overgrown with grass and bushes, and my attention was attracted first of all to the spot by the incessant chattering and uneasiness of the two old birds, one of which had a large grasshopper in its mouth. After hiding behind a bush for a few minutes, I saw the hen bird fly to the nest, which led to its discovery. The nest was dome-shaped, with an entrance upon one side, composed exteriorly of blades of rather coarse dry grass (green, however, as a rule when the nest is first built), and interiorly of similar, but finer, material. It is an easy nest to find when once the locality in which the birds breed is discovered, as it is a conspicuous ball of grass, smeared over, often more or less, exteriorly with a silky white vegetable-down or cobweb, and many of the blades of the tussock in which it is placed are often drawn down and woven into the nest, which at once attracts attention. Then, again, the cock bird is almost always to be found on the top of some low tree near the nest, uttering his peculiar ventriloquistic note ' *tissip, tissip, tissip,*' etc. All the above nests were exactly alike and in similar situations, viz. fixed in the centre of a tussock of coarse grass on the banks of some deep nullahs running through a large grass ' Beerh.' The eggs remind me more of the English Robin's eggs than those of any other species I know. The ground-colour is dull white, sometimes tinted with pale green, and the markings reddish fawn. In some cases the eggs are peppered all over with a conspicuous zone at the large end, sometimes a dense cap instead of a zone. In other cases the markings, though always present, are almost invisible, as also the zone or cap. They are about the size of the eggs of the Spotted Flycatcher. I found a few other nests besides those I have mentioned during July and August 1875."

Captain Cock informed me that this species is " common in the jungles around Seetapore. Nest is largish, dome-shaped, and placed low down in a thorny bush. The bird lays in August five eggs, the *fac-simile* of the eggs of *Pratincola ferrea*, perhaps of a more elongated type than the eggs of that bird."

Mr. H. Parker, writing on the birds of North-west Ceylon, refers to this bird under the titles *D. jerdoni* and *D. valida*, and informs us that it breeds from January to May.

The eggs of this species are somewhat elongated ovals. The ground-colour is a greenish or greyish stone-colour, and they are finely and often rather sparsely freckled all over with very faint reddish brown, or brownish pink in most eggs; these frecklings are gathered together into a more or less dense zone round the large end, forming a conspicuous ring there much darker-coloured than the frecklings over the rest of the surface. The eggs have a faint gloss.

In length they vary from 0·68 to 0·75, and in breadth from 0·49 to 0·52, but the average appears to be 0·7 by 0·5.

466. Prinia inornata, Sykes. *The Indian Wren-Warbler.*

Drymoipus inornatus (*Sykes*), Jerd. B. Ind. ii, p. 178; *Hume, Rough Draft N. & E.* no. 543.
Drymoipus longicaudatus (*Tick.*), Jerd. B. Ind. ii, p. 180.
Drymoipus terricolor, *Hume*; *Hume, Rough Draft N. & E.* no. 543 bis.

The breeding-season of this Wren-Warbler commences with the first fall of rain, and lasts through July and August to quite the middle of September.

The birds construct a very elegant nest, always closely and compactly woven, of very fine blades, or strips of blades, of grass, in no nests exceeding one-twentieth of an inch in width, and in many of not above half this breadth. The grass is always used when fresh and green, so as to be easily woven in and out. Both parents work at the nest, clinging at first to the neighbouring stems of grass or twigs, and later to the nest itself, while they push the ends of the grass backwards and forwards in and out; in fact, they work very much like the Baya (*P. baya*), and the nest, though much smaller, is in texture very like that of this latter species, the great difference being that the Baya, with us, more often uses *stems*, and *Prinia* strips of *blades* of grass. The nest varies in shape and in size, according to its situation: a very favourite locality is in amongst clumps of the *sarpatta*, or serpent-grass, in which case the bird builds a long and purse-like nest, attached above and all round to the surrounding grass-stems, with a small entrance near the top. Such nests are often 8 or 9 inches in length, and 3 inches or even more in external diameter, and with an internal cavity measuring 1½ inch in diameter, and having a depth of nearly 4 inches below the lower margin of the entrance-hole. At other times they are hung between bare twigs, often of some thorny bush, or are even placed in low herbaceous plants; in these cases they are usually nearly globular, with the entrance-hole near the top; they are then probably 3½ inches in external diameter in every direction. In other cases they are hung to or between two or more leaves to which the birds attach the nest, much as a Tailor-bird would do, using, however, fine grass instead of cobwebs or cotton-wool for ligaments. I have never found more than five eggs in any nest, and four is certainly the normal number.

Mr. R. M. Adam remarks:—" I had a nest brought me in
Oudh on the 17th April, containing four eggs. About Agra and
Muttra, where as you know the birds are *very* common, I have
always obtained the greatest number of eggs during August; four
is the regular number; in one taken on the 16th August I found
five eggs."

Mr. W. Blewitt writes:—" During July, August, and the early
part of September I found multitudes of nests of this species in
the neighbourhood of Hansie, almost exclusively in the Dhasapoor,
Dhana, and Secundapoor *Beerhs* or jungle-preserves.

" The nests, of which numerous specimens were sent to you,
were of the usual type, and were nearly all found in ber (*Z. jujuba*)
and hinse (*Capparis aphylla*) bushes, at heights of from 3 to 4 feet
from the ground. I did not meet with more than four eggs in
any one nest."

Colonel E. A. Butler says:—" The Indian Wren-Warbler is
very common in the plains, frequenting low scrub-jungle and long
grass studied with low bushes (*Calotropis, Zizyphus*, &c.). It
breeds during the monsoon, commencing to build in July, during
which month and August in the neighbourhood of Deesa I must
have examined some three or four dozen nests. There are two
distinct types of nests, and there may be two species of this genus
in this part of the country; but I must confess that after shooting
a large number of specimens of both sexes, and after examining an
immense series of the eggs, I have failed to make out more than one
species, and that Mr. Hume informs me is his *Drymoipus terricolor*.
The nests alluded to vary as follows:—One type is very closely
and compactly woven, as described of *D. terricolor* (' Nests and Eggs,
Rough Draft,' p. 349), with the entrance almost at the top. The
other type is built of the same material, with the exception that
the grass is rather coarser, but is more in shape like a Wren's nest,
and the grass is somewhat loosely put together instead of being
woven, and it has the entrance with a slight canopy over it upon one
side. The eggs four, and not uncommonly five, in number, were
exactly alike in both types, as also were the specimens of the birds
themselves that I obtained.

" Nearly all the nests I have seen have been built on the outside
of ber bushes (*Z. jujuba*), at heights varying from 2½ to 5 feet
from the ground."

Mr. B. Aitken says:—" I found this nest at Bombay on the
13th October, 1873, at the edge of a tank some 2 feet above the
ground. I have found four or five precisely similar ones before,
generally in similar situations. The nest was strongly attached to
the stems and leaves of four herbaceous plants growing close
together. In many cases the strips of grass had been passed
through and pierced the leaves. The nest is deep and purse-
shaped; the sides were prolonged upwards, except in front where
the entrance was, and joined above so as to form a canopy. The
nest has no lining, and none of the nests of this species that I ever
saw have ever had any lining. The whole nest inside and out is

composed of fine strips of blades of grass interwoven. The eggs, five in number, varied much in size. In colour they were bright blue, most irregularly blotched with various shades of purplish brown: some of the blotches very large, some mere specks. Each egg had also washed-out stains or blotches. The smaller eggs were by far the brighter.

"By reason of the roof and walls the entrance to the nest was at one side, but there was nothing that could be called a hole. The roof projected over the entrance, forming a porch.

"Six or eight nests which I have seen of this species were all over water. But the birds are by no means confined to marshy localities.

"Even in the middle of the rains the nests are invariably made of dry yellow grass.

"One nest found in Berar was in a babool bush, where of course there could have been no leaves pierced."

Mr. E. Aitken writes:—"I have found a good many nests in Bombay, and it breeds in Poona too. My notes only mention two nests with eggs, on the 22nd and 28th August, but I found some much later; and I am almost certain it begins to lay much earlier, if not actually at the beginning of the monsoon, like *Orthotomus* and *Prinia*.

"It builds in gardens and cultivated fields, especially in the vicinity of water, and often among plants growing in water.

"The nest is very firmly attached to the twigs of some plant where long grass or other plants completely surround and conceal it. It is usually about 3 feet from the ground. It varies much in size and shape, some being much deeper than others, and some having the top open; others an entrance somewhat to one side.

"I have always found three or four eggs—bright blue, with large irregular purplish-brown blotches and no hair-lines. I should have said that the nest is a bag, very uniformly woven, of fine grass, and *never with any lining*—at any rate in none that I have ever found. They never use the same nest twice, always building a fresh one even if you only rob without injuring the first. I think they have only one brood in the year, but, like *Orthotomus* and *Prinia*, one or two nests are generally deserted or destroyed by some accident before they succeed in rearing a brood."

Major C. T. Bingham informs us that this Wren-Warbler is "a common breeder both at Allahabad and at Delhi from March to September. Builds a neat bottle-shaped nest in clumps of surpat grass, of fine strips of the grass itself, which I have repeatedly watched the birds tearing off. The eggs are lovely little oval fragile shells of a deep blue, blotched and speckled and covered with fine hair-like lines, chiefly at the large end, of a deep chocolate-brown."

The eggs are a moderately long, and generally a pretty perfect, oval, often pointed towards one end, sometimes globular, seldom, if ever, much elongated. The shell is fine and glossy, and comparatively thick and strong. The ground-colour is normally a

beautiful pale greenish blue, most richly marked with various shades
of deep chocolate and reddish brown. Nothing can exceed the
beauty or variety of the markings, which are a combination of bold
blotches, clouds, and spots, with delicate, intricately interwoven
lines, recalling somewhat, but more elaborate and, I think, finer
than, those of our early favourite—the Yellow Ammer. The markings are invariably most conspicuous at the large end, where there
is very commonly a conspicuous confluent cap, and the delicate
lines are almost without exception confined to the broader half of
the egg.

Very commonly the smaller end of the egg is entirely spotless,
and I have a beautiful specimen now before me in which the only
markings consist of a ring of delicate lines round the large end.
Some idea of the delicacy and intricacy of these lines may be formed
when I mention that this zone is barely one tenth of an inch
broad, and yet in a good light between twenty and thirty interlaced
lines making up this zone may be counted.

The intricacy of the pattern is in some cases almost incredible,
and, what with the remarkable character of the patterns and the
rich and varying shades of their colours, these little eggs are, I
think, amongst the most beautiful known.

Occasionally the ground-colour of the eggs, instead of being a
bright greenish blue, is a pale, rather dull, olive-green, and still
more rarely it is a clear pinkish white. These latter eggs are so
rare that I have only seen six in about as many hundreds.

In size the eggs vary from 0·53 to 0·7 in length, and from 0·42
to 0·5 in breadth ; but the average of one hundred and twenty
eggs measured was 0·61 by 0·45.

467. Prinia jerdoni (Blyth).   *The Southern Wren-Warbler.*

Drymœca jerdoni (*Blyth*), *Hume, Cat.* no. 544 ter.

Mr. Davison says :—" The Southern Wren-Warbler breeds chiefly
on the slopes of the Nilgiris about the Badaga cultivation. The
nest is entirely composed of fine grass, and is generally placed
about 2 or 3 feet from the ground, either in a clump of long grass
or attached to the branch of a small bush. It is often suspended,
domed, and with the opening near the top. The eggs, generally
three, are blue, spotted and lined with deep red-brown."

From Kotagherry Miss Cockburn tells us that "the Common
Wren-Warbler has no song, but is loud and frequent in its repeti-
tion of a few notes during the breeding-season. Its nest, which is
globular, is built in the same shape as that of *P. socialis*, with the
entrance at one end, on some low bush, but it only uses *one* material, namely fine long grass, and does not add any soft lining. The
colour of its eggs, however, is totally different, of a light bluish
green, and having a number of spots and streaks like dark threads
carried round and through the spots, which are mostly at the thick
end. The breeding-season lasts from April to July."

Mr. C. J. W. Taylor, writing from Manzeerabad, Mysore, says:—"Fairly common throughout the district. Eggs taken on the 15th July, 1882."

Mr. Rhodes W. Morgan, writing from South India, remarks:—"It builds a neat pendent nest in long grass on the Nilgiris. The nest is composed entirely of short pieces of grass fitted together, and is very compact. The eggs are three in number, and are of a blue colour, with large blotches and hair-like streaks of a dark reddish brown at the upper end. An egg measured ·69 inch by ·5."

The eggs of this species do not differ materially in size, shape, or markings from those of *P. inornata*, which are very fully described above.

468. Prinia blanfordi (Walden). *The Burmese Wren-Warbler.*

Drymœca blanfordi, *Wald., Hume, Cat.* no. 543 ter.

Mr. Oates, who found this bird very common in Pegu, writes:—"The Burmese Wren-Warbler is perhaps the commonest bird of the Pegu plains. From Myitkyo on the Sittang, and possibly from further north, down to Rangoon, it is to be found in all the low tracts covered with grass.

"Where it occurs it is a constant resident and breeds from May to August. I have found the nest in the middle of May, but it is not till July that the bulk of the birds lay.

"The nest is never more than 4 feet from the ground, and is attached either to two or more stalks of elephant-grass or to the stem of a low weed, or to the blades of certain tender grasses which grow in thick tufts. There is little or no attempt at concealment. The materials forming the nest are entirely fine grasses, of equal coarseness or fineness throughout, gathered green, and so beautifully woven together that it is almost impossible to destroy a nest by tearing it asunder, although it may be looked through. In shape it is somewhat of a cylinder, with a tendency to swell out at the middle. Its length, or rather height (for its longer axis, being invariably parallel to the stalks to which the nest is attached, is generally upright), is from 6 to 8 inches, and its extreme width 4. The entrance is placed at the top of the nest, the sides of which are produced an inch or two above the lower edge of the entrance. The thickness of the walls is very small, seldom reaching half, and generally being only a quarter, of an inch. Occasionally the nest is almost globular, but the back of the entrance is in every case produced upwards some inches. There is no lining at all.

"The eggs never exceed four, and frequently are only three, in number, and the female does not commence sitting till the full number is laid. She deserts the nest on the slightest provocation; and if a nest with only one or two eggs is found, and the fingers inserted, it is useless to leave the eggs in hopes of getting more. She will lay no more. I have tested this in at least ten cases."

Major C. T. Bingham tells us:—"About Kaukarit, on the

Houndraw river in Tenasserim, I found this species, in June 1878, very common. They were then breeding, and I found several nests, all, however, unfinished; these were, in material and make, very like the nests of *P. inornata* which I had taken years ago in India."

The eggs of this species recall in many respects those of *P. inornata*, but the ground-colour is much more variable, and the markings are more blotchy and less intricate in shape. They are pretty regular ovals, and while some are very glossy others exhibit but little of this. The ground-colour is perhaps typically pale greenish blue, but in a great many specimens this is more or less obliterated by a reddish or pinkish tinge, as if the colour of the markings had run; in some the ground is a sort of reddish olive, in some pinky white. The markings are large blotches and spots, often forming zones or caps about the larger end, where they seem almost always to be most conspicuous, as they vary in colour from an intense burnt-sienna which is almost black, through a dingy maroon, and again to a dull, somewhat pale reddish brown; here and there individual eggs exhibit a hair-line or two, or a hieroglyphic-like mark, but these are the exceptions.

The eggs vary in length from 0·53 to 0·64 inch, and in breadth from 0·42 to 0·45; but the average of fourteen eggs is 0·58 by 0·44.

Very constantly smears or clouds of a paler shade than the blotches cover large portions of the surface between these. Occasionally all the markings are smeared and ill-defined, and in some eggs they are almost entirely wanting, and nothing but a scratch or two about the large end is to be seen.

Family LANIIDÆ.

Subfamily LANIINÆ.

469. Lanius lahtora (Sykes). *The Indian Grey Shrike.*

Lanius lahtora (*Sykes*), Jerd. B. Ind. i, p. 400.
Collyrio lahtora, *Sykes, Hume, Rough Draft N. & E.* no. 256.

The Indian Grey Shrike lays from January to August, and occasionally up to October, but the majority of my eggs have been obtained during March or April.

It builds, generally, a very compact and heavy, deep, cup-shaped nest, which it places at heights of from 4 to 10 or 12 feet from the ground in a fork, towards the centre of some densely growing thorny bush or moderate-sized tree, the various carounders, capers, plums, and acacias being those most commonly selected.

As a rule it builds a new nest every year, but it not unfrequently only repairs one that has served it in the previous season, and even at times takes possession of those of other species.

The nest is composed of very various materials, so much so that it is difficult to generalize in regard to them. I have found them built entirely of grass-roots, with much sheep's wool, lined with hair and feathers, or solidly woven of silky vegetable fibre, mostly that of the putsun (*Hibiscus cannabinus*), in which were incorporated little pieces of rag and strips of the bark of the wild plum (*Zizyphus jujuba*); but I think that most commonly thorny twigs, coarse grass, and grass-roots form the body of the nest, while the cavity is lined with feathers, hair, soft grass, and the like.

Generally the nests are very compact and solid, 6 or 7 inches in diameter, and the egg-cavity 3 to 4 in diameter, and 2 to $2\frac{1}{2}$ in depth, but I have come across very loosely built and straggling ones.

They have at times two broods in the year (but I do not think that this is always the case), and lay from three to six eggs, four or five being the usual number.

Mr. F. R. Blewitt, writing from Jhansie and Saugor, and detailing his experiences there and in the Delhi Districts, says :—

" The Common Indian Grey Shrike breeds from February to July; it builds on trees ; if it has a preference, it is for the close-growing roonj tree (*Acacia leucophlœa*). I have particularly noticed this fact both here and at Gurhi Hursroo. The nest in structure is neat and compact (though I have occasionally seen some very roughly put together), and generally well fixed into the forks of an off-shooting branch. In shape it is circular, varying from 5 to $7\frac{1}{2}$ inches in diameter, and from $1\frac{1}{2}$ to $3\frac{1}{2}$ inches in thickness; thorn twigs, coarse grass, grass-roots, old rags, &c. form the outer materials of the nest, and closely interwoven fine grass and roots the border-rim. The egg-cavity is deeply cup-shaped, from $3\frac{1}{2}$ to 5 inches in diameter, and lined with fine grass and khus; exceptionally shreds of cloth are interwoven with the khus and grass.

" On one occasion I got a nest with the cup interior entirely lined with old cloth pieces, very cleverly and ingeniously worked into the exterior framework. Five is the regular number of eggs, though at times six have been obtained in one nest. The birds often make their own nests each year, but this is not invariably the case. When at Gurhi Hursroo in February last, I found on an isolated roonj tree four nests within a foot of each other. The under centre one, an *old* Shrike nest (the other three were of other birds), was occupied by a Shrike sitting on five eggs. I very carefully examined it, and my impression at the time was that the parent birds had returned, to rear a second progeny, to the nest constructed by them the year previous.

" I do not know whether you have noticed the fact, but both *L. lahtora* and *L. erythronotus* often lay in old nests, of which they first carefully repair the egg-cavity with new materials. It is not only, however, in old nests of their own species that these birds make a home in the breeding-season. At times they take possession of fabrics clearly not the work of any Shrike. Quite recently I found a pair of *L. lahtora* with four eggs in a small nest entirely woven of hemp, the bottom of which was thickly coated with the droppings of former occupants. Again, on the 8th June, a nest with four eggs was found on a roonj tree. This wonderful nest, which I have kept, is entirely composed of what I take to be old felt and feathers, the bottom of the cavity of which, when found, was almost covered with the dung of young birds.

"Evidently this nest was not *originally* made by the Shrike, but, as would appear, was taken possession of by it, after the brood of some other species of birds had left it."

Mr. W. Theobald makes the following note of this bird's breeding in the neighbourhood of Pind Dadan Khan and Katas in the Salt Range :—" Lays in the last week of March to the end of April. Eggs five only, shape ovato-pyriform, size 1·06 inch by 0·8 inch ; colour pale greenish white, blotched and tinged with yellowish grey and neutral markings ; vary much in intensity and colour. Nest of twigs, lined with cotton or wool, and usually placed in stiff thorny bushes."

Lieut. H. E. Barnes, writing from Chaman in Southern Afghanistan, remarks :—" The Grey-backed Shrike is extremely common, breeding about the end of March, in much the same situations as in India. I have collected many specimens, and failed to detect any difference between the Indian bird and the one found here. The average of twelve eggs is ·97 by ·75."

He adds subsequently :—" This is the commonest Shrike in the country ; it breeds in March and April, and the young are easily reared in captivity."

Mr. W. Blewitt says that he " took four nests of this bird near Hansee on the 28th–30th March ; they contained, one 5, two 4, and one 3 eggs ; all but the latter (which, curiously enough, were a good deal incubated) quite fresh. The nests were placed in acacia and caper bushes, at heights of from 6 to 14 feet from the ground ; they were from 6 to 7 inches in diameter exteriorly, rather loosely constructed of thorny twigs, with egg-cavities from 2 to 2½ inches deep, lined with fine straw and leaves." Again he writes : " Took numerous nests in the neighbourhood of Hansee during the month of July ; most of the eggs were much incubated, and four was the largest number found in any one nest.

"The nests were all placed upon keekur trees at an average height of some 10 feet from the ground ; they were composed of thorny twigs, some with and some without a lining of fine grass and feathers, and averaged some 5 or 6 inches in diameter by 2 to 4 inches in depth."

Major C. T. Bingham says that " this bird is excessively common about Delhi, far more so than at Allahabad. At the latter place I only found it breeding in March and April, but at Delhi I have found nests in every month from March to August. One evening in June I remember counting in my walk thirteen nests within the radius of a mile ; some of these contained fresh eggs, some hard-set, some young. One nest I robbed in April of eggs contained young in the latter end of May, and I believe many of them have two if not more broods in the year. All nests that I have seen have been well made, firm, deep cups of babool branches, lined with grass-roots, and occasionally with bits of rag and tow. The eggs are broad ovals of a dead chalky bluish-white colour, spotted, chiefly at the large end, with purple and brown. Five is the greatest number of eggs I have found in a nest."

Mr. George Reid informs us that this Shrike breeds from March to July in the Lucknow Division, making a massive nest in babool trees, generally in solitary ones on open plains.

Colonel Butler writes :—" The Indian Grey Shrike breeds in the neighbourhood of Deesa in February, March, April, May, June, and July. I have taken nests on the following dates :—

" Feb.   19.   A nest containing 4 slightly incubated eggs.
March 13.   ,,   ,,   4 fresh eggs.
   ,,   16.   ,,   ,,   4   ,,
   ,,   19.   ,,   ,,   4   ,,
   ,,   20.   ,,   ,,   3   ,,
   ,,   20.   ,,   ,,   4   ,,
   ,,   28.   ,,   ,,   4 incubated eggs.
April   9.   ,,   ,,   4   ,,   ,,
June   1.   ,,   ,,   2 fresh eggs.
   ,,   7.   ,,   ,,   4 young birds.
   ,,   7.   ,,   ,,   2 incubated eggs.
July   9.   ,,   ,,   4   ,,   ,,

" The nest is usually placed in some low, isolated leafless thorny tree (*Acacia, Zizyphus*, &c.), from six to ten feet from the ground. It is solidly built of small dry thorny twigs, old rags, &c. externally, with a thick felt lining of the silky fibre of *Calotropis gigantea*. The eggs vary a good deal in shape, some being much more pointed at the small end than others; some I have are almost perfect peg-tops. They vary in number from three to five; and as a rule the colour is a dingy white, spotted and speckled sparingly all over with olive-brown and inky purple, which together form a well-marked zone at the large end."

Messrs. Davidson and Wenden remark :—" Common, and breeds abundantly in the Poona and Sholapoor Collectorates at the end of the hot weather. W. has noticed it breeding at Nulwar and Raichore. Davidson observed that it was very rare in the Satara Districts."

Mr. J. Davidson further informs us that *L. lahtora* is a permanent resident in Western Khandeish, and breeds in every month from January to July.

My friend Mr. Benjamin Aitken furnishes me with the following interesting note :—" You say that the Indian Grey Shrike lays from February to July. Now, in Berar, where this bird is very common, I have found their eggs frequently in the first week of January, and on not only to July, but to September ; and I once found a nest in October. I was never able to satisfy myself that the same pair had two broods in the year, but I scarcely think there can be any doubt about the matter. I once found, like your correspondent Mr. Blewitt, four nests in a small babool tree, and only one of them occupied. This was at Poona. My brother first pointed out to me that this species affects the dusty barren plain, whereas *L. erythronotus* prefers the cool and shaded country. This difference in the habits of the two birds is very observable at Poona, where both species are exceedingly common. Where a *jungly* or watered piece of country borders upon the open plain, you may see half a dozen of each kind within an area of half a mile radius, and yet never find the one trespassing upon the domain of the other. When you say you have never found a nest more than 1500 feet above the level of the sea, I would remind you that although *L. lahtora* never ascends the hills, it is yet very abundant in the Deccan, which is 2000 feet above the sea-level.

" I think I have written to you before that during a residence of twelve years I never saw *L. lahtora* in Bombay."

This Shrike is, however, essentially a plains bird, and never seems to ascend the Himalayas to any elevation. I have never myself found a nest more 1500 feet above the level of the sea.

Typically, the eggs are of a broad oval shape, more or less pointed towards one end, of a delicate greenish-white ground, pretty thickly blotched and spotted with various shades of brown and purple markings, which, always most numerous towards the large end, exhibit a strong tendency to form there an ill-defined zone or irregular mottled cap. The variations, however, in shape, size, colour, extent, and intensity of markings are very great; and yet, in the huge series before me, there is not one that an oologist would not at once unhesitatingly set down as a Shrike's. In some the ground-colour is a delicate pale sea-green. In some it is pale stone-colour; in others creamy, and in a few it has almost a pink tinge. The markings, commonly somewhat dull and ill-defined, are occasionally bold and bright; and in colour they vary through every shade of yellowish, reddish, olive, and purplish brown, while sub-surface-looking pale purple clouds are intermingled with the darker and more defined markings. In one egg the markings may be almost exclusively confined to a broad, very irregular zone of bold blotches near the large end. In others the whole surface is more or less thickly dotted with blotches and spots, so closely crowded towards the large end as almost wholly to obscure the ground-colour there. As a rule, the markings are irregular blotches of greater or less extent, but occasionally these blotches form the exceptions, and the majority of the markings are mere spots and specks. In some eggs the purple cloudings greatly predominate; in others scarcely a trace of them is observable. Some eggs are comparatively long and narrow, while some are pyriform and blunt at both ends; and yet, notwithstanding all these great differences, there is a strong family likeness between all the eggs. In size they are, I think, somewhat smaller than those of *L. excubitor*.

They vary in length from 0·9 to 1·17 inch, and in width from 0·75 to 0·83 inch; but the average of more than fifty eggs is 1·03 by 0·79 inch.

473. Lanius vittatus. *The Bay-backed Shrike.*

Lanius hardwickii (*Vigors*), *Jerd. B. Ind.* i, p. 405.
Lanius vittatus, *Dum., Hume, Rough Draft N. & E.* no. 260.

The Bay-backed Shrike breeds throughout the plains of India and in the Sub-Himalayan Ranges up to an elevation of fully 4000 feet.

The laying-season lasts from April to September, but the great majority of eggs are found during the latter half of June and July; in fact, according to my experience, the great body of the birds do not lay until the rains set in.

The nests are placed indifferently on all kinds of trees (I have notes of finding them on mango, plum, orange, tamarind, toon, &c.), never at any great elevation from the ground, and usually in *small* trees, be the kind chosen what it may. Sometimes a high hedge-row, such as our great Customs hedge, is chosen, and occasionally a solitary caper or stunted acacia-bush.

The nests (almost invariably fixed in forks of slender boughs) are neat, compactly and solidly built cups, the cavities being deep and rather more than hemispherical, from 2·25 to fully 3·5 inches in diameter, and from 1·5 to 2 inches in depth. The nest-walls vary from 0·5 to 1·25 inch in thickness. The composition of the nest is various. The following are brief descriptions which I have noted from time to time:—

" Compactly woven of grass-stems and a few fine twigs, but with more or less wool, rag, cotton, or feathers incorporated ; there *is no lining.*

" The nest was rather massive, externally composed of wool, rags, cotton, thread, and feathers, and a little grass ; the cavity rather neatly lined with fine grass.

" Composed almost entirely of cobweb, with a few soft feathers, wool, string, rags, and a few pieces of very fine twigs compactly woven. The interior was lined with fine straw and fibrous roots."

Elsewhere I have recorded the following note on the nidification of this species :—

" This bird, or rather birds of this species, have been laying ever since the middle of April, but nests were then few and far between, and now in July they are common enough. The nest that we had just found was precisely like twenty others that we had found during the past two months. Rather deep, with a nearly hemispherical cavity ; very compactly and firmly woven of fine grass, rags, feathers, soft twine, wool, and a few fine twigs, the whole entwined exteriorly with lots of cobwebs ; and the interior cavity about 1¾ inch deep by 2¼ in diameter, neatly lined with very fine grass, one or two horsehairs, shreds of string, and one or two soft feathers. The walls were a good inch in thickness. The nest was placed in a fork of a thorny jujube or ber tree (*Zizyphus jujuba*), near the centre of the tree, and some 15 feet from the ground. It contained four fresh eggs, feebly coloured miniatures of the eggs of *L. lahtora*, which latter so closely resemble those of *L. excubitor* that if you mixed the eggs, you could never, I think, certainly separate them again. The eggs exhibit the zone so characteristic of those of all Shrikes. They have a dull pale ground, not white, and yet it is difficult to say what colour it is that tinges it; in these four eggs it is a yellowish stone-colour, but in others it is greenish, and in some grey; near the middle, towards the large end, there is a broad and conspicuous, but broken and irregular zone of feeble, more or less confluent spots and small blotches of pale yellowish brown and very pale washed-out purple. There are a few faint specks and spots of the same colour here and there about the rest of the egg. In some eggs previously obtained the zone is quite in the middle, and in others close round the large end. In some the colours of the markings are clear and bright, in others they are as faint and feeble as one of our modern Manchester warranted-fast-coloured muslins, after its third visit to a native washerman. In size, too, the eggs vary a good deal.

"The little Shrike had a great mind to fight for his *penates*, and twice made a vehement demonstration of attack; but his heart failed him, and he retreated to a neighbouring mango branch, whence a few minutes after we saw him making short dashes after his insect prey, apparently oblivious of the domestic calamity that had so recently befallen him."

Mr. F. R. Blewitt, then at Gurhi Hursroo, near Delhi, sent me some years ago the following interesting note :—

"Breeds from March to at least the middle of August. It builds its nest in low trees and high hedgerows, preferring the former.

"In shape the nest is circular, with a diameter, outside, of from 5½ to 6½ inches, and from 1·5 to 2 in thickness.

"For the exterior framework thorny twigs, old rags, hemp, thread-pieces, and coarse grass are more or less used, and compactly worked together. The egg-cavity is deep and cup-shaped, lined with fine grass and khus; pieces of rag or cotton are sometimes worked up with the former.

"Five to six is the regular number of eggs. In colour they are a light greenish white, with blotches and spots generally of a light, but sometimes of a darker, reddish brown. The spots and blotches vary much in size, and they are mostly confined to the broad end of the eggs.

"I had frequently noticed on a tree in the garden an *old* Shrike's nest. It was in the beginning of May that a male bird suddenly made his appearance and established himself in the garden, and morning and evening without fail did he sit and alternately chatter and warble away for hours. His perfect imitation of the notes of other birds was remarkable.

"In the beginning of June his singing suddenly ceased, the

secret of which I soon discovered. He had secured a mate, and daily did I watch for the nest, which I thought they would prepare. Late on the evening of the 23rd June, happening to look up at the *old* nest, to my surprise I found it occupied by the female, the male the while sitting on a branch near her. Next morning on searching the nest I found four eggs. Whether this nest was prepared the year previous by these birds or by another pair I cannot tell.

" That day, the day of the robbery, the female disappeared. The male followed next day, but only to return after two or three days and recommence with renewed energy his chattering and warbling. This he continued daily till near the end of July, when, as before, he suddenly ceased to sing. I then found that he had again secured a mate, whether the old female or a new bride I am not certain ; they soon set about making a nest on a neighbouring tree, very cunningly, as I thought, selected ; and now the young birds reared are nearly full-fledged. An old nest, evidently of last year's make, was brought me the other day with five eggs, but the *lining*, as by the way was done in the one in the garden, had been wholly removed and *new* grass and khus substituted.''

Major C. T. Bingham writes :—" Breeds both at Allahabad and at Delhi in May, June, and July. At the former place I never got the eggs, but have seen some that were taken ; but at Delhi I found numbers of their nests in June and July, and one in May. It makes a much softer nest than either of the two above-mentioned Shrikes. One nest I took on the 15th June was composed wholly of tow, but generally they have an outer foundation of twigs, and are lined with tow, bits of cotton, human hair, or rags. Some eggs are a yellow-white, with very faint marks, others are miniatures of the eggs of *L. lahtora.*

" Five is the greatest number I have found in one nest.''

Mr. W. Theobald makes the following note of this bird's breeding in the neighbourhood of Pind Dadan Khan and Katas in the Salt Range :—

" Lays from the commencement of May to the middle of June. Eggs three or four in number ; shape varies from ovato-pyriform to blunt ovato-pyriform, and measuring from 0·73 to 0·87 inch in length and from 0·55 to 0·65* inch in breadth. Colour, same as *L. erythronotus,* also creamy or yellowish white, spotted with darker. Nest compact, in forks of thorny trees ; outside fibrous stalks, bound with silk or spider-web, and covered with lichens or cocoons, imitating a weathered structure ; inside lined with fine grass and vegetable down.''

Colonel C. H. T. Marshall, writing from Murree, says :— "These little Shrikes breed in the hills, as well as the plains, up to 5000 feet high.''

---

* I think that there must be some error in these dimensions, for mine are taken from forty-five specimens, the largest and smallest, out of some hundreds of eggs.—A. O. H.

Colonel Butler has the following notes on the breeding of this Shrike in Sind :—

" Kurrachi, 7th May, 1877.—I found two nests on this date, one in the fork of a babool tree, the other on the stump of a broken-off branch of a tree between the stump and the trunk of the tree.   The former contained four incubated eggs, exact miniatures of many eggs I have of *L. erythronotus*, the latter two small chicks.—May 12th, same locality, a nest containing two fresh eggs, and another containing two fully fledged young ones.— June 20th, same locality, one nest containing three fresh eggs, another containing four young birds.   Eggs most typical are those which have a well-marked zone near the centre."

" Hydrabad, Sind, 19th June, 1878.—A nest on the outer bough of a babool tree about ten feet from the ground, containing three fresh eggs."

And he further notes :—" The Bay-backed Shrike breeds in the neighbourhood of Deesa at the end of the hot weather.   The nest is a very firm and compactly built cup, usually placed in the fork of some low thorny tree at heights varying from seven to ten feet from the ground.

" June 15th, 1875.   A nest containing 3 fresh eggs.
" July 1st, 1876.   „   „   4 „   „
" July 15th, „   „   „   5 incubated eggs.
" July 29th, „   „   „   4 young birds.
"These birds always retire from the more open parts of the country to low thorny tree-jungle to breed."

Mr. R. M. Adam says :—" This species breeds about Sambhur in July.   On the 1st August I saw numbers of nests and fledglings in the Marot jungle."

Messrs. Davidson and Wenden, writing of the Deccan, say :— " Abundant, and breeds all over the Deccan."

And the former gentleman informs us that this species is also very common in Western Khandeish, and that it breeds in the plains in June and July, and in the Satpuras in March.

Mr. Benjamin Aitken writes :—" This is a very familiar bird, and builds readily in some roadside tree, where men and carts are passing all day long.   I have the following notes of its nests :—

" 1st–8th May, 1869.   Nest and three eggs taken at Khandalla, above the Bhore Ghat.

" 12th May, 1871.   Nest and four eggs at Poona.

" 16th–18th May, 1871.   Nest and four eggs at Khandalla. This nest was in a corinda bush, placed about 4½ feet from the ground.

" 13th May, 1873.   A clutch of young birds left the nest this morning at Poona.

" 19th May, 1873.   I found a nest of half-fledged young birds this day at Poona.   The tree was almost denuded of leaves, and the heat of the sun being very intense, the parent bird was nevertheless sitting close.   Its eyes were closed, and it was gasping hard.   One of the young ones had crawled out from under the parent, and was sitting on the edge of the nest, also gasping hard.

"I do not exactly gather from your notes in the 'Rough Draft' what form the spots usually take. In my nest taken on the 12th May all four eggs had the zone quite as distinct as the eggs of a Fan-tailed Flycatcher. The seven eggs taken from two nests at Khandalla, on the other hand, had not the least appearance of a zone, but were spotted, after the manner of Sparrows' eggs. In both the latter cases I saw the old bird fly off the nest and alight on a tree a few yards off.

"I remember one little Shrike of this species which used to come down every day to pick up crumbs of bread and pieces of potatoe put out for the Sparrows. (Being a true naturalist I love Sparrows.)

"My brother on one occasion saw one of these Shrikes trying to catch a garden lizard—not a gecko.

"Of course you know that the young of this handsome and brightly coloured Shrike have a plain and curiously marked plumage, reminding one a little of the *pateela* Partridge. I never saw this Shrike in Bombay."

The eggs of this, the smallest of all our Indian Shrikes, differ in no particular, so far as shape, colour, and markings go, from those of its larger congeners; that is to say, for every egg of this species an exactly similar one might be picked out from a large series of *L. lahtora* or *L. erythronotus*; but at the same time there is no doubt that pale-creamy and pale-brownish stone-coloured grounds predominate more amongst the eggs of this species than in those of the two above-named. The markings are also, as a rule, more minute and less well-defined; indeed, in the large series I possess there is not one which exhibits the bold sharp blotches common in the eggs of *L. lahtora*, and not uncommon in those of *L. erythronotus*.

In length they vary from 0·75 to 0·95 inch, and in breadth from 0·62 to 0·71 inch; but the average of forty-five eggs is 0·83 by 0·66 inch nearly.

475. Lanius nigriceps (Franklin). *The Black-headed Shrike.*

Lanius nigriceps (*Frankl.*), *Jerd. B. Ind.* i, p. 404.
Collyrio nigriceps, *Frankl., Hume, Rough Draft N. & E.* no. 259.

I have never myself taken the eggs or nests of the Black-headed Shrike.

Mr. R. Thompson says:—"This Shrike breeds all along the south-western termination of the Kumaon and Gurhwal forests, and is usually found in swampy, high grassy lands. It lays in July, August, and September, building a large cup-shaped nest, composed of roots and fine grasses, in small trees or shrubs in low, open grass-covered country.

"I found this the Common Shrike in the hilly jungly tracts in

Southern Mirzapore, but I do not know whether it breeds there.
The cry is quite like that of *L. erythronotus.*

" The southern limit of *Lanius nigriceps* is interesting and remarkable. It disappears after you go south-west of the Mykle
Range, and on the Range itself it is found only near marshy
places. This Mykle Range extends as far east as Ummerkuntuk,
with a spur going off north of that, and joining on with the
Kymore Range, parts of which I explored in March last in Per-
gunnahs Agrore and Singrowlee. Down in those places this
Lanius was the Common Shrike, but south and west of Ummerkuntuk all the Shrikes disappear more or less, and *L. nigriceps*
entirely."

According to Mr. Hodgson's notes and figures this species breeds
in the Valley of Nepal, laying in April and May, and building in
thorny bushes, hedges, and trees, often in the immediate neighbourhood of villages. The following are two of Mr. Hodgson's
notes :—

" Valley, May 18th.—Nest near the top of a fir of mean size,
fixed securely in the midst of several diverging branches, made
compactly of dry grasses, of which the inner ones, which constitute the lining, are hard and elastic, and well fitted to preserve
the shape, which is a deep cup with an internal cavity 3·5 inches
in diameter and nearly 3 deep. It contained six eggs, milk-and-
water white, with pale olive spots, chiefly at the large end,
measuring 0·95 by 0·68 inch.

" Jabar Powah, May 16th.—Ascent of Sheopoori, skirts of large
forests ; nest on lateral branches of a large tree made of downy
tops of plants, of moss and thick grasses strongly compacted, and
lined with fine elastic hair-like grass ; the cavity is circular, 3 inches
in diameter by more than 2 inches in depth ; the whole nest is a
solid deep cup ; it contained four eggs, bluish white, with grey-
brown remote spots."

Of another nest he gives the dimensions as :—external diameter
4·25 inches ; external height 3·87 ; internal diameter 2·87 ;. depth
of cavity 2·75. He figures it as a very compact and deep cup
resting on a horizontal fir branch between four or five upright
sprays. He states that the young are ready to fly towards the
end of June, and that it breeds only once a year.

Dr. Scully, also writing of Nepal, says :—" This Shrike breeds
on the hillsides of the valley, usually in places where there is no
tree-forest, and not uncommonly in the neighbourhood of hamlets.
Several nests were obtained in May and June ; these were large
cup-shaped structures, composed of grass-roots, fibres, and fine
seed-down intermixed. The egg-cavity was circular, lined with
fine grass-stems, about 4 inches in diameter, and 2 inches
deep in the middle. The usual number of eggs is five ; the
ground-colour pale greenish white, boldly blotched and spotted
with olive marks in an irregular zone round the large end. A
clutch of five eggs taken on the 14th June gave the following
dimensions :—0·94 to 0·97 in length, and 0·65 to 0·7 in breadth."

Mr. Gammie found a nest of this species on the 17th May at
Mongfoo, near Darjeeling, at an elevation of 3500 feet. The
nest was placed in a wormwood bush, and was supported between
several slender upright shoots, to which the exterior of the nest
was more or less attached. The nest was a deep compact cup,
externally composed of fine twigs, scraps of roots, and stems of
herbaceous plants, intermingled with a great deal of flowering
grass. Internally it was lined with very fine grass and moss-roots.
The cavity measured about 3 inches in diameter, and was fully
2 inches deep. The external diameter was about 5 inches, and
height 3½ or thereabout.

Subsequently he sent me the following full account of the nidification of this Shrike :—

" I have found this Strike breeding abundantly in the Cinchona
reserves in May and June, at elevations of from 3000 to 4500 feet
above the sea. It affects open, cultivated places, and builds, from
6 to 20 feet from the ground, in shrubs, bamboos, or small trees.
The nest is often suspended between several upright shoots, to which
it is firmly attached by fibres twisted round the stems and the
ends worked into the body of the nest; sometimes against a
bamboo-stem seated on, and attached to, the bunch of twigs given
out at a node ; or in a fork of a small tree, or end of an upright cut
branch where several shoots have sprung away from under the cut
and keep the nest in position, when it has a large pad of an everlasting plant or of the downy beads of a large flowering grass to
rest on—when the former material is handy it is preferred. The
nest is sometimes exposed to view, but generally is tolerably well
concealed. It is of a deep cup-shape, very compactly built of
flowering grass and stems of herbaceous plants intermixed with
fibry twigs, and lined with the small fibry-looking branchlets of
grass-panicles. Externally it measures 5 inches across by 3½ inches
in depth ; internally the cavity is 3½ inches in diameter by nearly
2 inches deep. Usually the eggs are either four or five in number.
On one occasion only have I seen so many as six. The coloration
is of two distinct types, but one type only is found in the same
nest. I suspect that the age of the bird has something to do with
the variation of colour in the eggs. In a nest containing four eggs
one had the majority of the spots collected on the small, instead
of the thick end as usual, and, strange to say, it was addled white.
The other three were hard-set. The parents get very much excited when their young are approached, and, as long as the intruder
is in the vicinity, keep up an incessant volley of their harsh grating
cries, at the same time stretching out their necks and jerking about
their tails violently."

Mr. J. R. Cripps, writing from Furreedpore in Eastern Bengal,
says :—" Excessively common and a permanent resident. Prefers
open plains interspersed with bushes, also the small bushes on road-
sides are a favourite haunt of theirs. Breeds in the district. I
took ten nests this season from the 11th April to 4th June, with
from one to five eggs in each. Four nests were placed in bamboo

clumps from 9 to 30 feet high; one 40 feet from the ground on a casuarina-tree, one 20 feet up in a but-tree, and the rest in babool-trees at from 6 to 15 feet high from the ground. There is no attempt at concealment. The nest is a deep cup fixed in a fork, and is made of grasses with a deal of the downy tops of the same for an outside lining; this peculiarity at once distinguishes the nest of this species. The description given by Mr. Hodgson of a nest found by him on the 16th May at Jahar Powah, in 'Nests and Eggs,' p. 172, correctly describes the nests I have found. This species imitates the call of several kinds of small birds, as Sparrows, King-Crows, &c., and I have often been deceived by it."

The eggs of this species, of which, thanks to Mr. Gammie, I now possess a noble series, vary very much in shape and size. Typically they are very broad ovals, a little compressed towards one end, but moderately elongated ovals are not uncommon. The shell is very fine and smooth, and often has a more or less perceptible gloss; in no case, however, very pronounced.

There are two distinct types of colouring. In the one, the ground-colour is a delicate very pale green or greenish white, in some few pale, still faintly greenish, stone-colour; and the markings consist as a rule of specks and spots of brownish olive, mostly gathered into a broad zone about the large end, intermingled with specks and spots of pale inky purple. In some eggs the whole of the markings are very pale and washed-out, but in the majority the brownish-olive or olive-brown spots, as the case may be, are rather bright, especially in the zone. In the other type (and out of 42 eggs, 12 belong to this type) the ground-colour varies from pinky white to a warm salmon-pink, and the markings, distributed and arranged as in the first type, are a rather dull red and pale purple. In fact the two types differ as markedly as do those of *Dicrurus ater*; and though I have as yet received none such, I doubt not that with a couple of hundred eggs before one intermediate varieties, as in the case of *D. ater*, would be found to exist—as it is, two more different looking eggs than the two types of this species could hardly be conceived. I may add that in eggs of both types it sometimes, though very rarely, happens that the zone is round the small end.

In length they vary from 0·82 to 1·01, and in breadth from 0·68 to 0·79; but the average of forty-two eggs measured is 0·92 by 0·75.

476. Lanius erythronotus (Vigors). *The Rufous-backed Shrike.*

Lanius erythronotus (*Vig.*), *Jerd. B. Ind.* i, p. 402.
Collyrio erythronotus, *Vigors, Hume, Rough Draft N. & E.* no. 257.
Collyrio caniceps * (*Blyth*), *Hume, Rough Draft N. & E.* no. 257 bis.

Lanius erythronotus.

The Rufous-backed Shrike lays from March to August; the first half of this period being that in which the majority of these

---

* Mr. Hume may probably still consider *L. caniceps* separable from *L. ery-thronotus.* I therefore keep the notes on the two races distinct as they appeared in the 'Rough Draft,' merely adding a few later notes.—ED.

birds lay in the Himalayas, which they ascend to elevations of 6000 feet; and the latter half being that in which we find most eggs in the plains; but in both hills and plains some eggs may be found throughout the whole period above indicated.

The nests of this species are almost invariably placed on forks of trees or of their branches at no great height from the ground; indeed, of all the many nests that I have myself taken, I do not think that one was above 15 feet from the ground. By preference they build, I think, in thorny trees, the various species of acacia, so common throughout the plains of India, being apparently their favourite nesting-haunts, but I have found them breeding on toon (*Cedrela toona*) and other trees. Internally the nest is always a deep cup, from 3 to 3¼ inches in diameter, and from 1¾ to 2½ deep. The cavity is always circular and regular, and lined with fine grass. Externally the nests vary greatly : they are always massive, but some are compact and of moderate dimensions externally, say not exceeding 5½ inches in diameter, while others are loose and straggling, with a diameter of fully 8 inches. Grass-stems, fine twigs, cotton-wool, old rags, dead leaves, pieces of snake's skin, and all kinds of odds and ends are incorporated in the structure, which is generally more or less strongly bound together by fine tow-like vegetable fibre. Some nests indeed are so closely put together that they might almost be rolled about without injury, while others again are so loose that it is scarcely possible to move them from the fork in which they are wedged without pulling them to pieces.

I have innumerable notes about the nests of this Shrike, of which I reproduce two or three.

"*Etawah, March 18th.*—The nest was on a babool tree, some 10 feet from the ground, on one of the outside branches ; an exterior framework of very thorny babool twigs, and within a very warm deep circular nest made almost entirely of sun (*Crotalaria juncea*) fibre, a sort of fine tow, and flocks of cotton-wool, there being fully as much of this latter as of the former ; a few fine grass-stems are interwoven ; there are a few human and a few sheep's wool hairs at the bottom as a sort of lining. The cavity of the nest is about 3 inches in diameter by 2 deep, and the side walls and bottom are from 1½ to 2 inches thick."

"*Bareilly, May 27th,* 1867.—Found a nest containing two fresh eggs. The nest was in a small mango tree, rather massive, nearly 2 inches in thickness at the sides and 3 inches thick at the bottom. It was rather stoutly and closely put together, though externally very ragged. The interior neatly made of fine grass-stems, the exterior of coarser grass-stems and roots, with a quantity of cotton-wool, rags, tow string and thread intermingled. The cavity was oval, about 3½ by 3 inches and 2 inches deep."

"*Agra, August 21st.*—Mr. Munro sent in from Bitchpoorie a beautiful nest which he took from the fork of a mango tree about 40 feet from the ground, a very compact and massive cup-shaped nest, not very deep."

Mr. F. R. Blewitt records the following note :—

"Breeds from March to August, on low trees, and, as would appear, without preference for any one kind.

"The nest in shape much resembles that of *Lanius lahtora*; but judging from the half-dozen or so I have seen, *L. erythronotus* certainly displays more skill and ingenuity in preparing its nest, which in structure is more neat and compact than that of *L. lahtora*. In shape it is circular, ordinarily varying from 5½ to 7 inches in diameter, and from 2 to 2½ inches in thickness. Hemp, old rags, and thorny twigs are freely used in the formation of the outer portion of the nest, but the Shrike shows a decided predilection for the former. In one nest I observed the cast skin of a snake worked in with the outer materials; in two others some kind of vegetable fibre was used to bind and secure the thorn twigs, and one had the margin made of fine neem-tree twigs and leaves. The egg-cavity is deeply cup-shaped, from 3 to 4 inches in diameter, and *lined* usually with fine grass. Five appears to be the regular number of eggs; but on this score I cannot be very certain, seeing that my experience is confined to some half-dozen or so of nests.

"I have recently reared three young birds, and it is very amusing to witness their many antics, shrewdness, and intelligence. They are very tame, flying in and out of the bungalow at pleasure; when irritated, which is rather a failing with them, they show every sign of resentment. If one is inclined to be rebellious, not coming to call, the show of a piece of meat at once secures its submission and capture. Singular how partial they are to raw meat, and more singular to see the expert way in which they catch up the meat with the claws of either leg, and hold it from them while they devour it piecemeal. I saw the other evening an old bird pounce on a field-mouse, kill it, and then bring and cleverly fix the victim firmly between the two forks of a branch and pull it in pieces. It consumed but a part of the mouse."

Mr. W. Theobald makes the following note on this bird's breeding in the neighbourhood of Pind Dadan Khan and Katas in the Salt Range :—"Lay in May; eggs five to six; shape blunt, ovato-pyriform: size varies from 0·88 to 0·93 of an inch in length, and from 0·68 to 0·81 of an inch in breadth. Colour white or pale greenish white, slightly ringed and spotted with yellowish grey and neutral tint. Nest of roots, coarse grass, rags, cotton, &c., lined with fine grass, and placed in forks of trees."

Captain Hutton, who recognizes the distinctions between this species and *L. caniceps*, says :—

"This is an abundant species in the Doon, but is found also within the mountains up to about 5000 feet. In the Doon I took a nest on the 28th June containing four eggs. It is composed of grass and fine stalks of small plants roughly put together, bits of rag, shreds of fine bark, and lined with very fine grass-seed stalks; internal diameter 3 inches, external 6 inches; depth 2½ inches."

Sir E. C. Buck notes having taken a nest containing four hard-set eggs on the 22nd of June, far in the interior of the Himalayas, at Niratu, north-east of Kotgurh. The nest was in a tuhar tree, and was composed externally of grass-seed ears, internally of finer grass; a very different-looking nest from any I have elsewhere seen, but he forwarded the bird and eggs, so that there could be no mistake.

From Murree, Colonel C. H. T. Marshall writes:—"Found numerous nests in the valleys in May and June, between 4000 and 5000 feet up."

From four to six eggs are laid, and in regard to this Shrike I have had no reason to think that it rears more than one brood in the year.

Major Wardlaw Ramsay says, writing of Afghanistan:—"I found a great many nests in May and June. The first (27th May) was situated in the centre of a dense thorny creeper, and contained six eggs, white, faintly washed with pale green, and spotted and blotched with purplish stone-colour and pale brown. The nest was composed of green grass, moss, cotton-wool, thistle-down, rags, cows' hair, mules' hair, shreds of juniper-bark, &c., &c. Other nests were found in willows by the river-bank and in apricot-trees. In a large orchard at Shalofyan, in the Kurrum valley, I found three nests within a few yards of one another."

Major C. T. Bingham writes:—"I have only found one nest of this Shrike, which is, however, common enough both at Allahabad and at Delhi. This nest I found on the 3rd June in the Nicholson gardens at Delhi. It was placed high up in the fork of a babool tree, and though more straggling and loosely built was very like that of L. lahtora; the two eggs it contained, except that they are a trifle smaller, are very like those of L. lahtora."

Colonel Butler has furnished me with the following note:—
"The Rufous-backed Shrike commences nidification at Mt. Aboo about the end of May. I took a nest on the 11th June containing five fresh eggs. It was placed in the fork of one of the outer branches of a mango-tree about 15 feet from the ground. The hen bird sat very close, allowing the native I sent up the tree to put his hand almost on to her back before she moved, and then she only flew to a bough close by, remaining there chattering and scolding angrily the whole time the nest was being robbed. The nest, which is coarse and somewhat large for the size of the bird, is composed externally of dry grass-roots, twigs, rags, raw cotton, string, and other miscellaneous articles all woven together. The interior is neatly lined with dry grass and horsehair. The eggs, five in number, are of a pale greenish-white colour, spotted all over with olivaceous inky-brown spots and specks, increasing in size and forming a zone at the large end. They vary much in shape, some being pyriform, and others blunt and similar in shape at both ends. I took another nest on the 19th June near the same place containing five fresh eggs, similar in every respect to the one already described, except that it was built on a thorn-tree about 10 feet from the ground. I took a nest at Deesa on the 8th July, 1875, containing four fresh eggs; these eggs are smaller and rounder than those from Aboo, and the blotches are larger and more distinct.

The same pair of birds built another nest a few days later, on 18th July, within ten yards of the tree from which the other nest was taken, laying five eggs.

"I found other nests at Deesa on the following dates:—

"July 2nd. A nest containing 4 incubated eggs.

„	7th.	„	„	2 fresh eggs.
„	8th.	„	„	4 „
„	9th.	„	„	2 „
„	10th.	„	„	5 „
„	10th.	„	„	4 „
Aug.	9th.	„	„	3 „

"I found many other nests in the same neighbourhood containing young birds during the last week of July."

Regarding the Rufous-backed Shrike, Mr. Benjamin Aitken has sent me the subjoined interesting note:—"This Shrike makes its appearance in Bombay regularly during the last week of September, and announces its arrival by loud cries for the first few days, till it has made itself at home in the new neighbourhood; after which it spends nearly the whole of its days on a favourite perch, darting down on every insect that appears within a radius of thirty yards. It pursues this occupation with a system and perseverance to which *L. lahtora* makes but a small approach. When its stomach is full, it enlivens the weary hours with the nearest semblance to a song of which its vocal organs are capable; for while many human bipeds have a good voice but no ear, the *L. erythronotus* has an excellent ear but a voice that no modulation will make tolerable. It remains in Bombay till towards the end of February, and then suddenly becomes restless and quarrelsome, making as much ado as the *Koel* in June, and then taking its departure, for what part of the world I do not know. This I know, that from March to August there is never a Rufous-backed Shrike in Bombay.

"The Rufous-backed Shrike, though not so large as the Grey Shrike, is a much bolder and fiercer bird. It will come down at once to a cage of small birds exposed at a window, and I once had an Amadavat killed and partly eaten through the wires by one of these Shrikes, which I saw in the act with my own eyes. The next day I caught the Shrike in a large basket which I set over the cage of Amadavats. On another occasion I exposed a rat in a cage for the purpose of attracting a Hawk, and in a few minutes found a *L. erythronotus* fiercely attacking the cage on all sides. I once caught one alive and kept it for some time. As soon as it found itself safely enclosed in the cage, it scorned to show any fear, and the third day took food from my hand. It was very fond of bathing, and was a handsome and interesting pet."

Messrs. Davidson and Wenden remark:—"Very common in Satara; breeding freely in beginning of the rains; observed at Lanoli. Rare in the Sholapoor District and does not appear to breed there." And the former gentleman, writing of Western

Khandeish, says :—" A few pairs breed about Dhulia in June and July."

Mr. C. J. W. Taylor records the following note from Manzeera-bad in Mysore :—" Plentiful all over the district. Breeding in May ; eggs taken on the 7th."

I have so fully described the eggs of *L. lahtora,* of which the eggs of this present species are almost miniatures, that I need say but little in regard to these. On the whole, the markings in this species are, I think, feebler and less numerous than in *L. lahtora ;* and though this would not strike one in the comparison of a few eggs in each, it is apparent enough when several hundreds of each are laid side by side, four or five abreast, in broad parallel rows. The ground-colour, too, in the egg of *L. erythronotus* has seldom, if ever, as much green in it, and has commonly more of the pale creamy or pinky stone-colour than in the case of *L. lahtora.*

In size the eggs of *L. erythronotus* appear to approach those of the English Red-backed Shrike, though they average perhaps somewhat smaller.

In length they vary from 0·85 to 1·05 inch, and in breadth from 0·65 to 0·77 inch, but the average of more than one hundred eggs measured is 0·92 by 0·71 inch.

Lanius caniceps.

This closely allied species, the Pale Rufous-backed Shrike, breeds only, so far as I yet know, in the Nilghiris, Palanis, &c.

It lays from March to July, the majority, I think, breeding in June.

Its nest is very similar and is similarly placed to that of the preceding, from which, if it differs at all, it only differs in being somewhat smaller.

It lays from four to six eggs, slightly more elongated ovals than those of *L. erythronotus,* taken as a body, but not, in my opinion, separable from these when mixed with a large number.

Captain Hutton, however, does not concur in this : he remarks :—" This species, which is very common in Afghanistan, occurs also in the Doon and on the hills up to about 6000 feet. At Jeripanee I took a nest on the 21st June containing five eggs, of a pale livid white colour, sprinkled with brown spots, chiefly collected at the larger end, where, however, they cannot be said to form a ring ; interspersed with these are other dull sepia spots appearing beneath the shell. Diameter 0·94 by 0·69 inch, or in some rather more. Shape rather tapering ovate.

" The differences perceptible between this and the last are the much smaller size of the spots and blotches, the latter, indeed, scarcely existing, while in *L. erythronotus* they are large and numerous ; there is great difference likewise in the shape of the egg, those of the present species being less globular or more tapering. The nest was found in a thick bush about 5 feet from the ground, and was far more neatly made than that of the fore-

going species; it is likewise less deep internally. It was composed
of the dry stalks of 'forget-me-not,' compactly held together by
the intermixture of a quantity of moss interwoven with fine flax
and seed-down, and lined with fine grass-stalks. Internal diameter
3¼ inches; external 6 inches; depth 1½ inch, forming a flattish
cup, of which the sides are about 1½ inch thick. The depth,
therefore, is less by 1 inch than in that of the last-mentioned
nest."

Mr. H. R. P. Carter tells me that "at Coonoor, on the Nilghiris,
this species breeds in April and May, placing its nest in large
shrubs, orange-trees, and other low trees which are thick and leafy.
The nest is externally irregular in shape, and is composed of fibres
and roots mixed with cotton-wool and rags; in one nest I found a .
piece of lace, 6 or 8 inches long; internally it is a deep cup, some
4 inches in diameter and 2 in depth. The eggs are sometimes
three in number, sometimes four."

Mr. Wait says that "the breeding-season extends from March
to July in the Nilghiris; the nest, cup-shaped and neatly built, is
placed in low trees, shrubs, and bushes, generally thorny ones;
the outside of the nest is chiefly composed of weeds (a white downy
species is invariably present), fibres, and hay, and it is lined with
grass and hair; there is often a good deal of earth built in, with
roots and fibres in the foundation of this nest; four appears to be
the usual number of eggs laid."

Miss Cockburn, from Kotagherry, also on the Nilghiris, tells me
that "the Pale Rufous-backed Shrike builds in the months of
February and March and forms a large nest, the foundation of
which is occasionally laid with large pieces of rags, or (as I have
once or twice found) pieces of carpet. To these they add sticks,
moss, and fine grass as a lining, and lay four eggs, which are white,
but have a circle of ash-coloured streaks and blotches at the thick
end, resembling those on Flycatchers' eggs. They are exceedingly
watchful of their nests while they contain eggs or young, and
never go out of sight of the bush which contains the precious
abode."

Mr. Davison remarks that "this species builds in bushes or
trees at about 6 to 20 feet from the ground; a thorny thick bush
is generally preferred, Berberis asiatica being a favourite. The
nest is a large deep cup-shaped structure, rather neatly made of
grass, mingled with odd pieces of rag, paper, &c., and lined with
fine grass. The eggs, four or five in number, are white, spotted
with blackish brown, chiefly at the thicker end, where the spots
generally form a zone. The usual breeding-season is May and the
early part of June, though sometimes nests are found in April and
even as late as the last week in June, by which time the south-
west monsoon has generally burst on the Nilghiris."

Dr. Fairbank writes:—"This bird lives through the year on the
Palanis and breeds there. I found a nest with five eggs when
there in 1867, but have not the notes then made about it."

Captain Horace Terry informs us that this Shrike is a most

common bird in the Palani hills, found everywhere and breeding freely.

Mr. H. Parker, writing from Ceylon, says :—" A pair of these Shrikes reared three clutches of young in my compound (two of them out of one nest) from December to May, inclusive; but this must be abnormal breeding."

Colonel Legge writes in his ' Birds of Ceylon ':—" This bird breeds in the Jaffna district and on the north-west coast from February until May. Mr. Holdsworth found its nest in a thorn-bush about 6 feet high, near the compound of his bungalow, in the beginning of February . . . . Layard speaks of the young being fledged in June at Point Pedro, and says that it builds in *Euphorbia*-trees in that district."

The eggs of this species, sent me by Captain Hutton from the Doon and by numerous correspondents from the Nilghiris, are un-distinguishable from many types of *L. erythronotus*, and indeed the birds are so closely allied that this was only to be expected. It is unnecessary to describe these at length, as my description of the eggs of *L. erythronotus* applies equally to these.

In size the eggs, however, vary less and *average* longer than those of this latter species. In length they range from 0·93 to 1 inch, and in breadth from 0·7 to 0·72 inch, but the average of twenty was 0·95 by 0·7 inch.

477. Lanius tephronotus (Vigors). *The Grey-backed Shrike.*

Lanius tephronotus (*Vig.*), *Jerd. B. Ind.* i, p. 403.
Collyrio tephronotus, *Vigors, Hume, Rough Draft N. & E.* no. 258.

As far as I yet know, the Grey-backed Shrike breeds, within our limits, only in the Himalayas, and chiefly in the interior, at heights of from 5000 to 8000 feet above the sea-level. In the interior of Sikhim, in the Sutlej Valley near Chini, in Lahoul, and well up the valley of the Beas, they are pretty common during the summer; they lay from May to July, and the young are about by the end of July or the early part of August. I have never seen a nest, although I have had eggs and birds sent me from both Sik-him and the Sutlej Valley. There were only two eggs in each case, but doubtless, like other Shrikes, they lay from four to six.

Mr. Blanford remarks that *L. tephronotus* was " common at Láchúng, in Sikhim, 8000 to 9000 feet, in the beginning of September, but three weeks later all had disappeared. Many of those seen were in young plumage, with hair on the breast, back, and scapulars."

Colonel C. H. T. Marshall records from Murree :—" This species much resembles *L. erythronotus*, but the eggs differ considerably, being more creamy white, blotched and spotted (more particularly at the larger end) with pale red and grey. They are the same size as those of the preceding species. Lays in the beginning of July at the same elevation as *L. erythronotus*."

As to the size I cannot concur with the above.

Colonel Marshall has since kindly sent me two of the eggs above referred to; they are clearly, it seems to me, eggs of *Dicrurus longicaudatus*, or the slightly smaller hill-form named *himalayanus*, Tytler.

Colonel G. F. L. Marshall writes:—" A nest found at about three feet from the ground in a thick bush at Bheem Tal, at the edge of the lake, contained five fresh eggs on the 28th May : the nest was a coarsely built massive cup; the eggs were about the same size as those of *L. erythronotus*, but the spots were larger and less closely gathered than is usual with that species."

Dr. Scully says :—" The Grey-backed Shrike is common in the Valley of Nepal from about the end of September to the middle of March ; it is the only Shrike found in the Valley during the winter season, but it migrates further north to breed. In December it was fairly common about Chitlang, which is higher than Kathmandu, but seemed to be entirely replaced in the Hetoura Dun by *L. nigriceps*. It frequents gardens, groves, and cultivated ground, perching on bushes and hedges and small bare trees. It has a very harsh chattering note, louder than that of *L. nigriceps*, and appears to be most noisy towards sunset, when its cry would often lead one to suppose that the bird was being strangled in the clutches of a raptor."

Mr. O. Möller has kindly furnished me with the following note :—" On the 7th June, 1879, my men brought a nest containing four fresh eggs, together with a bird of the present species ; I send two of the eggs : perhaps you recollect the eggs of *L. tephronotus*, in which case you of course will be able to see at a glance if I am correct. I have never come across such large eggs of *L. nigriceps*, the eggs of which also as a rule have well-defined spots and no blotches ; the two other eggs the nest contained measure 1 by 0·74, and 1·01 by 0·76 inch."

The eggs of this species are of the ordinary Shrike type, moderately elongated ovals, a little compressed towards the small end. The shell extremely smooth and compact, but with scarcely any perceptible gloss. The ground-colour pale greenish or yellowish white ; the markings chiefly confined to a broad irregular ill-defined zone round the large end—blotches, spots, specks, and smears of pale yellowish brown more or less intermingled with small clouds and spots of pale sepia-grey or inky purple. In some eggs a good number of the smaller markings and occasionally one or two larger ones are scattered over the entire surface of the egg, but typically the bulk of the markings are comprised within the zone above referred to.

In length four eggs vary from 0·97 to 1·06 inch, and in breadth from 0·76 to 0·81 inch.

481. Lanius cristatus, Linn. *The Brown Shrike.*

Lanius cristatus, *Linn.*, *Jerd. B. Ind.* i, p. 406; *Hume, Rough Draft N. & E.* no. 261.

I am induced to notice this species, the Brown Shrike, although I possess no detailed information as to its nidification, in consequence of Lord Walden's remarks on this subject in 'The Ibis' of 1867. He says, "Does it, then, cross the vast ranges of the Himalaya in its northern migration? or does it not rather find on the southern slopes and in the valleys of those mountains all the conditions suitable for nesting?"; and he adds in a note, "It is extremely doubtful whether any passerine bird which frequents the plains of India during the cooler months crosses to the north of the snowy ranges of the Himalaya after quitting the plains to escape the rainy season or the intense heat of summer."

Now, it is quite certain, as I have shown in 'Lahore to Yarkand,' that several of our Indian passerine birds do cross the entire succession of Snowy Ranges which divide the plains of India from Central Asia, and it is tolerably certain from my researches and those of numerous contributors that *L. cristatus* breeds *only* north of these ranges. True, Tickell gives the following account of the nidification of this species in the plains of India :—

"Nest found in large bushes or thickets, shallow, circular, 4 inches in diameter, rather coarsely made of fine twigs and grass. Eggs three, ordinary ; $\frac{?9}{}$ by $\frac{?4}{}$: pale rose-colour, thickly sprinkled with blood-red spots, with a darkish livid zone at the larger end.— *June*." But Tickell, though he warns us at the commencement of his paper (Journal As. Soc. 1848, p. 297) of the "attempts at duplicity of which the wary oologist must take good heed," gives the egg of the Sarus as plain white, and says he has seen upwards of a dozen like this, those of the Roller as full deep Antwerp blue, those of *Cypselus palmarum* as white with large spots of deep claret-brown, and so on, and it is quite clear that his supposed eggs and nest of *L. cristatus* belonged to one of the Bulbuls.

Of more than fifty oologists who have collected for me at different times in hills and plains, from the Nilghiris to Huzára on the one side, and to Sikhim on the other, not one has ever met with a nest of *L. cristatus*. This is doubtless purely negative evidence, but it is still entitled to considerable weight.

From the valleys of the Beas and the Sutlej, as also from Kumaon and Gurhwal, these Shrikes seem to disappear entirely during the summer, and they are then, as we also know, found breeding in Yarkand. It is only in the latter part of the autumn that they reappear in the former named localities, finding their way by the commencement of the cold season to the foot of the hills.

Mr. R. Thompson, to quote one of many close observers, remarks :—"This bird appears regularly at Huldwanee and Rumnugger at the foot of the Kumaon Hills during the cold weather, confining itself to thick hedges and deep groves of trees. Where it goes to in summer I cannot say, it certainly does not remain in our hills."

484. **Hemipus picatus** (Sykes). *The Black-backed Pied Shrike.*

Hemipus picatus (*Sykes*), *Jerd. B. Ind.* i, p. 412 ; *Hume, Rough Draft N. & E.* no 267.

I quite agree with Mr. Gray that this bird is a Flycatcher and

not a Shrike; no one in fact who has watched it in life can have
any doubt on this subject; but yet, except for their being more
strongly marked, its eggs have no doubt a very Shrike-like character,
at the same time that they exhibit many affinities to those of
Rhipidura albifrontata and other undoubted Flycatchers.

Mr. W. Davison says:—"About the first week in March 1871,
I found at Ootacamund a nest of this bird placed in the fork of
one of the topmost branches of a rather tall *Berberis leschenaulti.*
For the size of the bird this was an exceedingly small shallow nest,
and from its position between the fork, its size, and the materials
of which it was composed externally, might very easily have passed
unnoticed; the bird sitting on it appeared to be sitting only on a
small lump of moss and lichen, the whole of the bird's tail, and as
low down as the lower part of the breast, being visible. The nest
was composed of grass and fine roots covered externally with cobweb and pieces of a grey lichen, and bits of moss taken apparently
from the same tree on which the nest was built; the eggs were
three in number. The tree on which this nest was built was
opposite my window, and I watched the birds building for nearly
a week; and, again, when having the nest taken, the birds sat till
the native lad I had sent up put out his hand to take the nest. I
am *absolutely* certain as to the identity of this nest and these eggs."

The eggs brought me by Mr. Davison, of the authenticity of
which he is positive, are very Shrike-like in their appearance;
they are rather elongated ovals, somewhat obtuse at both ends,
and entirely devoid of gloss. The ground-colour is a pale greenish
or greyish white, and they are profusely blotched, spotted, and
streaked with darker and lighter shades of umber-brown; in both
eggs these markings are more or less confluent along a broad
zone, which in one egg encircles the larger, in the other the smaller
end: these eggs measure 0·7 by 0·5 inch and 0·69 by 0·49 inch.

Captain Horace Terry writes from the Palani Hills:—"Pittur
Valley. I had a nest brought me which from the description of
the bird must, I think, have belonged to this species. Nest
rather a shallow cup placed in a thorny tree about ten feet from
the ground, neatly made of grass and moss, lined with fine grass
and a few feathers, covered a great deal on the outside with
dusky-coloured cobwebs, 2·5 inches across and 1·5 inch deep inside,
and 3·25 inches to 3·5 inches across, and 2·25 inches deep outside: contained five very much incubated eggs; shape and marking
exactly like those of *L. caniceps,* having a well-defined zone round
the larger end; size about the same or rather smaller than those of
Pratincola bicolor."

485. Hemipus capitalis (McClelland).   *The Brown-backed Pied Shrike.*

Hemipus capitalis (*McClell.*), *Hume, Cat.* no. 267 A.

I must premise that to the best of my belief there is no such thing as *H. capitalis*, McClell., in India, or, in other words, that this latter name is a mere synonym of *H. picatus* *.

Mr. Blyth remarks, Ibis, 1866:—" *Hemipus picatus*. Under this name two very distinct species are brought together by Dr. Jerdon: *H. capitalis* (McClell., 1839; *H. picæcolor*, Hodgson, 1845) of the Himalaya, which is larger, with proportionally longer tail, and has a brown back; and *H. picatus* (Sykes) of Southern India and Ceylon, which has a black back. Mr. Wallace has good series of both of them.

" *Hemipus capitalis* has accordingly to be added to the birds of India."

Now, out of India, Mr. Wallace may have got hold of some brown-backed *Hemipus*, which is really distinct, but nothing is more certain (I speak after comparison of a large series from Southern India with a still larger, gathered from all parts of the Himalayas) than that the Southern and Northern Indian birds are identical, and that in both localities the males have black and the females brown backs.

Capt. T. Hutton says:—" On the 12th of May I procured a nest of this bird in the Dehra Doon; it was placed on the ground at the base of an overhanging rock, and was composed entirely of the hair of horses and cows and other cattle, which had doubtless been collected from the bushes and pasture-lands in the vicinity. There were four eggs of a pale sea-green, spotted with rufous-brown, and forming an indistinct and nearly confluent ring at the larger end. The bird had begun to sit.

" This curious little species is not uncommon in the outer hills up to 5000 feet in the summer months."

The three eggs sent me by Captain Hutton appear to differ somewhat conspicuously from any other eggs of the *Laniidæ* that I have yet seen. The ground-colour is a very pale greenish white, and they are moderately thickly freckled and mottled all over, but most densely towards the large end (where, in one egg, there is a well-marked, though somewhat irregular, zone), with pale brownish pink and very pale purple. In shape the eggs are very regular, rather broad ovals, and appear to have but little or no gloss. They vary in length from 0·66 to 0·7 inch, and in breadth from 0·53 to 0·55 inch.

Dr. Jerdon's evidence, so far as it goes, tallies with Captain Hutton's account. He says:—" I obtained its nest once at Darjeeling, made of roots and grasses, with three greenish-white eggs, having a few rusty-red spots."

---

* Mr. Hume would probably now agree with me that *H. picatus* and *H. capitalis* are distinct species. *H. picatus*, however, is not confined to Southern India, but occurs along the Terais of Sikhim and Nepal, and throughout Burma. *H. capitalis* occurs on the Himalayas from Gurwhal to Assam. There is little doubt that Captain Hutton's nest did not really belong to a Pied Shrike.—ED.

From Sikhim, Mr. Gammie writes:—" At page 178 of 'Nests and Eggs of Indian Birds' (Rough Draft), Captain T. Hutton's description of the nest and eggs of *Hemipus picatus* is given, and at page 179 that of Mr. W. Davison. The two descriptions differ so radically that, as there remarked, one of the two must be in error. Permit me to record my limited experience of the nesting of this bird.

" Common as it is in Sikhim I have but once taken its nest, and that in the first week of May, at 4000 feet elevation. The nest, which is well described by Mr. Davison, is made of black, fibry roots, sparingly lined with fine grass-stalks, and covered outwardly with small pieces of lichens bound to the sides with cobwebs. It is a very neat diminutive cup, measuring externally 1·9 inch across by an inch deep; internally 1·5 by half an inch.

" The whole nest, although quite a substantially built structure, is barely the eighth part of an ounce in weight. It was placed on the upper side of a horizontal branch close to its broken end, about fifteen feet from the ground, and contained two fresh eggs. I send you the nest and an egg, both of which will, I think, be found on comparison to agree exactly with those taken by Mr. Davison."

Mr. Mandelli has sent me two nests of this species, found on the 15th August above Namtchu in Native Sikhim. They were placed about two feet from each other, each in a small fork of the branches of a small tree which was situated in heavy forest. Each contained two fresh eggs. The nests are very similar, but one is rather larger and less tidily finished-off than the other. Both are shallow cups, miniatures of some of the nests of *Dicrurus*, composed of excessively fine grass-stems, coated exteriorly all round the sides with cobwebs, and, in the case of one of them, plastered exteriorly with tiny films of bark and dry leaves like some of the nests of the *Pericrocoti*. Both have a little soft silky vegetable down at the bottom of the cavity. The one nest is about two inches, the other about two and a half inches in diameter exteriorly, and both are a little less than three quarters of an inch high outside. The cavity in the one is about an inch and a half, in the other about an inch and three quarters in diameter, and both are about half an inch deep.

Eggs received from Sikhim are broad ovals, glossless, with greenish-white grounds, profusely speckled and mottled with slightly varying shades of brown, here and there intermingled with dull, pale inky purple. The markings are densest generally round the broadest part of the egg. They measured from 0·61 to 0·7 in length, and from 0·51 to 0·55 in breadth.

486. Tephrodornis pelvicus (Hodgs.). *The Nepal Wood-Shrike.*

Tephrodornis pelvica (*Hodgs.*), *Jerd. B. Ind.* i, p. 409 ; *Hume, Cat.* no. 263.

The Nepal Wood-Shrike is a permanent resident throughout

Burma, Assam, Cachar, and the sub-Himalayan Terais and Ranges to which the typical Indo-Burmese fauna extends. Still we have no information as to its nidification, and the only egg of the species that I possess was extracted from the oviduct of a female shot by Mr. Davison on the 26th of March, 1874, near Tavoy in Tenasserim. The egg is rather a handsome one—very Shrike-like in its character, but rather small for the size of the bird. In shape it is a broad oval, very slightly compressed towards one end. The shell is fine and compact, but has no gloss. The ground is white, with the faintest possible greenish tinge only noticeable when the egg is placed alongside a pure white one, such as a Bee-eater's for instance. The markings are bold, but except at the large end not very dense—spots and blotches of a light clear brown, and (chiefly at the large end) somewhat pale inky grey. Where the two colours overlap each other, there the result of the mixture is a dark dusky brown, so that the markings appear to be of three colours. Fully half the markings are gathered into a broad conspicuous but very broken and irregular zone about the broad end. The egg measured only 0·86 by 0·69.

Subsequently to writing the above Mr. Mandelli sent me a nest of this species found at Ging near Darjeeling on the 27th April. It contained four fresh eggs, and was placed on branches of a very large tree about 22 feet from the ground. The tree was situated at an elevation of about 3000 feet. The nest is a large massive cup, 5 inches in exterior diameter and rather more than 3 in height. It is composed of tendrils of creepers and stems of herbaceous plants, to many of which the bright yellow amaranth flowers remain attached; and all over the sides and bottom masses of flower-stems of grass with the white silky down attached are thickly plastered, which, intermingled as this white down is with the glistening yellow flowers, produces a very ornamental effect, and looks as if the bird had really had an eye to decoration.

Inside the nest is entirely lined with very fine grass-stems. The nest is everywhere about an inch thick, and the cavity about 3 inches in diameter by nearly 2 deep.

Eggs said to belong to this species kindly sent me by Mr. Mandelli, whose men obtained them on the 27th April, are very Shrike-like in their appearance. In shape they vary from broad to ordinary ovals, generally somewhat compressed towards the small end. The shell is white but almost glossless. The ground-colour is a dead white, and they are profusely speckled and spotted with yellowish brown, paler in some eggs, darker in others. In all the eggs the markings are by far the most numerous towards the large end. Two eggs measure 0·95 and 0·91 in length by 0·74 and 0·72 in breadth respectively.

487. **Tephrodornis sylvicola**, Jerdon. *The Malabar Wood-Shrike.*

Tephrodornis sylvicola, *Jerd., Jerd. B. Ind.* i, p. 409; *Hume, Cat.* no. 264.

Major M. Forbes Coussmaker has furnished me with the following note on the nidification of the Malabar Wood-Shrike :—" I took the nest of this bird on April 13th, 1875. It was composed of fine roots and fibres, neatly woven into a shallow cup-like nest, secured to the fork of a horizontal bough and fixed in its place with cobweb, and covered externally with lichen corresponding to that on the bough. It measured 4·2 inches in diameter externally, and 2·4 internally and ·7 deep. Both parent birds were shot. The eggs two in number, rather round, coloured white with faint inky and brown spots."

One of these eggs is a very regular oval, the shell fine but glossless, the ground-colour white, with a faint greenish tinge ; round the large end is a pretty conspicuous zone of black or blackish-brown and pale inky purple spots and small blotches, and similar spots and blotches of the same colour are somewhat sparsely scattered over the rest of the surface of the egg. The egg measured 0·98 by 0·73.

488. **Tephrodornis pondicerianus** (Gm.). *The Common Wood-Shrike.*

Tephrodornis pondiceriana (*Gm.*), *Jerd. B. Ind.* i, p. 410; *Hume, Rough Draft N. & E.* no. 265.

The Common Wood-Shrike lays during the latter half of March and April. This at least is, I think, the normal season, but Mr. W. Blewitt found a nest at Hansee on the 2nd of June containing two fresh eggs.

I have only taken one nest myself (though I have had many others sent me), and that was on the 2nd of April at Chundowah in Jodpoor, Rajpootana. The nest was in the fork of a ber tree (*Zizyphus jujuba*), on a small horizontal bough, about 5 feet from the ground. It was a broad shallow cup, somewhat oval interiorly, with the materials very compactly and closely put together. The basal portion and framework of the sides consisted of very fine stems of some herbaceous plant about the thickness of an ordinary pin. It was lined with a little wool and a quantity of silky fibre ; exteriorly it was bound round with a good deal of the same fibre and pretty thickly felted with cobwebs. The egg-cavity measured 2·5 inches in diameter one way and only 2 the other way, while in depth it was barely ·86. The exterior diameter of the nest was about 4 inches and the height nearly 2 inches. It contained three fresh eggs, of a slightly greyish-white ground, very thickly spotted and speckled with yellowish brown, dark umber-brown, and a pale washed-out inky-purple. In all, the spots were thickest in a zone round the large end, where they became more or less confluent. I have, however, a large series of these nests, and taking them as a whole, although much more massive, they remind one no little of those of *Rhipidura albifrontata* and *Terpsiphone paradisi* and even *Ægithina tiphia*. They are broad shallow cups, measuring internally $2\frac{1}{4}$ inches across and about $\frac{7}{8}$ inch in depth. They are placed in a horizontal fork of a branch, and are composed of vegetable fibre and fine grass-roots, thickly coated externally with cobwebs, by which also they are fixed on to branches, and lined internally with silky vegetable down or fibre. Externally their colour always approximates closely to the bark of the branch on which they are placed; they are not thin basket-like structures like those of *Ægithina* or *Rhipidura*, but are fully ½ inch thick at the sides and probably ¾ inch thick at the bottom.

Colonel G. F. L. Marshall writes :—" The Common Wood-Shrike builds in the Saharunpoor district in the latter half of March, the young being hatched early in April. The bird is common ; but owing to the small size and bark-like colour of its nest, the latter is very difficult to find. On the 8th April I fired at a specimen and missed it ; it then flew off and settled in a fork of another tree about 30 feet from the ground. On looking carefully with an opera-glass, I found that it was sitting on its nest. I drove it off and shot it. The nest was very small and shallow, cup-shaped, and wedged in between two small boughs at their junction, and not appearing either above or below. The egg-receptacle was 2¼ inches in diameter. The nest was made of grass and bits of bark, beautifully woven together and bound with cobwebs, and exactly resembling the boughs between which it was placed, or, I might say, wedged in. The eggs, four in number, were slightly set ; they were small for the bird, and of a rather round oval shape ; the colour was a creamy-yellow ground, thickly spotted and blotched with the different shades of brown and sienna, the bulk of the spots tending to form a zone near the thick end, as in the typical form of the eggs of the *Laniidæ*, and a number of faint purple blotches underlying the zone."

Major C. T. Bingham says :—" I have only found three nests of this bird, and these at Delhi. At Allahabad it was not very common. It is a difficult nest to find, being generally well hidden in the forks of leafy trees. All three nests I got were of one type—shallow saucers, made of vegetable fibre matted together into a soft felt-like substance. In two of the nests I found three and in the third one egg. These are thickly spotted and blotched with brown and a washed-out purple, on a pale greyish-yellow ground. The average measurements of the seven eggs are —length 0·77, breadth 0·61."

Colonel E. A. Butler writes from Sind :—

"*Hyderabad, 19th April,* 1878.—Noticed two young birds scarcely able to fly ; fresh eggs were laid, therefore, about the beginning of March. On the 20th April near the same place I found a nest containing young birds. It consisted of a neat little cup composed of dry grass smeared all over exteriorly with cobwebs, and fixed in a fork of one of the outer branches of a large babool-tree about 10 feet from the ground. The nest was very small for the size of the bird, and had I not seen the old bird on it I should have taken it for a nest of *Rhipidura albifrontata*."

The late Captain Beavan remarked that this bird " appears to come to the Maunbhoom District for the purpose of breeding. I procured the nest and eggs early in April, and the young were nearly fledged by the 20th of that month ; they appear to come year after year to particular localities to breed.

" Several nests were brought me from the neighbourhood of Kashurghur both in 1864 and 1865, whereas none were seen elsewhere. The nest is very small for the size of the bird, and the material of which it is composed closely resembles the bird's plumage in colour. The nest is round and very shallow, something like a Chaffinch's, being very neatly made ; diameter inside 2 inches, depth 1 inch : composed of grey fibres, bits of bark, grass, and the like, cemented with spider's web. The eggs are two in number, greenish white, spotted with brown and slate-coloured dots, which in most specimens form a well-defined zone round the thickest part of the egg, leaving both ends without marks. Length of the egg ·75 inch ; breadth ·59 inch. This bird was not observed in Maunbhoom except during the breeding-season."

Mr. G. W. Vidal, writing from the South Konkan, remarks :— " Common, as also at Sávant Vádi. Nest found with three hard-set eggs on the 18th February, low down in a mango-tree. Nest a very neat compact cup of grasses and fibres, woven throughout with spiders' webs. Eggs greyish white, with brown and inky-purple spots."

Dr. Jerdon remarks :—" The nest has been brought to me in August at Nellore, chiefly made of roots and lined with hair ; and the eggs, three in number, were greenish white with large brown blotches."

Major M. F. Coussmaker sends me the following note from Mysore :—" I took the nest of this bird on April 16th. It was composed of fine roots and fibres closely woven into a compact nest, secured to a horizontal bough with cobweb and covered externally with lichen to match the tree. It measured in diameter 4·1 inches externally and 2·2 internally and ·8 deep. The parent bird was shot from the nest.

" The nest contained two eggs, white with brown spots and markings. They were so broken when I got them that no reliable measurements could be taken."

Lastly, Mr. Oates writes from Pegu :—" Nest with three fresh eggs on the 3rd March near Pegu."

The eggs are very Shrike-like in appearance, and many of them are perfect miniatures of the eggs of *Lanius lahtora*, but some of them have a more uniformly brown tint than any of this latter species that I have yet met with. The ground-colour is generally either a very pale greenish white or a creamy-stone colour, and more or less thickly spotted and blotched with different shades of yellowish and reddish brown ; many of the markings are almost invariably gathered into a conspicuous, but irregular and ill-defined, zone near the large end, in which zone clouds of subsurface-looking, pale, and dingy purple, not usually observable on any other portion of the egg, are thickly intermingled. The texture of the shell is fine and close, but scarcely any gloss is ever perceptible. Occasionally the eggs are very faintly coloured, and have a dull white ground, while the markings consist of only a few spots and specks of very pale purple and pale rust-colour confined to a zone near the large end.

In length the eggs vary from 0·69 to 0·8 inch, and in breadth from 0·57 to 0·65 inch; but the average of a dozen eggs is 0·75 by 0·61 inch nearly.

490. Pericrocotus speciosus (Lath.). *The Indian Scarlet Minivet.*

Pericrocotus speciosus (*Lath.*), *Jerd. B. Ind.* i, p. 419; *Hume, Rough Draft N. & E.* no. 271.

Captain Hutton records that the Indian Scarlet Minivet breeds both on the Doon and in the hills overlooking it, to an elevation of about 5000 feet. He says:—"The nest is generally placed high up on the branch of some tall tree, often overhanging the side of a fearful precipice. On the 6th and 17th of June I procured two nests in ravines opening upon the Doon, one of which contained four, and the other five eggs, of a dull-white colour, sparingly spotted and blotched with earthy brown, more thickly so at the larger end, where they form an open ring of spots; other small blotches of a fainter colour are seen beneath the shell.

"It is a curious fact that in the latter nest, out of the five eggs *three* were ringed at the larger end, and the other two *at the smaller end.* The nest is rather coarsely made, being very thick at the sides, and the materials not neatly interwoven; it is composed externally of dried grasses and the fine stalks of various small plants, interspersed with bits of cotton and grass-roots, and lined with the fine seed-stalks of small grasses."

I am not at all sure that there is not some mistake here. The nest described is rather that of *L. erythronotus* than of any of the *Pericrocoti*, and but for the excellent authority on which the above rests, I should certainly not have accepted it.

This species breeds in the forests of the central hills of Nepal: according to Mr. Hodgson's notes and drawings they begin laying about April, and lay three or four eggs, which are neither described nor figured. The nest is a beautiful deep cup externally about 3·25 inches in diameter, and rather more than 2 inches high, composed of moss and moss-roots lined internally with the latter, and entirely coated exteriorly with lichen and a few stray pieces of green moss firmly secured in their places by spiders' webs. The nest is placed in some slender branch between three or four upright sprays. This, I may note, is just the kind of nest one would have expected this Large Minivet to build.

The only specimens, supposed to be the eggs of this species, that I possess I owe to Captain Hutton. They closely resemble the eggs of *L. erythronotus,* but are perhaps shorter, and hence *look* broader than those of this latter. They are slightly bigger than the eggs of *L. vittatus.* In shape they seem to be typically a slightly broader oval than those of any of our true Shrikes, but elongated and pointed examples occur. Their ground-colour is a very pale greyish white, thickly spotted all over the large end, and thickly dotted elsewhere, with specks, spots, and tiny blotches of pale yellowish brown and pale inky-purple. Compared with the eggs of the other *Pericrocoti*, they are very dingily coloured. The eggs are devoid of gloss. I am doubtful about these eggs.

In length they vary from 0·88 to 0·93 inch, and in breadth from 0·72 to 0·75 inch; but the average of five eggs is 0·9 by 0·72 inch.

494. Pericrocotus flammeus (Forst.). *The Orange Minivet.*

Pericrocotus flammeus (*Forst.*), *Jerd. B. Ind.* i, p. 420; *Hume, Rough Draft N. & E.* no. 272.

The Orange Minivet lays, I believe, in June and July on the Nilghiris. I have never taken a nest myself, but I have received several, with a few words in regard to them, from Miss Cockburn.

The nests are comparatively massive little cups placed on, or sometimes in, the forks of slender boughs. They are usually composed of excessively fine twigs, the size of fir-needles, and they are densely plastered over the whole exterior surface with greenish-grey lichen, so closely and cleverly put together that the side of the nest looks exactly like a piece of a lichen-covered branch. There appears to be no lining, and the eggs are laid on the fine little twigs which compose the body of the nest.

The nests are externally from 3 to 3¼ inches in diameter, and about 1½ inch deep, with an egg-cavity about 2 inches in diameter and about ¾ inch in depth. Some, however, when placed in a fork are much deeper and narrower, say externally 2¼ inches in diameter and the same height; the egg-cavity about 1¾ inch in diameter and 1¼ inch in depth.

Miss Cockburn notes that one nest was found on the 24th of June on a high tree, the nest being placed on a thin branch between 30 or 40 feet from the ground. It contained a single fresh egg, which was broken in the fall of the branch, which had to be cut. This egg, the remains of which were sent me, had a pale greenish ground, and was pretty thickly streaked and spotted, most thickly so at the large end, with pale yellowish brown and pale rather dingy-purple, the latter colour predominating.

Another egg which she subsequently sent me, obtained on the 17th of July, is a 'regular, moderately elongated oval, a little pointed towards one end. The shell is fine, but glossless. The ground is a delicate pale sea-green or greenish white, and it is rather sparsely spotted and speckled with pale yellowish brown. Only one or two purplish-grey specks are to be detected on this egg; it measures 0·9 by 0·67.

Mr. J. Darling, junior, sends me the following note:—"I had the good fortune to find a nest of the Orange Minivet at Neddivattum, about 6000 feet above the level of the sea, on the 5th September, 1870. It was placed on a tall tree near the edge of a jungle and was built in a fork, about 30 feet from the ground.

"The nest was built of small twigs and grasses, and covered on the outside with lichens, moss, and cobwebs, making it appear as part and parcel of the tree. I noticed it merely from the fact of seeing the bird sitting on her nest, and even then could not make up my mind, and came away. Being of an inquisitive nature, next day I went again and saw the bird in the same place, so I climbed up and managed to pull the nest towards me with a hook, and took two eggs, one of which I send you.

"In August 1874 at Vythery I saw a bird sitting on her nest, and watched her rear and take away her brood, but could not get at the nest."

An egg sent me by Mr. Darling is very similar to the eggs sent me by Miss Cockburn, except that the brown markings are rather more numerous, especially in a broad zone round the large end, and that with these a good many pale purple or lilac spots or specks are intermingled. It measures 0·88 by 0·68 inch.

495. Pericrocotus brevirostris (Vigors). *The Short-billed Minivet.*

Pericrocotus brevirostris (*Vig.*), *Jerd. B. Ind.* i, p. 421; *Hume, Rough Draft N. & E.* no. 273.

The Short-billed Minivet breeds in the Himalayas at elevations of from 3000 to 6000 feet in Kumaon, and again in Kulu and the valley of the Sutlej. It lays in May and June, building a compact and delicate cup-shaped nest on a hoizontal bough pretty high up in some oak, rhododendron, or other forest tree. I have never seen one on any kind of fir-tree.

Sometimes the nest is merely placed on, and attached firmly to, the upper surface of the branch; but, more commonly, the place where two smallish branches fork horizontally is chosen, and the nest is placed just at the fork. I got one nest at Kotgurh, however, wedged in between two upright shoots from a horizontal oak-branch. The nests are composed of fine twigs, fir-needles, grass-roots, fine grass, slender dry stems of herbaceous plants, as the case may be, generally loosely, but occasionally compactly interlaced, intermingled and densely coated over the whole exterior with cobwebs and pieces of lichen, the latter so neatly put on that they appear to have grown where they are. Sometimes, especially at the base of the nest, a little moss is attached exteriorly, but, as a rule, there is nothing but lichen. The nest has no lining. The external diameter is about 2½ inches, and the usual height of the nest from 1½ to 2 inches; but this varies a good deal according to situation, and the bottom of the nest, which in some may be at most ¼ inch thick, in another is a full inch. The sides rarely exceed ⅛ inch in thickness. The egg-cavity has a diameter of about 2 inches, and a depth of from 1 to 1·25 inch.

Five seems to be the maximum number of eggs laid, but I have now twice met with three, more or less incubated, eggs.

Mr. Hodgson notes:—"May 16th: At the top of the great forest of Sheopoori, secured a nest built near the top of a kaiphul tree, and laid on a thick branch amongst smaller twigs. The nest is about 2 inches deep and the same in diameter: inside it is 1·5 inch deep; it is made of paper-like bits of lichen welded together with spiders' webs, and with a lining of elastic fibres. It is the shape of a deep soap-stand, open at the top of course. It contained two eggs of a bluish or greenish-white ground, much spotted with liver colour, especially near the large end, where the spots are clustered into a zone."

Dr. Scully, writing also from Nepal, says:—"During the breeding-season (May and June) this Minivet is found in forests on the hills up to an elevation of 7500 feet. A nest was found in the Sheopoori forest on the 17th June, which contained two very young birds and one egg."

The eggs of this species that I have seen are moderately broad ovals, as a rule, very regular in their shape, and scarcely compressed at all towards the lesser end. The shell is fine and satiny, but the eggs have little or no real gloss. The ground-colour is a dull white, sometimes slightly tinged with pink, sometimes with green, and they are richly and profusely blotched, spotted, and streaked, most densely, as a rule, towards the large end, with brownish red and pale purple. Most eggs exhibit a more or less conspicuous, though irregular, zone round the larger end.

The eggs vary in length from 0·71 to 0·8 inch, and in breadth from 0·54 to 0·6 inch.

499. **Pericrocotus roseus** (Vieill.). *The Rosy Minivet.*

Pericrocotus roseus (*Vieill.*), *Jerd. B. Ind.* i, p. 422; *Hume, Rough Draft N. & E.* no. 275.

The only one of my contributors who appears to have taken the eggs of the Rosy Minivet is Colonel C. H. T. Marshall. Mr. R. Thompson says:—"They breed in the warmer valleys of Kumaon, up to an elevation of some 5000 feet, in May and June;" but he adds: "I have never got down the nests."

Colonel Marshall, writing from Murree, says:—"The Rosy Minivet builds a beautifully little shallow cup-shaped nest, the outer edge being quite narrow and pointed. The external covering of the nest is fine pieces of lichen fastened on with cobwebs. It was found on the 12th of June, and contained three fresh eggs, white, with greyish-brown spots and blotches sparsely scattered about the larger end; the length is 0·8 by 0·55 inch; 5000 feet up."

The nest, which I owe to this gentleman, is externally a short section of a cylinder, rather than a cup, the walls standing up outside almost perpendicularly. It is 2·5 inches in diameter and nearly 1·75 in height. The rim of the nest is ¼ inch wide, and the cavity, a shallow cup, 2 inches wide by scarcely an inch deep ; the walls of the nest increase in thickness as they approach the base.

Externally the whole surface is *entirely* covered by small scales of lichen, firmly bound into their respective places by gossamer threads ; internally the nest is a very loosely put together basket-work of excessively fine twigs and grass-stems not thicker than common needles. A morsel or two of moss have become involved in the fabric, as well as two fine blades of grass ; but there is no lining, and the eggs are obviously laid upon the soft loose basket frame of the nest.

The egg which accompanied the nest is a regular oval, slightly compressed towards one end. The ground-colour is pale greenish white entirely devoid of gloss. The egg is richly blotched, spotted, and speckled (most densely so towards the larger end) with reddish brown and greenish purple, there being two conspicuously different shades (a much darker and a much lighter, the latter of which appears like subsurface tints) of each of these colours. This egg measures 0·82 by 0·6 inch nearly.

Another egg of the same clutch was less richly coloured, the markings being merely brown, with scarcely a perceptible reddish tinge, and dull mostly inky, but here and there somewhat reddish, purple. The markings, too, were fewer in number, but there was a more marked tendency for these to form a zone about the larger end.

In another clutch the markings were almost entirely confined to a dense zone round the larger end about a third of the way up from the middle of the egg. In this zone they were so densely set as to be quite confluent, and they consisted of yellowish brown and inky purple.

Mr. J. R. Cripps found the nest of this Minivet in the Bhaman tea-garden, in the Dibrugarh District of Assam, on the 31st May, 1879. The nest contained three eggs, and was placed on the upper side of a large lateral branch of a tree that grew on the main garden road, about 15 feet from the ground.

Seven eggs of this bird vary in length from 0·75 to 0·86, and in breadth from 0·58 to 0·6.

500. Pericrocotus peregrinus (Linn.). *The Small Minivet.*

Pericrocotus peregrinus (*Linn.*), *Jerd. B. Ind.* i, p. 423 ; *Hume, Rough Draft N. & E.* no. 276.

Our Small Minivet lays during the latter half of June (as soon, in fact, as the rains set in), and throughout July and August. I believe it breeds pretty well all over India and Burma.

The nest is small and neat, and done up generally like a Chaffinch's, to resemble the bark of the tree on which it is placed.

22*

The nests that I have seen have been invariably placed at a considerable height from the ground in the fork of a branch, most commonly, I think, a mango-tree, though I have occasionally noticed them in other trees.

The nest is a small moderately deep cup, with an internal cavity about 1·7 inch to 1·9 in diameter, and nearly an inch in depth. The sides of the nest are about $\frac{3}{8}$ inch thick, and the thickness of the bottom of the nest varies according to the shape of the fork chosen, whether obtuse or acute-angled. In the former case the bottom of the nest is sometimes not above $\frac{1}{4}$ inch in depth. In the latter case, it is sometimes as much as an inch in thickness. It is composed of very fine, needle-like twigs (with at times here and there a few feathers) carefully bound together externally with cobwebs, and coated with small pieces of bark or dead leaves, or both, so that looked at from below with the naked eye it is impossible to distinguish it from one of the many little excrescences so common, especially on mango-trees. There appears to be rarely any regular lining, a very little down and cobwebs forming the only bed for the eggs, and even this is often wanting. Sometimes a few tiny dead leaves or a little lichen will be found incorporated in the nest, and occasionally, but rarely, fine grass-stems take the place of very slender twigs.

Three is, I believe, the normal number of the eggs. I extract a couple of old notes I made in regard to the nests of this species :— "*August 5th.*—Took three eggs of this bird, shooting the two old birds at the same time. The tree was a mango, the nest was in the fork of a branch, some 40 feet from the ground, built interiorly with very small twigs, with here and there a very few feathers intermixed, and was exteriorly coated with fine flakes of bark held in their place by gossamer threads. It was cup-shaped, with an interior diameter of $1\frac{7}{8}$ by $\frac{3}{4}$ inch.

"The eggs had a slightly greenish-white ground, thickly spotted and speckled, and towards the larger end blotched, with somewhat brownish red ; the markings showing a decided tendency to form a zone round, or cap at the larger end."

"*Allygurh, August 27th.*—Another beautiful little nest in a mango-tree high up, a tiny cup about $1\frac{1}{2}$ inch internal diameter by $\frac{3}{4}$ inch deep, woven with very fine twigs, and exteriorly coated with tiny fragments of bark and dead leaves firmly secured in their places with gossamer threads and cobwebs. It contained two fresh eggs ; a pale slightly greenish-white ground, richly speckled and spotted and sparsely blotched with a purplish and a brownish red, the markings greatly predominating towards the larger end."

Mr. F. R. Blewitt, detailing his experiences in Jhansie and Saugor, says :—"Breeds in June and July. The tamarind-tree is by preference chosen by this bird for its nest ; at least the three I saw were all on tamarind-trees. The nest, cup-shaped, is a compactly made structure ; the exterior appeared to be composed of the very fine petioles of leaves, with a thick coating all over of what looked like spider's web; attached to this web-like substance here and there, for better disguise, were the dry leaves of the tamarind-tree; the lining of very fine grass. The outer diameter of a nest may fairly be given at 2·2 inches, inner at 1·8, depth of nest 0·9. Two is the regular number of eggs, at least that was the number in the three nests I took. In colour they are of a pale greenish white, sparingly speckled on the narrower half of the egg with brownish spots, but they have on the broader half the spots more dense, and forming at the end a more or less complete cap. The feat of securing a nest is a most hazardous one, for it is always fixed close in between two delicate forks at the extreme end of a slight side-branch near to the top of the tree. On each occasion that the nest was detected the male bird was found flitting about near to it, the female all the while sitting on the eggs. On the last two occasions of finding the nests, it was this flitting to and fro of the male that attracted us; otherwise the nest is so small that from the ground the eye can scarcely distinguish it from the branch. The bird appears to be migratory, for since the termination of the breeding-season it has disappeared from these parts."

Major C. T. Bingham writes to me:—"Although this bird is common enough both at Allahabad and at Delhi, I have found it difficult to find its nest, from the fact that it is placed at the very extreme tip of leafy branches. However, with careful watching and patience, I managed to find one nest at Allahabad and five at Delhi. The first I found on the 3rd July at Chupree near Allahabad. It contained two well-fledged young ones, that hopped out as soon as the nest was touched. Out of the five at Delhi I managed to get six eggs; three of the nests when found being empty, were afterwards deserted by the birds. Of the two nests with eggs, one contained four and the other two. The nests are tiny little cups, made of very fine grass, and coated externally with cobwebs, to which are attached bits of bark and dry leaves. The eggs are a greenish stone-colour, thickly speckled with light purple and brownish red. The earliest nest I have found was on the 21st March, on the banks of the canal at Delhi, so that the bird occasionally, at Delhi at least, lays in spring. The average of eggs I have is 0·68 in length, and 0·55 in breadth."

Colonel E. A. Butler furnishes us with the following interesting note:—" Found a nest at Belgaum, containing two fresh eggs, on the 3rd September, 1879. It was situated in the fork of one of the small outer top branches of a tall mango-tree, and was on the whole about the prettiest nest I have seen in India. It consisted of a tiny cup about $1\frac{1}{4} \times 2$ inches measured interiorly, and $1\frac{7}{8} \times 2\frac{1}{4}$ inches exteriorly. Depth inside 1 inch, outside $1\frac{1}{2}$ inches from rim to proper base, excluding about an inch of lichen continued down one side of the bough below the fork in which the nest was built. It was composed, so far as I could judge after a very minute examination, almost entirely of the white lichen which grows so freely on the bark of every tree during the rains, with a few cobwebs incorporated and wound round the outside to keep it together, assimilating so perfectly with the branch upon which it was placed, which was also overgrown with the same kind of lichen, that without watching the old birds closely it never could have been discovered.

"It contained no regular lining, though a few coarse dry leaf-stems of a dark colour were encircled within. I observed the birds building first on the 21st August, and the nest from below looked then almost finished. The cock and hen worked together, flying to and fro very busily with bits of lichen picked off the branches of another tree adjoining. On the 25th I watched the nest for some time, but the birds only came to it once, and then the hen bird went on and smeared some cobwebs round the outside, at least that is what she seemed to me to be doing. On the 28th I watched it again, and although both birds were in the adjoining tree, I did not see them go to the nest. On the 31st, about 10 A.M., I found the hen on the nest, and she remained on till about 10.30, when she flew off and joined the cock, who was sitting pluming himself on a branch of the next tree the whole time she was on the nest. Immediately she joined him, he commenced catching flies and feeding her, as if she were a young bird, and eventually they both flew away together. Arriving at the conclusion that she only went on the nest to lay, I decided on taking the nest three days later, and accordingly returned for that purpose with a small boy on the 3rd Sept., and found, as I expected, the hen sitting and the cock in another tree close by.

"I sent the boy up the tree, and as he approached the nest, which was some 30 or 35 feet from the ground, the hen bird became very uneasy, moving her head from side to side, and looking down to see what was going on below. When the boy was within about 10 feet of the nest she flew off and joined the cock, after which I saw her no more. The eggs were then secured with difficulty, as the branches surrounding the nest were very thin and blown about a good deal by the wind.

"After breaking off the bough, nest and all, the boy descended. One branch of the fork in which the nest was placed was rotten, and broke off at the junction at the base of the nest as the boy was descending the tree; but the nest, which was firmly bound to it with cobwebs, remained in its place and was not injured, and I had the nest and bough beautifully painted for me by a lady friend the same day. The eggs were pale bluish green, speckled and spotted, most densely at the large end, with two shades of dusky purple, the markings of the lighter shade appearing to underlie those of the darker. On the 6th Sept., the same pair of birds commenced a new nest on another mango-tree about 20 yards off. This time it was placed in a fork of one of the small outside lateral branches about 25 feet from the ground, and resembled in every respect the first nest. On the 15th Sept., the hen bird began to sit, and on the 18th I sent a boy up the tree by means of a ladder, and secured two more fresh eggs, similar to those already described.

On this occasion the two old birds evinced signs of the greatest anxiety, the hen remaining on the nest till the boy was close to her, and, joined by the cock immediately she left it, the pair kept flying from bough to bough in the greatest possible state of excitement the whole time the nest was being taken, the hen actually once or twice going on to the nest again after she had left it, when the boy was within 3 feet of her. On examining the nest I found that one of the branches of the fork consisted of a small rotten stump, similar to the one described in the first nest, and in the bottom of both nests there were three or four small black downy feathers, intermingled with the dead leaf-stems that constituted the lining."

In his recent "Notes on Birds'-nesting in Rajpootana," Lieut. H. E. Barnes writes, "The Small Minivet breeds during July and August."

Mr. Benjamin Aitken writes:—"You say that the Small Minivet lays during the latter half of June and throughout July and August. I would therefore remark that on the 11th November, 1871, I saw several newly-fledged young ones at Poona. There could be no mistake about this, as I stood under the tree, which was a small one, and saw the young ones being fed."

Messrs. Davidson and Wenden remark that in the Deccan it is "common, and breeds in the rains."

The latter gentleman subsequently added the following note:—
"In July, my men found a nest with two eggs at Nulwar, Deccan. It was built on a small branch of a tamarind-tree, 20 feet from the ground. The nest is similar to that described in the 'Rough Draft' as being found at Allyghur. The whole of the bark used on the outer coating is that of tamarind-tree, and there are a good many feathers and much down incorporated into the structure, inside and out. The eggs differ considerably in colouring. In both the ground-colour is greenish white. One is profusely speckled all over, but more thickly at the smaller end, with brownish red and a few purple blotches, whilst the other egg has the specks less numerous but larger, and chiefly on the larger end, with little or no purple, and the small end almost unsullied."

Finally, Mr. Oates records that "in Lower Pegu nests of this bird may be found from the end of April to the middle of June."

The eggs are of a rather broad oval shape, and, as is often the case even in the typical Shrikes, very blunt at both ends. The ground-colour is a pale delicate greenish white, and they are more or less richly marked with bright, slightly brownish-red specks, spots, and blotches, which, always more numerous at the large end, have a tendency there to form a mottled irregular cap. In many eggs, besides these primary markings, a number of small faint patches and blotches of pale inky purple are observable, almost exclusively at the large end. The eggs appear to be quite devoid of gloss. I have eggs both of *Copsychus saularis* and *Thamnobia cambaiensis*, strange as it may seem, closely resembling, except in size, some types of this bird's egg; and I have one egg of *Merula simillima* from the Nilghiris, which, though immensely larger, so far as tint, colour, and character of ground and markings go, is positively identical with eggs that I have of this species.

In length the eggs vary from 0·6 to 0·7 inch, and in breadth from 0·5 to 0·56 inch, but the average of twenty-eight eggs is 0·67 nearly by 0·53 inch.

501. Pericrocotus erythropygius (Jerd.).   *The White-bellied Minivet.*

Pericrocotus erythropygius (*Jerd.*), *Jerd. B. Ind.* i, p. 424; *Hume, Cat.* no. 277. ,

Mr. J. Davidson, C.S., is apparently the only ornithologist who has discovered the nest of the White-bellied Minivet. Writing on the 25th August, from Khandeish, he says :—" Yesterday I took two nests of *Pericrocotus erythropygius.* Both nests were like those of *P. peregrinus*, and were placed about 2½ feet from the ground in a fork of a straggling thorn-bush among thin scrub-jungle. One contained 3 young birds, and one 3 hard-set eggs. I watched the nest, and found the cock sitting on the eggs, and watched him for a minute, so there is no possibility of mistake ; but the eggs are not the least what I expected. They are fairly glossy, one being very much elongated, of a greenish-grey ground, with long longitudinal dashes of dark brown, as unlike Minivets' eggs as they can possibly be. They were the only two pairs I saw in a long morning walk, and the nests were easily found by watching the birds. I wish I had known the birds were breeding where they were, as by going three weeks ago I should probably have found many nests, as there are miles and miles of similar jungle, and it is barely 12 miles from Dhulia. It is very provoking. I have had great trouble trying to make the Bhils work for me. They will bring in eggs but not mark them down."

Later on, Mr. Davidson wrote :—" I happened to be staying a few days at Arvee, in the extreme south of Dhulia, and found this bird breeding there in considerable numbers. This was in the end of August (26th to 31st), and I was rather late, most of the nests containing young, and in some cases the young were able to fly. I, however, found eight nests with eggs (most of them hard-set). All the nests, which are small and less ornamented than those of *P. peregrinus*, were placed from 3 to 4 feet from the ground, in a small common thorny scrub. They were all placed in low thin jungle, and never where the jungle was thick and difficult to walk through. A great deal of the jungle round Arvee is full of anjan-trees, but none of the birds seem to breed in these."

The nests are elegant little cups, reminding one of those of *Rhipidura albifrontata*, measuring internally about 1·75 inch in diameter and 1 inch in depth, the thickness of the walls of the nest being usually somewhat less than a quarter of an inch. Interiorly the nest is composed of excessively fine flowering-stems of grasses, and externally and on the upper edge it is densely coated with fine,

rather silky greyish-white vegetable fibres, in places more or less
felted together. It is not ornamented externally with moss and
lichen, as those of so many of the *Pericrocoti* commonly are, only
occasionally one or two little ornamental brown patches of withered
glossy vegetable scales are worked into the exterior of the nest.

The eggs are not at all like those of the other *Pericrocoti* with
which we are best acquainted; though less densely, and even more
streakily marked, they most remind me of the egg of *Volvocivora*,
and in a lesser degree of that of *Hemipus picatus*.

The eggs vary in shape from rather broad to rather elongated
ovals. The shell is very fine and smooth, but has scarcely any
perceptible gloss. The ground-colour is greenish or greyish white,
and they are profusely marked with comparatively fine longitudinal
streaks of a moderately dark brown, which in some lines is more
of a chocolate, in others perhaps more umber. At both ends of
the egg, but especially the smaller end, the markings often become
spotty or speckly, but the fine longitudinal streaking of the sides
of the egg is very conspicuous.

In size the eggs vary from 0·69 to 0·71 in length, by 0·51 to
0·58 in breadth. I have measured too few eggs to be able to
give a reliable average.

505. Campophaga melanoschista (Hodgs.). *The Dark-grey Cuckoo-Shrike.*

Volvocivora melaschistos, *Hodgs., Jerd. B. Ind.* i, p. 415; *Hume, Rough Draft N. & E.* no. 269.

I have never found the nest of the Dark-grey Cuckoo-Shrike.
Captain Hutton tells us :—

" This, too, is a mere summer visitor in the hills, arriving up to
7000 feet about the end of March, and breeding early in May.
The nest is small and shallow, placed in the bifurcation of a horizontal bough of some tall oak tree, and always high up; it is
composed externally almost entirely of grey lichens picked from
the tree, and lined with bits of very fine roots or thin stalks of
leaves. Seen from beneath the tree the nest appears like a bunch
of moss or lichens, and the smallness and frailty would lead one
to suppose it incapable of holding two young birds of such size.
Externally the nest is compactly held together by being thickly
pasted over with cobwebs. The eggs, two in number, of a dull
grey-green, closely and in part confluently dashed with streaks of
dusky brown."

This species, according to Mr. Hodgson's notes and drawings,
breeds in Nepal in the central districts of the hills from April to
July, laying three or four eggs. The nest is a broad shallow
saucer, some 4 inches in external diameter and 1·75 inch in height;
it is placed in a fork where two or three slender branches divide,
to one or more of which it is firmly bound with vegetable fibres
and grass-roots, and is composed of fine roots and vegetable fibres,

and plastered over externally with pieces of lichen and moss. The eggs are regular ovals, with a pale-greenish ground, blotched and spotted with a somewhat olivaceous brown.

A nest of this species found at Mongphoo (elevation 5500 feet) on the 15th June contained three eggs nearly ready to hatch off. The nest was placed on a nearly horizontal fork of a small branch. It is composed of very fine twigs loosely twisted together and coated everywhere exteriorly with cobwebs and scraps of grey lichen. At the lower part, which, owing to the slope of the branch, had to be thicker, it is exteriorly about an inch and a half in height. At the upper end it is only about half an inch high. The shallow saucer-like cavity is about two and a half inches in diameter and about half an inch in depth.

The eggs of this species, sent me by Captain Hutton from Mussoorie, much resemble those of *Graucalus macii* and *C. sykesi*, but they are decidedly longer than the latter, and the general tone of their colouring is somewhat duller. In shape they are somewhat elongated ovals, more or less compressed towards one end; the general colour is greenish white, very thickly blotched and streaked with dull brown and very pale purple. The markings are very closely set, leaving but little of the ground-colour visible. They have little or no gloss.

They measure 1·03 by 0·72 inch, and 0·95 by 0·68 inch.

Other eggs that I have since obtained have been quite similar, but have not had the markings quite so densely set ; the secondary markings have been greyer and less purple, and several eggs have exhibited an appreciable gloss ; others, again, were quite like those first described and entirely devoid of gloss. They measured 0·9 to 0·98 in length by 0·65 to 0·71 in breadth.

508. Campophaga sykesi (Strickl.). *The Black-headed Cuckoo-Shrike.*

Volvocivora sykesii (*Strickl.*), *Jerd. B. Ind.* i, p. 414; *Hume, Rough Draft N. & E.* no. 268.

Mr. F. R. Blewitt took the eggs of Sykes's Cuckoo-Shrike many years ago. He furnishes the following note :—

" I first met with this bird in the southern part of Bundlekund. Nowhere here is it common, and I have never seen more than a pair together. It is to be found in wooded tracts of country, but more frequently among thin large trees surrounding villages. Dr. Jerdon has correctly described its restless habits, and its careful examination of the foliage and branches of trees for food. It is usually a silent bird, but during the earlier portion of the breeding-season the male bird may frequently be heard repeating for minutes together his clear plaintive notes. Each time, as it flies from one tree to another, the song is repeated. The flight is easy, slightly undulating, and the strokes of the wing somewhat rapid. In the latter end of July I procured one nest. It was found on a mowa-

CAMPOPHAGA. 347

tree (*Bassia latifolia*), placed on and at the end of two small out-shooting branches. When my man, mounting the tree, approached the nest the parent birds evinced the greatest anxiety, flew just above his head, uttering all the while a sharply repeated cry. Even when one of the birds was shot the other would not leave the spot, but remained hovering about and uttering its shrill cry. The nest is slightly made, and constructed of thin twigs and roots; the exterior is covered slightly with spider's web. If we except the size, the formation of this Cuckoo-Shrike's nest is almost identical with that of *Graucalus macii*. I secured two eggs in the nest. In colour they are, when fresh, of a deepish green, mottled with dark brown spots; indeed the eggs, when first taken, a good deal resemble those of *Copsychus saularis*. The maximum number of eggs, no doubt, is three, as those I secured were fresh-laid. The bird breeds from June to August."

The nest above referred to, and now in my museum, was a very shallow, rather broad cup. The egg-cavity about 2¼ inches in diameter and about ¾ inch deep, and the nest very loosely put together of very fine twigs, and exteriorly coated and bound together with cobwebs. The sides of the nest are about 0·6 inch thick, but the bottom is a mere network of slender twigs, not above ¼ inch thick, and can be readily looked through.

Mr. I. Macpherson writes:—"This bird is found in the open scrub-forests of the Mysore district, but is nowhere common.

"14th May, 1880.—While passing a small sandal-wood tree a bird flew out, and on looking into the tree I found a very shallow nest at the junction of two small branches about 10 feet from the ground; the nest contained three eggs.

"Returned again in a quarter of an hour and shot the bird (the male) as it flew out of the tree. The eggs were within a few days of being hatched off.

"20th May, 1880.—While out driving this morning saw a male bird of this species fly out of a small sandal-wood tree close to the roadside. Pulled up to watch, and shortly saw the female bird fly into the tree. Got out and shot her and took the nest, which was beautifully fixed in a fork with three branches only eight feet from the ground.

"The nest contained three eggs very hard-set."

Mr. J. Davidson, C.S., remarks:—"This pretty little Cuckoo-Shrike is one of the earliest migrants in the rains, arriving about the 8th of June, and breeding all along the scrub-jungles which stretch between the Nasik and Khandeish Collectorates. It appears particularly partial to the Angan forest, and, as far as I remember, all the many nests I have seen have been in forks of angan trees. The nest is a pretty firm platform composed of fine roots; and the eggs, which much resemble those of the Magpie-Robin, are three in number."

Colonel Legge writes, in his 'Birds of Ceylon':—"With us this Cuckoo-Shrike breeds in April in the Western Province. Mr. MacVicar writes me of the discovery, by himself, of two nests

last year near Colombo. One was built on the topmost branch of
a young jack-tree about 40 feet high. It was very small and
shallow, measuring 2·8 inches in breadth and only 0·8 inch in
depth, and the old bird could be seen plainly from beneath sitting
across it. The other was situated on the top of a tree about
20 feet from the ground, and was built in the same manner. The
materials are not mentioned."

I have only seen two eggs of this species, sent me with the nest
and parent bird by Mr. F. R. Blewitt. They are oval eggs,
moderately broad and obtuse at both ends, about the same size as
average eggs of *Lanius vittatus*. They are slightly glossy, have a
pale greenish-white ground, and are thickly blotched and streaked
throughout, but most densely so towards the large end, with somewhat pale brown, much the same colour as the markings on typical
eggs of *L. erythronotus*. They measure 0·85 inch in length by
0·65 and 0·68 inch in breadth respectively. Other eggs since
received from Calcutta and Mysore measure from 0·87 to 0·81 in
length, and from 0·68 to 0·62 in breadth.

509. Campophaga terat (Bodd.)*. *The Pied Cuckoo-Shrike.*

Lalage terat (*Bodd.*), *Hume, Cat.* no. 269 ter.

The eggs are quite of the *Graucalus* and *Campophaga* type, but
perhaps a little more elongated in shape. Very regular, slightly
elongated ovals, with scarcely any gloss on them, the ground
greenish white, but everywhere thickly streaked and mottled and
freckled over, most thickly about the large end, with a dull pale
slightly olivaceous brown intermingled with brownish, or in some
specimens faintly purplish grey. The two eggs I possess measure
0·85 and 0·87 in length, by 0·61 and 0·62 respectively in breadth.

510. Graucalus macii, Lesson. *The Large Cuckoo-Shrike.*

Graucalus macei, *Less., Jerd. B. Ind.* i, p. 417 ; *Hume, Rough Draft
N. & E.* no. 270.

My friend Mr. F. R. Blewitt seems to be the only ornithologist
who has taken many nests of the Large Grey Cuckoo-Shrike. I
never was so fortunate as to find one. He says :—" This Shrike
begins to pair about May, and in June the work of nidification
commences. The place selected for the nest is the most lofty
branch of a tree, and is built near the fork of two outlying twigs.
If this bird has a preference it would appear to be for mango and
mowa trees, on which I found most of the nests. The nest is in
form circular, and its exterior is somewhat thickly made ; the

* I cannot find any note among Mr. Hume's papers regarding the discovery
of the nest of this bird. The nest may possibly have been found at Camorta
(Nicobar Islands), where this species is not uncommon.—ED.

interior is moderately cup-shaped. Thin twigs and grass-roots are freely used in its construction, while the outer part of the nest is somewhat thickly covered with what appears to be spider's web. Altogether the nest, considering the size of the birds, is of light structure. I am sorry I did not take the dimensions of each nest secured, but I sent you two very perfect ones. I found the first eggs in the beginning of July. They are of a dull lightish green, with brown spots of all sizes, more dense towards the large end. The maximum number of eggs is three. The bird breeds from June to August."

The nests which Mr. Blewitt sent me remind one a good deal of those of the *Dicruri*. They are broad shallow saucers, with an egg-cavity about 3 inches in diameter, and ¾ inch in depth, composed in the only two specimens that I possess of very fine twigs, chiefly those of the furash (*Tamarix orientalis*). Exteriorly they are bound round with cobwebs, in which a quantity of lichen is incorporated. The nests are loose flimsy fabrics, which but for the exterior coating of cobwebs would certainly never have borne removal.

Dr. Jerdon remarks :—" I once obtained its nest and eggs. The nest was built in a lofty casuarina tree, close to my house at Tellicherry; it was composed of small twigs and roots merely, of moderate size, and rather deeply cup-shaped, and contained three eggs, of a greenish-fawn colour, with large blotches of purplish brown."

Professor H. Littledale writing from Baroda says :—" The Large Cuckoo-Shrike is a permanent resident here. I found six nests last August near Baroda, each with one egg ; and my men found a nest building in the Police Lines at Khaira on the 10th October."

Mr. J. Davidson informs us that " a pair of *Graucalus macii* were apparently breeding near this place (the Kondabhari Ghât). He found a nest with two young in the previous September near the same place."

Mr. G. W. Vidal, referring to the South Konkan, says :— " Common ; breeds in February and March."

A nest that was placed in the fork of a bough was composed entirely of slender twigs, the petioles of some pennated-leaved tree, bound together all round the outside with abundance of cobwebs, so that notwithstanding the incoherent nature of the materials the nest was extremely firm. It is a shallow saucer quite of the Dicrurine type, with a cavity 3 inches in diameter and barely 0·75 in depth.

The eggs are typically of a somewhat elongated oval, a good deal pointed towards one end, but some are broader and more of a typical Shrike shape. The eggs are of course considerably larger than those of *Lanius lahtora*. The shell is compact and fine, and faintly glossy. The ground-colour is a palish-green stone-colour, greener in some, and somewhat more creamy in others. The markings are very Shrike-like, and consist of brown blotches, streaks, and spots, with numerous clouds and blotches of pale inky-purple, which appear to underlie the brown markings. The markings in some eggs are all very faint, and, as it were, half washed out, while in others they are very bright and clear. In some these are comparatively sparse and few; in others close-set and numerous, especially in a broad zone near the large end; but this zone is by no means invariably present; in fact, not above one in five eggs exhibit it. There is something in these eggs which reminds one of some of the Terns' eggs; and although, when compared with a large series of *L. lahtora*, individuals of this latter species may be found resembling them to a certain extent, I do not think that at first sight any zoologist would have felt sure that they *were* Shrike's eggs.

They vary in length from 1·12 to 1·41 inch, and in breadth from 0·8 to 0·95 inch, but the average of eight eggs is 1·26 by 0·9 inch nearly.

Subfamily ARTAMINÆ.

512. Artamus fuscus, Vieill. *The Ashy Swallow-Shrike.*

Artamus fuscus, *V., Jerd. B. Ind.* i, p. 441; *Hume, Rough Draft N. & E.* no. 287.

Mr. R. Thompson says:—"I have frequently found the nests of the Ashy Swallow-Shrike, and have watched the old birds constructing them, but never took down their eggs. Two or three pairs may always be found nesting on the long-leaved pine, as one comes up from Kaladoongee to Nyneetal and passes halfway up from the first dâk chokee at Ghutgurh. They lay in May and June, constructing their nest on the horizontal extension of a main branch of some lofty tree, generally *Pinus longifolia*. The nest, composed of fine grasses, roots, and fibres, is a loose, only slightly cup-shaped structure, some 5 inches in diameter."

Dr. Jerdon says on the other hand:—" I have procured the nest of this bird situated on a palmyra tree on the stem of the leaf. It was a deep cup-shaped nest, made of grass, leaves, and numerous feathers, and contained two eggs, white with a greenish tinge, and with light brown spots, chiefly at the larger end. I see that Mr. Layard procured the nest in Ceylon, where this bird is common, in the heads of cocoanut trees, made of fibres and grasses, and it was probably the nest of this bird that was brought to Tickell as that of the Palm-Swift."

According to Mr. Hodgson this species begins to lay in March, the young being fledged in June; the nest is a broad shallow saucer, from 6 to 8 inches in diameter, composed of grass and roots, together with a little lichen, loosely put together, a green leaf or two being sometimes found as a lining to the nest. The nest is placed on some broad horizontal branch, where two or three slender twigs or shoots grow out of it, or on the top of some stump of a tree, or broken end of a branch, generally, at a considerable height from the ground. The eggs are *figured* as white, spotted and blotched almost exclusively at the large end with yellowish brown, and measuring 0·8 by 0·52 inch, but no actual measurements are recorded.

Mr. Gammie, however, himself found, and kindly sent me, a nest and eggs of this species, at Mongpho near Darjeeling, at an elevation of about 3500 feet, on the 13th May, 1873. It was placed in the hole of a trunk of a dead tree at a height of about 40 feet from the ground, and it contained three hard-set eggs. The nest was a loose shallow saucer of coarse roots devoid of lining. The eggs were rather narrow ovals, a good deal pointed towards one end; the shell fine and with a slight gloss. The ground-colour was creamy white, and the markings, which are almost entirely confined to a broad ring round the large end and the space within it, consisted of spots and clouds of very pale yellowish brown, intermingled with clouds and specks of excessively pale, nearly washed out, lilac.

He subsequently furnished me with the following note from Sikhim :—" In the hills this bird is migratory, coming about the last week in February and leaving in the last week of October. It is exceedingly abundant on the outer ridges running in from the Teesta Valley, and most numerous about the elevation of 3000 feet, but stragglers get up as high as 5000 feet. It prefers dry ridges on which there are a few scattered tall trees, from the tops of which it can make short flights, over the open country, after insects. It goes very little abroad in the height of the day, and feeds principally in the evenings. It rarely keeps on the wing for more than a minute or two at a time, but occasionally will fly for ten minutes on end. It is quite as bold and persevering in its habit of attacking and driving off hawks and kites as the king-crow. Towards the end of September it begins to congregate in rows along dead branches in the tops of trees.

" It begins to lay in April and, I think, has only one brood in the year. It builds in holes of trees, on surfaces of large horizontal branches 30 or 40 feet up, or in depressions in ends of lofty stumps. The nest is a shallow saucer, made entirely of light-coloured roots and twigs loosely put together. The usual number of eggs appears to be three."

Mr. J. R. Cripps informs us that at Furreedpore in Eastern Bengal this species is " common, and a permanent resident, very partial to perching on the tips of bamboos, and I have seen as many as 13 sitting side by side on a bamboo tip. I took seven nests this season, all from date-trees (*Phœnix sylvestris*), which trees are very common in the district. The nest is generally built at the junction of the leaf-stem and the trunk of the tree, though in two instances the nest was placed on a ledge from which all leaves had been removed to enable the tree to be tapped for its juice. In every instance the nest was exposed, and if any bird, even a hawk, came near, these courageous little fellows would drive it off. My nests were found from the 5th April to 6th June ; shallow saucers made of fine twigs and grasses with a lining of the same, and contained two to four eggs in each. Height of nest from ground about 12 to 15 feet. On the 17th April I took two fresh eggs from a nest, and the birds laying again, I, on the 8th May, again took three fresh eggs. When on the wing they utter their note, generally returning to the same perch."

And he adds :—

"16th April, 1878.—Took two perfectly fresh eggs from a nest built on a date-tree. The date-trees in this district are tapped annually for the juice, from which sugar is manufactured. The leaves and the bark for a depth of 3 inches are sliced away from one half of the trunk, the leaves on the other half remaining, and at the root of one of these the nest was built, wedged in between the trunk and the leaves; the external diameter was $4\frac{1}{2}$ inches, depth 3 inches, thickness of sides of nest $\frac{3}{4}$ inch ; a rather shallow cup, composed exclusively of fine grasses with no attempt at a lining.

"17th April, 1878.—Secured two fresh eggs from another nest on a date-tree. In size and shape they were similar and the materials were the same grasses with no lining. The trees these nests were on formed a small clump alongside a ryot's house. People were passing under them all day, but the birds never noticed them. Any bird, from a Kite to a Bulbul, coming near received a warm welcome. The nests are at all times exposed, and the natives believe that two males and one female are found occupying one nest. The birds being gregarious build on adjoining trees, and while the ladies are engaged with their domestic affairs their lords keep each other company, so the natives put them down as polyandrous. I have found over a dozen nests, and every one has been the counterpart of the other, and only on date-trees."

Miss Cockburn writes from the Nilghiris :—" On the 17th May, 1873, a nest of this bird was found. It was formed in a perpendicular hole in a dried stump of a tree, about 15 feet in height. The nest consisted entirely of slight sticks lined with fine grass, no soft material being added as a finish, and the whole structure went to pieces when removed. This nest contained three eggs, their colour white, with a few dark and light brown spots and blotches all over, and a strongly marked ring round the thick end.

"The birds frequently returned to the place while the eggs were being taken, till one of them was shot."

Mr. J. Davidson remarks :—" This bird is very local in the Tumkur districts in Mysore, and I have only found it in three or four gardens. I knew it had been breeding (from dissection) since March, but till to-day (May 9th) I could not find its nest. To-day, however, I saw four or five birds perpetually flying round and round a very ragged old cocoanut-tree, the highest in that part of the garden, and determined to send a man up. Two birds, however, at that moment lit on one branch and I shot them both, and they proved to be fully-fledged young ones. I sent the man up, however, and was rewarded by his announcing two old nests and a new one containing one egg. The nests were near the trunk of the tree on the horizontal leaves, and were formed of thin roots and a little grass and were very slight. The egg, which is large for the size of the bird, is creamy white, with a broad ring round the larger end formed of blotches of orange, brown, and purple, and in the cap within the ring there are a number of faint purple spots. The egg was perfectly fresh, and the old birds defended it by swooping down upon the man; and I can't help thinking that both the young birds and the new nest belonged to one pair of birds, and that as soon as their first brood was fledged they had commenced to lay again."

A nest taken by Mr. Gammie on the 24th April, at an elevation of about 3500 feet in Sikhim, was placed on a dead horizontal limb near the top of a large tree. It contained four eggs slightly set; it is a somewhat shallow cup, interiorly 3 inches in diameter by nearly 1½ in depth, and composed almost entirely of fine roots, pretty firmly interwoven. It has no lining, but at the bottom exteriorly it is coated partially with a sort of plaster, composed apparently of strips of bark and vegetable fibre partially cemented together in some way.

The egg sent me by Miss Cockburn is of quite the same type as those found by Mr. Gammie, but it is a trifle longer, measuring 1·0 by 0·7, and the colouring is much brighter. The ground is a sort of creamy white. There is a strongly marked though irregular zone round the large end of more or less confluent brownish rusty patches (amongst which a few pale grey spots may be detected), and a good many spots and small blotches of the same are scattered about the whole of the rest of the surface of the egg.

Numerous eggs subsequently obtained by Mr. Gammie correspond well with those already described as procured by himself and Miss Cockburn.

In length the eggs vary from 0·82 to 1·0, and in breadth from 0·6 to 0·72, but the average is 0·94 by 0·68.

513. Artamus leucogaster (Valenc.). *The White-rumped Swallow-Shrike.*

Artamus leucorhynchus (*Gm.*), *Hume, Rough Draft N. & E.* no. 287 bis.

The White-rumped Swallow-Shrike breeds, we know, in the Andamans and Great Cocos, and that is nearly all we do know. Mr. Davison says:—" On the 2nd of May I saw a bird of this species fly into a hollow at the top of a rotten mangrove stump about 20 feet high. The next day I went, but did not like to climb the stump, as it appeared unsafe, so I determined to cut it down, and after giving about six strokes that made the stump shake from end to end, the bird flew out. I made sure that as the bird sat so close the nest must contain eggs, so I ceased cutting and managed to get a very light native, who volunteered to climb it; but on his reaching the top, he found, to my astonishment,

that the nest, although apparently finished, was empty. The nest
was built entirely of grass, somewhat coarse on the exterior, finer
on the inside ; it was a shallow saucer-shaped structure, and was
placed in a hollow at the top of the stump."

Family ORIOLIDÆ.

518. Oriolus kundoo, Sykes. *The Indian Oriole.*

Oriolus kundoo, *Sykes, Jerd. B. Ind.* ii, p. 107 ; *Hume, Rough Draft
N. & E.* no. 470.

The Indian Oriole breeds from May to August (the great
majority, however, laying in June and July) almost throughout
the plains country of India and in the lower ranges of the Himalayas to an elevation of 4000 feet. In Southern and Eastern
Bengal it only, so far as I know, occurs as a straggler during the
cold season, and I have no information of its breeding there. It
does not apparently ascend the Nilghiris, and throughout the
southern portion of the peninsula it breeds very sparingly, if at
all ; indeed, it is just at the commencement of the breeding-season,
when the mangoes are ripening, that Upper India is suddenly
visited by vast numbers of this species migrating from the south.
    The nest is placed on some large tree, I do not think the bird
has any special preference, and is a moderately deep purse or
pocket, suspended between some slender fork towards the extremity of one of the higher boughs. From below it looks like a
round ball of grass wedged into the fork, and the sitting bird is completely hidden within it ; but when in the hand it proves to be a
most beautifully woven purse, shallower or deeper as the case may
be, hung from the fork of two twigs, made of fine grass and slender
strips of some tenacious bark and bound round and round the
twigs, and secured to them much as a prawn-net is to its wooden
framework. Some nests contain no extraneous matters, but
others have all kinds of odds and ends—scraps of newspaper or
cloth, shavings, rags, snake-skins, thread, &c.—interwoven in the
exterior. The interior is always neatly lined with fine grass-
stems.
    Very commonly the bird so selects the site for its nest that
the leaves of the twigs it uses as a framework form more or less
of a shady canopy overhead ; in fact, as a rule, it is from very few
points of view that even a passing bird of prey can catch sight of
the female on her eggs. Possibly the brilliant plumage of the
bird (which has endowed it amongst the natives with the name of
Peeluk, or " The Yellow One ") may have had something to do
with the concealment it so generally affects.
    The nests vary a good deal in size. I have seen one with an

internal cavity $3\frac{1}{2}$ inches in diameter and over $2\frac{1}{2}$ deep. I have seen others scarcely over $2\frac{1}{2}$ inches in diameter and not 2 in depth, which you could have put bodily, twigs and all, inside the former. As a rule, the purse is strong and compact, the material closely matted and firmly bound together; but I have seen very flimsy structures, through which it was quite possible to see the eggs.

Four is the greatest number of eggs I have ever found in one nest, but it is quite common to find only three well-incubated ones.

Colonel C. H. T. Marshall reports having found several nests of this species about Murree at low elevations.

Mr. W. Blewitt tells me that he obtained two nests near Hansie on the 1st and 14th July respectively. The nests (which he kindly sent) were of the usual type, and were placed, the one on an acacia, the other on a loquat tree, at heights of 10 and 12 feet from the ground. Each contained three eggs, the one clutch much incubated, the other perfectly fresh.

Dr. Scully writes :—"The Indian Oriole is a seasonal visitant to the valley of Nepal, arriving about the 1st of April and departing in August. It frequents some of the central woods, gardens, and groves, and breeds in May and June."

Colonel J. Biddulph remarks regarding the nidification of this Oriole in Gilgit :—"A summer visitant and common. Appears about the 1st of May. Nest with three eggs hard-set, taken 8th of June; several other nests taken later on."

Writing from near Rohtuk, Mr. F. R. Blewitt says :—"The breeding-season is from the middle of May to July. The nest is made on large trees, and always suspended between the fork of a branch. I have certainly obtained more nests from the tamarind than any other kind of tree.

"The nest is cup-shaped, light, neat, and compact. The average outer diameter is 4·8 inches ; the inner or cup-cavity about 3·6. Hemp-like fibre is almost exclusively used in the exterior structure of the nest, and by this it is firmly secured to the two limbs of the fork. Cleverly indeed is this work performed, the hemp being well wrapped round the stems and then brought again into the outer framework. Occasionally bits of cloth, thread pieces, vegetable fibres, &c. are introduced. On one occasion I got a nest with a cast-off snake-skin neatly worked into the outer material.

"The lining of the egg-cavity is simply fine grass, if we except the occasional capricious addition of a feather or two, an odd piece of cotton or rag, &c. Three appears to be the regular number of eggs. This bird is to be found in small numbers all over the country here ; its habits are well described by Jerdon. It is, as I have observed, hard to please in its choice of a nest site. I have watched it for days going backwards and forwards, from tree to tree and from fork to fork, before it made up its mind where to commence work."

23*

Capt. Hutton records that " this is a common bird in the Dhoon, and arrives at Jerripanee, elevation 4500 feet, in the summer months to breed. Its beautiful cradle-like nest was taken in the Dhoon on the 29th of May, at which time it contained three pure white eggs, sparingly sprinkled over with variously sized spots of deep purplish-brown, giving the egg the appearance of having been splashed with dark mud. The spots are chiefly at the larger end, but there is no indication of a ring. The nest is a slight, somewhat cup-shaped cradle, rather longer than wide, and is so placed, between the fork of a thin branch, as to be suspended between the limbs by having the materials of the two sides bound round them. It is composed of fine dry grasses, both blade and stalk, intermixed with silky and cottony seed-down, especially at that part where the materials are wound round the two supporting twigs ; and in the specimen before me there are several small silky cocoons of a diminutive *Bombyx* attached to the outside, the silk of which has been interwoven with the fibres of the external nest. It is so slightly constructed as to be seen through, and it appears quite surprising that so large a bird, to say nothing of the weight of the three or four young ones, does not entirely destroy it."

From Futtehgurh, the late Mr. A. Anderson remarked :—" The nest and eggs of this bird so closely resemble those of its European congener (*O. galbula*) that little or no description is necessary. The Mango-bird lays throughout the rains, July being the principal month. One very beautifully constructed nest was taken by me on the 9th July, 1872, containing four eggs, which, according to my experience, is in excess of the number usually laid. I have frequently taken only a pair of well-incubated eggs.

" Two of the four eggs above alluded to were quite fresh, while the other two were tolerably well incubated. The nest is fitted outwardly with tow, which I have never before seen. One of the pieces of cloth used in the construction of this nest was 6 inches long."

" At Lucknow," writes Mr. R. M. Adam, " I found this species on the 20th May building a nest in a neem-tree, and on the 24th I took two eggs from the nest. On the 10th June I saw another pair, only making love, so they probably did not lay till the end of that month."

Dr. Jerdon notes that he " procured a nest at Saugor from a high branch of a banian tree in cantonments. It was situated between the forks of a branch, made of fine roots and grass, with some hair and a feather or two internally, and suspended by a long roll of cloth about three quarters of an inch wide, which it must have pilfered from a neighbouring verandah where a tailor worked. This strip was wound round each limb of the fork, then passed round the nest beneath, fixed to the other limb, and again brought round the nest to the opposite side ; there were four or five of these supports on either side. It was indeed a most curious nest, and so securely fixed that it could not have been removed till the supporting bands had been cut or rotted away. The eggs were white, with a few dark claret-coloured spots."

Major Wardlaw Ramsay says, writing from Afghanistan :—"At Shalofyan, in the Kurrum valley, in June, I found them in great numbers : some were breeding ; but as I saw quite young birds, it is probable that the nesting-season was nearly over."

Colonel Butler contributes the following note:—"The Indian Oriole breeds in the neighbourhood of Deesa in the months of May, June, and July. I took nests on the following dates :—

" 24th May, 1876. A nest containing 1 fresh egg.
" 29th   ,,   ,,   ,,   ,,   3 fresh eggs.
" 12th June ,,   ,,   ,,   2 much incubated eggs.
" 12th   ,,   ,,   ,,   ,,   3 fresh eggs.
" 13th   ,,   ,,   ,,   ,,   2   ,,
" 19th   ,,   ,,   ,,   ,,   3   ,,
" 29th   ,,   ,,   ,,   ,,   2   ,,
" 29th   ,,   ,,   ,,   ,,   2   ,,
" 29th   ,,   ,,   ,,   ,,   3   ,,
" 3rd July   ,,   ,,   ,,   2   ,,
" 6th   ,,   ,,   ,,   ,,   3   ,,
" 30th   ,,   ,,   ,,   ,,   2   ,,

"The nest found on the 24th May was suspended from a small fork of a neem-tree about ten feet from the ground, and was very neatly built of dry grass (fine interiorly, coarse exteriorly), old rags, and cotton (woven, not raw). The rim was firmly bound to the branches of the fork with rags and coarse blades of dry grass. It is an easy nest to find when the birds are building, as both birds are always together and keep constantly flying to and from the nest with materials for building. The cock, as before mentioned, always accompanies the hen to and from the nest whilst she is building ; but I do not think he assists in its construction, as I never saw him carrying any of the materials, neither have I ever seen him on the nest. On the contrary, whilst the hen is at the nest building he is generally waiting for her, either on the same tree or else on another close by, occasionally uttering his well-known rich mellow note. On the 29th May I sent a boy up a tree to examine a nest. The hen bird had been sitting for a week, and was on the nest when the boy ascended the tree. The cock bird flew past, and being a brilliant specimen I shot him, thinking of course that the nest contained a full complement of eggs. To my astonishment, however, though the hen bird sat very close, there were no eggs in the nest, and although she returned to it once or twice afterwards, she eventually forsook it without laying. Possibly she may have laid, and that the eggs were destroyed by Crows. In addition to the materials already mentioned, this nest was also composed of tow, string, and strips of paper, all neatly woven into the exterior, and many of the other nests mentioned were exactly similar; sometimes I have found pieces of snake-skin woven into the exterior.

"On the 9th of July I observed a pair of Orioles building on a neem-tree in one of the compounds in Deesa. When the nest was nearly finished a gale of wind rose one night and scattered it all over the bough it was fixed to. The birds at once commenced to remove it, and in a couple of days carried off every particle of it to another tree about 100 yards off, upon which they built a new nest of the materials they had removed from the other tree. I ascended the tree on the 17th of July, and found it contained three fresh eggs.

"The eggs are pure white, sparingly spotted with moderately-sized blackish-looking spots, if washed the spots run. They vary a good deal in shape and size, some being very perfect ovals, others greatly elongated, &c."

Major C. T. Bingham writes :—"The Indian Oriole builds at Allahabad and at Delhi from the beginning of April to the end of July. In the cold weather this bird seems to migrate more or less, as but few are seen and none heard during that season. The nests are built generally at the top of mango-trees and well concealed; they are constructed of fine grass, beautifully soft, mixed with strips of plaintain-bark, with which, or with strips of cotton cloth purloined from somewhere, the nest is usually bound to a fork in the branch. The egg-cavity is pretty deep, that is to say from $1\frac{1}{2}$ to 3 inches."

Mr. George Reid records the following note from Lucknow :— "The Mango-bird, or Indian Oriole, though a permanent resident, is never so abundant during the cold weather as it is during the hot and rainy seasons from about the time the mango-trees begin to bloom to the end of September. It frequents gardens, avenues, mango-topes, and is frequently seen in open country, taking long flights between trees, principally the banian and other *Fici*, upon the berries and buds of which it feeds. I have the following record of its nests :—

"June 16th. Nest and no eggs (building).
"July 2nd. 2 eggs (fresh).
"July 2nd. 1 egg (fresh).
"July 5th. 3 eggs (fresh).
"July 25th. 3 young (just hatched).
"August 5th. 2 young (fledged)."

Messrs. Davidson and Wenden, writing of this bird in the Deccan, say :—"Common, and breeds in June and July."

Colonel A. C. McMaster informs us that he "found several nests of this bird at Kamptee during June and July; they corresponded exactly with Jerdon's admirable description. Has any writer mentioned that this bird has a faint, but very sweet and plaintive song, which he continues for a considerable time? I have only heard it when a family, old and young, were together, i. e. at the close of the breeding-season."

Lieut. H. E. Barnes, writing of Rajpootana in general, tells us that this Oriole breeds during July and August.

Mr. C. J. W. Taylor, speaking of Manzeerabad in Mysore, says:—" Abundant in the plains. Rare in the higher portions of the district. Breeding in June and July."

The eggs are typically a moderately elongated oval, tapering a good deal towards one end, but they vary much in shape as well as size. Some are pyriform, and some very long and cylindrical, quite the shape of the egg of a Cormorant or Solan Goose, or that of a Diver. They are always of a pure excessively glossy china-white, which, when they are fresh and unblown, appears suffused with a delicate salmon-pink, caused by the partial translucency of the shell. Well-defined spots and specks, typically black, are more or less thinly sprinkled over the surface of the egg, chiefly at the large end. Normally, as I said, the spots are black and sharply defined, and there are neither blotches nor splashes, but numerous variations occur. Sometimes, as in an egg sent me by Mr. Nunn, all the spots are pale yellowish brown. Sometimes, as in an egg I took at Bareilly, a few spots of this colour are mingled with the black ones. Deep reddish brown often takes the place of the typical black, and the spots are not very unfrequently surrounded by a more or less extensive brownish-pink nimbus, which in one egg I have is so extensive that the ground-colour of the whole of the large end appears to be a delicate pink. Occasionally several of the clear-cut spots appear to run together and form a coarse irregular blotch, and one egg I possess exhibits on one side a large splash. The eggs as a body, as might have been expected, closely resemble those of the Golden Oriole, to which the bird itself is so nearly related; and as observed by Professor Newton in regard to the eggs of that species, so in *my* large series, the prevalence of greatly elongated examples is remarkable.

The eggs vary in length from 1·03 to 1·32, and from 0·75 to 0·87 in breadth; but the average of fifty eggs measured was 1·11 by 0·81.

521. Oriolus melanocephalus (Linn.). *The Indian Black-headed Oriole.*

Oriolus melanocephalus, *Linn., Jerd. B. Ind.* ii, p. 110; *Hume, Rough Draft N. & E.* no. 472.
Oriolus ceylonensis, *Bonap., Jerd. B. Ind.* ii, p. 111.

I have already noticed ('Stray Feathers,' vol. i, p. 439) how impossible it is to draw any hard-and-fast line, in practice, between this the so-called " Bengal Black-headed Oriole " and the supposed distinct southern species, *O. ceylonensis*, Bp.

The present species certainly breeds in suitable (*i. e.* well-wooded and not too bare or arid) localities throughout Northern and Central India, Assam, and Burma, and I have specimens from Mahableshwar, from the Nilgiris, and even Anjango, that are

nearer to typical *O. melanocephalus* than to typical *O. ceylonensis.*
Of its nidification southwards I know nothing. I have only
myself taken its eggs in the neighbourhood of Calcutta.

It appears to lay from April to the end of August. The nest
of this species, though perhaps slightly deeper, is very much like
that of *O. kundoo;* it is a deep cup, carefully suspended between
two twigs, and is composed chiefly of tow-like vegetable fibres,
thin slips of bark and the like, and is internally lined with very
fine tamarisk twigs or fine grass, and is externally generally more
or less covered over with odds and ends, bits of lichen, thin flakes
of bark, &c. It is slightly smaller than the average run of the
nests of *O. kundoo.* The egg-cavity measures about 3 inches in
diameter and nearly 2 inches in depth. I myself have never found
more than three eggs, but I daresay that, like *O. kundoo,* it may
not unfrequently lay four.

The late Captain Beavan writes :—"A nest with three eggs,
brought to me in Manbhoom on 5th April, 1865, is cup-shaped ;
interior diameter 3·5, depth inside 2 inches. It is composed
outside of woolly fibres, flax, and bits of dried leaves, and inside of
bents and small dried twigs, the whole compact and neat. The
eggs are of a light pink ground (almost flesh-coloured), with a few
scattered spots of brownish pink, darker and more numerous at
the blunt end. They measure 1·125 by barely 0·8."

From Raipoor, Mr. F. R. Blewitt remarks :—"*Oriolus melanocephalus* indiscriminately selects the mango, mowah, or any other
kind of large tree for its nest, which is invariably firmly attached
to the extreme terminal twigs of an upper horizontal branch, varying from 20 to 35 feet from the ground, Owing to the position
it selects for the safety of its nest, it sometimes happens that the
latter cannot be secured without the destruction of the eggs. It
nidificates in June and July, and it would appear that both the
birds, male and female, engage in the construction of the nest.
Three is the normal number of the eggs, though on one occasion
my shikaree found four in a nest."

Buchanan Hamilton tells us that this species "frequents the
groves and gardens of Bengal during the whole year, and builds a
very rude nest of bamboo-leaves and the fibres that invest the top
of the cocoanut or other palms. In March I found a nest with
the young unfledged."

I confess that I believe this to be a mistake : neither season nor
nest correspond with what I have myself seen about Calcutta.
The nests, so far from being *rude,* are very neat.

Mr. J. R. Cripps writes from Furreedpore in Eastern Bengal :—
" Very common, and a permanent resident. On the 20th April I
found a nest containing two half-fledged young ones ; in the garden
was a clump of mango-trees, and attached to one of the outer twigs,
but overhung by a lot of leaves, and about 12 feet from the
ground, hung the nest, of the usual type."

Mr. J. Davidson met with this Oriole on the Kondabhari Ghât
in Khandeish. On the 16th August he saw a brood, while on an

adjoining tree there was a nest with two slightly-set eggs. He says :—" It was a very deep cup on the end of a thin branch, and though in cutting the branch to get at the nest, it got turned at right angles to its proper position, the eggs were uninjured. I do not think this nest belonged to the same pair as that which had young ones flying.

" These Orioles are very common here, and I found three nests ; one was new and empty ; from another the birds had just flown ; while the remaining one contained one fresh egg. The bird would no doubt have laid more ; but to get at the nest I had to cut the branch off, and it was only then I discovered that only one egg had been laid."

Major C. T. Bingham says :—" Plentiful at Allahabad across the Ganges, notwithstanding which I only found one nest, and that I have no note about, but I remember it was some time in June, and contained four half-fledged young ones ; the materials of the nest were the same as those used by *O. kundoo.*"

Writing of his experience in Tenasserim he adds :—" On the 5th March I found a nest of this bird in a small tree near the village of Hpamee. It, however, contained three unfledged young, so I left it alone.

" On the 21st April I found a second nest suspended from the tip of a bamboo that overhung the path from Shwaobah village to Hpamee. This contained two awfully hard-set eggs, white, with a few dark purple blotches and spots at the larger ends. Nest made of grass and dry bamboo-leaves, lined with the dry midribs of leaves, and firmly bound on to the fork of the bamboo with a strip of some bark."

Mr. Oates writes from Pegu :—" My nests of this Oriole have been found in March, April, and May, but I have no doubt they also breed in June. No details appear necessary."

Typically the eggs are somewhat elongated ovals, only slightly compressed towards one end, but pyriform as well as more pointed varieties may be met with. The shell is very fine and moderately glossy. The ground-colour varies from a creamy or pinky white to a decided but very pale salmon-colour. They are sparingly spotted and streaked with dark brown and pale inky purple. In most eggs the markings are more numerous towards the large end. Some have no markings elsewhere. The dark spots, especially towards the large end, are not unfrequently more or less enveloped in a reddish-pink nimbus. Though much larger and much more glossy, some of the eggs, so far as shape, colour, and markings go, exactly resemble some of the eggs of *Dicrurus ater*. The eggs of *O. kundoo* are typically excessively glossy china-white, with few well-defined black spots. The eggs of *O. melanocephalus* are typically somewhat less glossy, with a pinky ground and more numerous and less defined brownish-purple spots and streaks. I have not yet seen one egg of either species that could be mistaken for one of the other, although of course abnormal varieties of each approach each other more closely than do the typical forms.

The dozen eggs that I possess of this species vary from 1·1 to 1·2 in length, and from 0·78 to 0·87 in breadth, and the average is 1·14 by 0·82. Although the average is somewhat larger than that of the preceding species, and although none of the eggs are quite *as* small as many of those of *O. kundoo*, still none are nearly so large as the finest specimens of the latter's egg. Probably had I an equally large series of the eggs of the present species, we should find that as regards size there was no perceptible difference between the two.

522. Oriolus traillii (Vigors).   *The Maroon Oriole.*

Oriolus traillii (*Vig.*), *Jerd. B. Ind.* ii, p. 112; *Hume, Cat.* no. 474.

From Sikhim Mr. Gammie writes:—" I took a nest of this Oriole on the 24th April, at an elevation of about 2500 feet. It was suspended, within ten feet of the ground, from an outer fork of a branch of a small leafy tree, which grew in a patch of low dense jungle. It is a neat cup, composed of fibrous bark and strips of the outer part of dry grass-stems, intermixed with skeletonized leaves and green moss, and lined with fine grass. Besides being firmly bound by the rim of the cup to the horizontal forking branches by fibrous barks, several strings extended from one branch to the other, both under and in front of the nest, while other strings from the body of the nest were fastened to an upright twig that rose immediately behind the fork, thus most securely retaining it in its position.

" Externally the nest measured 5 inches wide by 2·75 in height; internally 3·25 wide by 2 deep. It contained three fresh eggs.

" The female came quite close, making loud complaints against the robbing of her nest."

The nest is that of a typical Oriole, usually very firmly and substantially built, and of course always suspended at a fork between two twigs. A nest taken by Mr. Gammie in Sikhim on the 20th April, at an elevation of about 2500 feet, is a deep substantial cup, nearly 4 inches in diameter and 2½ in depth internally. It is everywhere nearly an inch in thickness. The suspensory portion composed of vegetable fibres; towards the exterior dead leaves, bamboo-sheaths, green moss, and tendrils of creeping plants are profusely intermingled; interiorly, it is closely and regularly lined with very fine grass.

A nest sent me by Mr. Mandelli was found on the 3rd April at Namtchu, and contained three fresh eggs. It is precisely similar to the one above described, except that in the lining roots are mingled with the fine grass, and that instead of being suspended in a fork, it was partly wedged into and partly rested on a fork.

As a rule, however, as I know from other nests subsequently obtained, the nests are always suspended like those of the Common Oriole.

Two eggs of this species obtained by Mr. Gammie closely

resemble those of *O. melanocephalus*. In shape they are regular moderately elongated ovals; the shell is strong, firm, and moderately glossy. The ground is white with a creamy or brownish-pink tinge; the markings are blackish-brown spots and specks, almost confined to a zone about the large end, where they are all more or less enveloped in a brownish-red haze or *nimbus*. In length they measure 1·12 by 0·82, and 1·14 by 0·83.

Family EULABETIDÆ.

523. Eulabes religiosa (Linn.). *The Southern Grackle.*

Eulabes religiosa (*Linn.*), *Jerd. B. Ind.* ii, p. 337; *Hume, Rough Draft N. & E.* no. 692.

The Southern Grackle breeds in Southern India and Ceylon from March to October.

Mr. Frank Bourdillon, writing from Travancore, gives me the following account of the eggs. He says :—" This bird, an abundant resident, lays a blue egg pretty evenly marked with brown spots, some light and some darkish, in a nest of straw and feathers in a hole of a tree generally a considerable height from the ground.

" I have only taken one nest, which contained a single egg slightly set, on 23rd March, 1873, the egg measuring 1·37 long and 0·87 broad."

Later Mr. Bourdillon says :—" Since writing the foregoing I took on 21st April two fresh eggs from the nest of a Southern Hill-Mynah (*Eulabes religiosa*). The nest was of grass, feathers, and odds and ends in a hole in a nânga (*Mesua coromandeliana*) stump, about 25 feet from the ground. The eggs of this Mynah are blue, with purplish and more decided brown spots.

" I am *positive* as to the identity of the egg. Both the eggs taken last year and the two taken the other day were obtained under my personal supervision. In both instances I watched the birds building, and when we robbed the nests saw the female fly off them."

These two eggs sent me by Mr. Bourdillon are very beautiful. In shape they are very gracefully elongated ovals; the shell is very fine and smooth, but has only a rather faint gloss. The ground-colour is a delicate pale sea-green or greenish blue, and the eggs are more or less profusely spotted or splashed with purplish, or, in some spots, chocolate-brown and a very pale purple, which looks more like the stain that might be supposed to be left by one of the more decided coloured markings that had been partially washed out than anything else.

The eggs measure 1·37 by 0·9 and 1·35 by 0·87.

Mr. J. Darling, junior, writes :—" The Southern Grackle breeds in the S. Wynaad rather plentifully, and I have had numbers of

tame ones brought up from the nest, but have never succeeded
in getting a perfect egg owing to my having found all the nests in
very hard places to get at.

"I cut down a tree containing a nest and broke all the eggs,
which must have been very pretty—blue ground, very regularly
marked with purplish-brown spots. The nest was composed of
sticks, twigs, feathers, and some snake-skin. I have found them
in March, April, September, and October. I hope this year to
get a number of eggs, as Culputty is a very good place for
them."

Mr. C. J. W. Taylor notes from Manzeerabad in Mysore:—
"Common up in the wooded portions of the district. Breeding in
April and May."

Mr. T. Fulton Bourdillon, speaking of this Grackle in Travancore, says :—"This bird lays one or two light blue eggs beautifully
blotched with purple in the holes of trees. It does not like heavy
jungle, but after a clearing has been felled and burnt it is sure to
appear. During the fine weather it is very abundant on the hills,
descending to the low country at the foot when the rains have
fairly set in. The nest scarcely deserves the name, being only
a few dead leaves or some powdered wood at the bottom of the
hole, and there about the end of March the egg or eggs are laid.
The young birds, which can be taught to speak and become very
tame, are often taken by the natives, as they can sell them in the
low country. I have obtained on the following dates eggs and
young birds :—

"March 29th. One egg slightly set.
"April 20th. Two young birds.
"April 22nd.   „   „
"April 25th. Two eggs slightly set.
"May 2nd. One young bird.

"I also had three eggs, slightly set, brought me on May 21.
They are rather smaller and a deeper blue than the ones obtained
before, being 1·25 × 1, 1·19 × ·95, 1·21 × ·97 inch. They were all
out of the same nest, so that the bird sometimes lays three eggs,
though the usual number is two."

Colonel Legge writes in the 'Birds of Ceylon':—"The Black
Myna was breeding in the Pasdun Korale on the occasion of a
visit I made to that part in August, but I did not procure its
eggs."

Other eggs subsequently sent me by Mr. Bourdillon from
Mynall, in Southern Travancore, taken on the 9th and 13th April,
1875, are precisely similar to those already described. The eggs
that I have measured have only varied from 1·2 to 1·37 in length,
and from 0·86 to 0·9 in width.

524. **Eulabes intermedia** * (A. Hay). *The Indian Grackle.*

Eulabes intermedia (*A. Hay*), *Jerd. B. Ind.* ii, p. 339.
Eulabes javanensis (*Osbeck*), *Hume, Rough Draft N. & E.* no. 693.

The Indian Grackle, under which name I include *E. andamanensis*, Tytler, breeds, I know, in the Nepal Terai and in the Kumaon Bhabur; and many are the young birds that I have seen extracted by the natives out of holes, high up in large trees, in the old anti-mutiny days when we used to go tiger-shooting in these grand jungles. I never saw the eggs however, which, I think, must have all been hatched off in May, when we used to be out.

"In the Andamans," writes Davison, "they breed in April and May, building a nest of grass, dried leaves, &c. in holes of trees." He also, however, never took the eggs.

Mr. J. R. Cripps tells us that this species is "common during March to October in Dibrugarh, after which it retires to the hills which border the east and south of the district. About the tea-gardens of Dibrugarh there are always a number of dead trees standing, and in these the Grackles nest, choosing those that are rotten, in which they excavate a hole. I have seen numbers of nests, but as these were so high up and the tree so long dead and rotten, no native would risk going up."

Mr. J. Inglis notes from Cachar :—"This Hill-Mynah is common in the hilly district. It breeds in the holes of trees during April, May, and June."

Major C. T. Bingham writes from Tenasserim :—"I saw several nest-holes of this bird, which was very common in the Reserve, but none of them were accessible; and it wasn't till the 18th April that I chanced on one in a low tree, the nest being in the hollow of a stump of a broken branch. It was composed and loosely put together of grass, leaves, and twigs, and contained three half-fledged young and one addled egg of a light blue colour, spotted, chiefly at the large end with purplish brown."

The eggs very similar to those of *E. religiosa*, but, what is very surprising, it is very considerably *smaller*.

Of *E. religiosa* the eggs vary from 1·2 to 1·37 in length, and from 0·86 to 0·9 in breadth, and the average of eight is 1·31 by 0·88.

This present egg only measures 1·12 by 0·8, and it must, I should fancy, be abnormally small.

In shape it is an extremely regular oval. The ground is a pale greenish blue, and it is spotted and blotched pretty thickly at the

---

* Mr. Hume does not recognize *E. javanensis* and *E. intermedia* as distinct. The following account refers to the nidification of the latter, except perhaps Major Bingham's later note, in which he states that he procured two distinct sizes of eggs in the Meplay valley (Thoungyeen). It is very probable that Major Bingham found the nests of both species on this occasion. I have seen no specimen of *E. javanensis* from the Thoungyeen valley, but at Malewun, further south, it occurs along with *E. intermedia.*—ED.

large end (where all the larger markings are) and very thinly at the smaller end with purple and two shades (a darker and lighter one) of chocolate-brown, the latter colour much predominating. The shell is very fine and close, but has but little gloss.

And later on Major Bingham again wrote:—"One of the commonest and most widely spread birds in the province. The following is an account of its nidification:—

"This bird lays two distinct sizes of eggs, all, however, of the same type and coloration. Out of holes in neighbouring trees, on the bank of the Meplay, on the 13th March, 1880, I took two nests, one containing three, and the other two eggs. The first lot of eggs measured respectively 1.15×0.77, 1.15×0.80, and 1.16×0.79 inch; while those in the second nest 1.30×0.95, and 1.27×0.93 inch respectively. All the eggs, however, are a pale blue, spotted chiefly at the larger end with light chocolate. The nests were in natural hollows in the trees, and lined with grass and leaves loosely put together."

The eggs apparently vary extraordinarily in size; they are generally more or less elongated ovals, some slightly pyriform and slightly obtuse at both ends, some rather pointed towards the small end. The shell in all is very fine and compact and smooth, but some have scarcely any appreciable gloss, while others have a really fine gloss. The ground-colour is pretty uniform in all, a delicate pale greenish blue. The markings are always chiefly confined to one end, usually the broad end; even about the large end they are never very dense, and elsewhere they are commonly very sparse or almost or altogether wanting. In some eggs the markings are pretty large irregular blotches mingled with small spots and specks, but in many eggs again the largest spot does not exceed one twelfth of an inch in diameter. In colour these markings are normally a chocolate, often with more or less of a brown tinge, in some of the small spots so thickly laid on as to be almost black, in many of the larger blotches becoming only a pale reddish purple, or here and there a pale purplish grey. In some eggs all the markings are pale and washed out, in others all are sharply defined and intense in colour. Occasionally some of the smaller spots become almost a yellowish brown.

526. Eulabes ptilogenys (Blyth). *The Ceylon Grackle.*

Eulabes ptilogenys (*Bl.*), *Hume, Cat.* no. 693 bis.

Colonel Legge writes in his 'Birds of Ceylon':—"This species breeds in June, July, and August, laying its eggs in a hole of a tree, or in one which has been previously excavated by the Yellow-fronted Barbet or Red Woodpecker. It often nests in the sugar- or kitool-palm, and in one of these trees in the Peak forest I took its eggs in the month of August. There was an absence of all nest or lining at the bottom of the hole, the eggs, which were two in number, being deposited on the bare wood. The female was sitting at the time, and was being brought fruit and berries by the male bird. While the eggs were being taken the birds flew round repeatedly, and settled on an adjacent tree, keeping up a loud whistling. The eggs are obtuse-ended ovals, of a pale greenish-blue ground-colour (one being much paler than the other), sparingly spotted with large and small spots of lilac-grey, and blotched over this with a few neutral-brown and sepia blots. They measure from 1·3 to 1·32 inch in length by 0·96 to 0·99 in breadth."

527. Calornis chalybeïus (Horsf.). *The Glossy Calornis.*

Calornis chalybaeus* (*Horsf.*), Hume, Cat. no. 690 bis.

Of the Glossy Calornis Mr. Davison remarks that "it is a permanent resident at the Nicobars, breeding in holes in trees and in the decayed stumps of old cocoanut-palms, apparently from December to March. At the Andamans it is much less numerous, and is only met with in pairs or in small parties, frequenting the same situations as it does in the Nicobars."

Mr. J. Inglis writes from Cachar :—" This Tree-Stare is rather rare. It breeds about April in the holes of dead trees ; when the young are able to fly it departs. It again returns about the middle of February."

In Tenasserim this species was observed nesting by Mr. J. Darling, junior, who says :—" 22nd March. Noticed several pairs of *Calornis*, with nests, in the big wooden bridge over the Kyouk-tyne Creek about 1½ mile out of Tavoy, and also a great number of their nests in the old wooden posts of an old bridge further down the Creek."

Mr. W. Davison, when in the Malay peninsula, took the eggs of this bird. He remarks :—" I found a few pairs frequenting some areca-palms at Laugat, and breeding in them, but only one nest contained eggs, three in number. The nest was a loose structure almost globular, but open at the top, composed externally of very coarse dry grass (lalung or elephant-grass), and lined with green durian leaves cut into small bits. The nest was too lightly put together to preserve. This nest and several other empty ones were placed at the base of the leaves where they meet the trunk.

" The three eggs obtained were slightly set, so that three is probably the normal number laid.

" I noticed several other pairs breeding at the same time in holes of a huge dead tree on Jugra Hill at Laugat, but I was unable to get at the nests."

The eggs are quite of the *Eulabes* type, moderately broad ovals, more or less compressed towards the small end, occasionally pyriform. The shell firm and strong, though fine, smooth to the touch in some cases, with but little, but generally with a fair amount of

---

* Mr. Hume considers the Andaman *Calornis* distinct from the *Calornis* inhabiting Cachar, Tenasserim, &c. I have united them in the 'Birds of India.'—ED.

gloss. The ground is a very pale greenish blue. A number of fairly large spots and blotches, intermingled with smaller specks and spots, are scattered about the large end, often forming an imperfect irregular zone, and a few similar specks and spots are scattered thinly about the central portion of the egg, occasionally extending to the small end. The colour of these spots varies; they are generally a brownish-reddish purple and a paler greyer purple, but in some eggs the spots are so thick in colour that they seem almost black. In some they are almost purely reddish brown without any purplish tinge, and some again, lying deep in the shell, are pale grey.

Six eggs measure from 0·92 to 1·1 in length, and from 0·71 to 0·76 in breadth, but the average of six eggs is 1 by 0·74.

Family STURNIDÆ.

528. Pastor roseus (Linn.). *The Rose-coloured Starling.*

Pastor roseus (*Linn.*), *Jerd. B. Ind.* ii, p. 333; *Hume, Cat.* no. 690.

The Rose-coloured Starling has not yet been discovered breeding in India, but Mr. Doig has written the following note on the subject, which is one of great interest. He writes from the Eastern Narra, in Sind :—

"Though I have not as yet discovered the breeding-place of this bird, I think it as well to put on record what little I have noticed, in the hope that it may be of assistance in eventually finding out where it goes to breed. I began watching the birds in the middle of April, and every week shot one or two and dissected them, but did not perceive any decisive signs of their breeding until the 10th May, when I shot two males, both of which showed signs of being about to breed at an early date. Again, on the 15th May, out of seven that I shot in a flock, six were males with the generative organs fully developed; the seventh was a young female in immature plumage, the ovaries being quite undeveloped. The birds were feeding in the bed of a dried-up swamp, along with flocks of *Sturnus minor*, and were constantly flying in flocks, backwards and forwards, in one direction. Unfortunately, important work called me to another part of the district, and when I returned in a fortnight's time I could not see one. Where can they have gone? And they remain away such a short time! I have seen the old birds return as early as the 7th July, accompanied by young birds barely fledged, and I should not be at all surprised if these birds are found to breed in some of the Native States on the *east* of Sind. That they could find time to migrate to the Caspian Sea and Central Asia to breed, and return again by the middle of July, I cannot believe, especially after having found them so thoroughly in breeding-time, while still in the east of Sind. Another suspicious circumstance is the absence of females in the flocks I met with. Perhaps some of my readers may have an opportunity of finding out whether *Pastor roseus* occurs in the districts lying to the east of Sind in the month of June, as there is no doubt that the breeding-time lies between the 20th May and the commencement of July."

529. Sturnus humii, Brooks. *The Himalayan Starling.*

Sturnus unicolor, *Marm., apud Jerd. B. Ind.* ii, p. 322.
Sturnus nitens, *Hume* ; *Hume, Rough Draft N. & E.* no. 682.

The Himalayan Starling breeds in Candahar, Cashmere, and the extreme north-west of the Punjab. It is the bird which Dr. Jerdon includes in his work as *S. unicolor* (a very different bird, which does not occur within our limits), and which Mr. Theobald referred to as breeding in Cashmere as *Sturnus vulgaris*, which bird does not, as far as I can learn, occur in the Valley of Cashmere, though it may in Yarkand.

This Starling lays towards the end of April at Peshawur, where I found it nesting in holes in willow-trees in the cantonment compounds. In Candahar it lays somewhat earlier, and in the Valley of Cashmere somewhat later, viz. in the month of May.

It builds in holes of trees, in river-banks, and in old buildings and bridges, constructing a loose nest of grass and grass-roots, with sometimes a few thin sticks ; it is perhaps more of a lining to the hole than a true nest. It lays five or six eggs.

Mr. Brooks says :—" It is like *S. unicolor*, but smaller, with shorter wing and more beautiful reflections. It is excessively abundant in Cashmere, at moderate elevations, and in the Valley, and breeds in holes of trees and in river-banks. The eggs are like those of *S. vulgaris*, but rather smaller. The latter bird * occurs plentifully in the plains of India in the cold weather, and is as profusely spotted as English specimens. The bills vary in length, and are not longer, as a rule, than those of British birds. I did not meet with *S. vulgaris* in Cashmere. It appears to migrate more to the west, for it is said to be common in Afghanistan. *S. nitens* also occurs in the plains in the cold season. I have Etawah specimens. They are at that time slightly spotted, but can always be very easily distinguished from *S. vulgaris*."

Mr. W. Theobald makes the following remark on its nidification in the Valley of Cashmere :—" Lays in the second and third weeks of May ; eggs ovato-pyriform ; size 1·15 by 0·85 ; colour, pale clear bluish green ; valley generally, in holes of bridges, tall trees, &c., in company with *Corvus monedula*."

Captain Hutton records that " *S. vulgaris* remains only during the coldest months, and departs as spring approaches ; whereas the present species builds in the spring at Candahar, laying seven or eight blue eggs, and the young are fledged about the first week in May."

---

* Mr. Brooks here refers to *S. menzbieri.*—ED.

The eggs of this species are generally somewhat elongated ovals, a good deal compressed towards one end, and not uncommonly more or less pyriform. They are glossy, but in a good light have the surface a good deal pitted. They are entirely devoid of markings, and seem to have the ground one uniform very pale sea-greenish blue. They appear to vary very little in colour, and to average generally a good deal smaller than those of the Common Starling.

They vary in length from 1·02 to 1·19, and in breadth from 0·78 to 0·87; but the average of twenty eggs is 1·13 by 0·83.*

531. **Sturnus minor**, Hume.   *The Small Indian Starling.*

Sturnus minor, *Hume; Hume, Cat.* no. 681 bis.

Mr. Scrope Doig furnishes us with the following interesting note on the breeding of *S. minor* in Sindh :—

" Last year I mentioned to my friend, Captain Butler, that I had noticed Starlings going in and out of holes in trees along the ' Narra ' in the month of March, and that I thought they must be breeding there ; he said that I must be mistaken, as *S. vulgaris* never bred so far south. As it happens we were both correct—he in saying *S. vulgaris* did not breed here, and I in saying that *Starlings* did. My Starling turns out to be the species originally described from Sindh as *Sturnus minor* by Mr. Hume; and as I have now sent Mr. Hume a series of skins and eggs, I trust he will give us a note on the subject of our Indian Starlings. In February I shot one of these birds, and on dissection found that they were beginning to breed; later on, early in March, I again dissected one and found that there was no doubt on the subject, and so began to look for their nests; these I found in holes in kundy trees growing along the banks of the Narra, and also situated in the middle of swamps. The eggs were laid on a pad of feathers of *Platalea leucorodia* and *Tantalus leucocephalus*, which were breeding on the same trees, the young birds being nearly fledged ; the greatest number of eggs in any one nest was five. The first date on which I took eggs was the 13th March, and the last was on the 15th May.

" The eggs are oval, broad at one end and elongated at the other; the texture is rather waxy, with a fine gloss, and they are of a pale delicate sea-green colour.

---

* STURNUS PORPHYRONOTUS, Sharpe.   *The Central-Asian Starling.*

This species breeds in Kashgharia, and visits India in winter. Dr. Scully writes :—"This Starling breeds in May and June, making its nest in the holes of trees and walls, and in gourds and pots placed near houses by the Yarkandis for the purpose. It seems to make only a simple lining for its hole, composed of grass and fibres. The eggs vary in shape from a broadish oval to an elongated oval compressed at one end ; they are glossy and, in a strong light, the surface looks pitted. The eggs are quite spotless, but the colour seems also to vary a good deal—from a deep greenish blue to a very pale light sea-blue. In size they vary from 1·1 to 1·22 in length, and from 0·80 to 0·86 in breadth ; but the average of nine eggs is 1·19 by 0·83."

" The birds during the breeding-time confine themselves closely to their breeding-ground, so much so, that except when close to their haunts none are ever seen.

" The size of the eggs varies from 1·00 to 1·10 in length, and from ·70 to ·80 in breadth. The average of twelve eggs is 1·03 in length and ·79 in breadth."

He subsequently wrote:—" I first noticed this bird breeding on the 11th March ; on the 10th, while marching, I saw some on the side of the road and shot one, and on opening it found it was breeding. Accordingly on the 11th, on searching, I found their breeding-ground, which was in the middle of a Dhund thickly studded over with kundy trees, in the holes of which they had their nests. The nest lay at the bottom of the hole, which was generally some 18 inches deep, and consists of a few bits of coarse sedge-grass and feathers of *T. leucocephalus* and *P. leucorodia* (which were breeding close by). Five was the maximum number of eggs, but four was the normal number in each nest.

" I afterwards found these birds breeding in great numbers all along the Eastern Narra wherever there were suitable trees (kundy trees). At the place I first found them in, the young ones are now many of them fledged and flying about, while in other places they are just beginning to lay.

" The total length of their breeding-ground in any district must be close on 200 miles, but entirely confined to the banks of the river. If you looked four miles from the river, one side or the other, you would not see one. Can *Pastor roseus* breed in India in some similar secluded spot ? I have been rather unlucky in getting their eggs, as at each place which I visited personally the birds had either young ones or were just going to lay."

The eggs of this species are moderately broad ovals, sometimes slightly elongated, always more or less appreciably pointed towards the small end. The shell is extremely smooth and has a fine gloss. The colour, which is extremely uniform in all the specimens, is an excessively delicate pale blue with a faint greenish tinge, a very beautiful colour. They vary from 1 to 1·18 in length, and from 0·71 to 0·82 in breadth.

537. Sturnia blythii (Jerdon). *Blyth's Myna.*

Temenuchus blythii *(Jerd.)*, *Jerd. B. Ind.* ii, p. 331.
Sturnia blythii *(Jerd.)*, *Hume, Cat.* no. 689.

Mr. Iver Macpherson sent me from Mysore three eggs and a skin of a Myna, which latter, although in very bad order, is undoubtedly *S. blythii.* He says :—" It is very possible that the bird now sent is *S. malabarica*, and it is such a bad specimen that I fear it will not be of much use to you for the purpose of identification. I think it is *Sturnia blythii*, as Jerdon says that *S. malabarica* is only a cold-weather visitant in the south of India.

" I will, however, try and procure you a good specimen of the
24*

bird. It is only found in our forests bordering the Wynaad, and as it is far from common, I am not well acquainted with it.

"I am also inclined to think that it is not a permanent resident with us, but that a few couples come to these forests only to breed.

"The only nest I have ever found was taken on the 24th April, 1880, and was in a hole of a dry standing tree in a clearing made for a teak plantation and contained three fresh eggs.

"A few days subsequently I saw a brood of young ones flying about a dry tree in the forest, so probably the breeding-season here extends through April and May."

The eggs are very similar to those of *Sturnia malabarica* and *S. nemoricola*, but perhaps slightly larger. They are moderately elongated ovals, generally decidedly pointed towards the small end. The shell is very fine and smooth, and has a fair amount of gloss. In colour they are a very delicate pale greenish blue. They measure 0·99 and 1 in length by 0·71 in breadth.

538. Sturnia malabarica (Gm.).   *The Grey-headed Myna.*

Temenuchus malabaricus (*Gm.*), *Jerd. B. Ind.* ii, p. 330; *Hume, Rough Draft N. & E.* no. 688.

I have never met with the nest of the Grey-headed Myna myself, but am indebted to Mr. Gammie for its eggs and nest. That gentleman says :—" I obtained a nest of this species near Mongphoo (14 miles from Darjeeling), at an elevation of about 3400 feet. The nest was in the hollow of a tree, and was a shallow pad of fine twigs, with long strips of bark intermingled in the base of the structure, and thinly lined with very fine grass-stems. The nest was about 4 inches in diameter and less than 1½ inch in height exteriorly, and interiorly the depression was perhaps half an inch deep. It contained four hard-set eggs."

This year he writes to me :—" The Grey-headed Myna breeds about Mongphoo, laying in May and June. I have taken several nests now, and I found that they prefer cleared tracts where only a ew trees have been left standing here and there, especially on low but breezy ridges, at elevations of from 2500 to 4000 feet. They always nest in natural holes of trees both dead and living, and at any height from 20 to 50 feet from the ground. The nest is shallow, principally composed of twigs put roughly together in the bottom of the hole. They lay four or five eggs.

"The Grey-headed Myna is not a winter resident in the hills. It arrives in early spring and leaves in autumn. It is very abundant on the outer ranges of the Teesta Valley, and is generally found in those places frequented by *Artamus fuscus*. It feeds about equally on trees and on the ground, and a flock of 40 or 50 feeding on the ground in the early morning is no unusual sight."

Mr. J. R. Cripps, writing from Fureedpore, Eastern Bengal, says :—"Very common from the end of April to October, after which a few birds may be seen at times. I cannot call to mind ever having seen these birds descend to the ground. They must nest here, though I failed to find one. In front of my verandah was a large *Poinciana regia*, in the trunk of which, and at about seven feet from the ground, was an old nest-hole of *Xantholæma* which a pair of these birds widened out. During all May and June I watched these birds pecking away at the rotten wood and throwing the bits out. They generally used to engage in this work during the heat of the day; and, although I several times searched the hole, no eggs were found; the pair were not pecking at the decayed wood for insects, for I watched them through a glass. Had I remained another month at the factory most likely they would have laid during that time; it was on this account their lives were spared. This species associates with its congeners on the peepul trees when they are in fruit, which they eat greedily."

Subsequently detailing his experiences at Dibrugarh in Assam, he adds:—" On the 27th May I found a nest with three callow young and one fresh egg. The birds had excavated a hole in a rotten and dead tree about 18 feet from the ground, and had placed a pad of leaves only at the bottom of the hole. They build both in forest as well as the open cultivated parts of the country."

Mr. Oates remarks:—" This Myna lays in Pegu in holes of trees at all heights above 20 feet. It selects a hole which is difficult of access, and I have only been able to take one nest. This was on the 13th May. This nest, a small pad of grass and leaves, contained three eggs, which were slightly incubated. They measured 0·86 by 0·7, 0·8 by 0·7, and 0·83 by 0·72."

Major C. T. Bingham writes from Tenasserim:—"I shot a Myna as she flew out of a hole in a zimbun tree (*Dillenia pentagyna*). I had nearly a fortnight before seen the birds; there was a pair of them, busy taking straw and grass-roots into the hole; and so on the 18th April, when I shot the birds, I made sure of finding the full complement of eggs, but to my regret on opening the hollow, I only found one egg resting in a loose and irregularly formed nest of roots and leaves. This solitary egg is of a pale blue colour."

The eggs vary a good deal in shape: some are broad and some are elongated ovals, but all are more or less pointed towards the small end; the shell is very fine and delicate, and rather glossy; the colour is a very delicate pale sea-green, without any markings of any kind. They vary from 0·89 to 1·0 in length, and from 0·69 to 0·72 in breadth; but the average of ten eggs is 0·93 by 0·7.

539. Sturnia nemoricola, Jerdon. *The White-winged Myna.*

Sturnia nemoricola, *Jerd., Hume*, Cat. no. 688 bis.

Mr. Oates writes from Lower Pegu:—" Of *S. nemoricola* I have taken two sets of eggs: one set of two eggs fresh, and one of three on the point of being hatched; the former on 12th May, the latter on 6th June. In size the two clutches vary extraordinarily.

The first two eggs measure ·82 × ·62 and ·85 × ·63 ; the second lot measure 1·01 × ·7, 1·0 × ·7, and 1·0 × ·7.

"The eggs are very glossy, and the colour is a uniform dark greenish blue, of much the same tint as the egg of *Acridotheres tristis*."

543. Ampeliceps coronatus, Blyth. *The Gold-crest Myna.*

Ampeliceps coronatus, *Bl.*, *Hume, Rough Draft N. & E.* no. 693 sex ; *id. Cat.* no. 693 ter.

Of the nidification of this beautiful species, the Gold-crest Myna, we possess but little information. My friend Mr. Davison, who has secured many specimens of the bird, writes :—" On the 13th April, 1874, two miles from the town of Tavoy, on a low range of hills about 200 feet above the sea-level, I found a nest of the Gold-crest Grakle. The nest was about 20 feet from the ground in a hole in the branch of a large tree. It was composed entirely of coarse dry grass, mixed with dried leaves, twigs, and bits of bark, but contained no feathers, rags, or such substances as are usually found in the nests of the other Mynas. The nest contained three young ones only a day or two old."

544. Temenuchus pagodarum (Gm.). *The Black-headed Myna.*

Temenuchus pagodarum (*Gm.*), *Jerd. B. Ind.* ii, p. 329; *Hume, Rough Draft N. & E.* no. 687.

The Pagoda or Black-headed Myna breeds throughout the more open, dry, and well-wooded or cultivated portions of India. In Sindh and in the more arid and barren parts of the Punjab and Rajpootana on the one hand, or in the more humid and jungly localities of Lower Bengal on the other, it occurs, if at all, merely as a seasonal straggler. How Adams, quoted by Jerdon (vol. ii, p. 330), could say that he never saw it in the plains of the North-West Provinces (where, as a matter of fact, it is one of our commonest resident species), altogether puzzles me.

Neither in the north nor in the south does it appear to ascend the hills or breed in them at any elevations exceeding 3000 or 4000 feet.

The breeding-season lasts from May to August, but in Upper India the great majority lay in June.

According to my experience in Northern India it nests exclusively in holes in trees. Dr. Jerdon says that "at Madras it breeds about large buildings, pagodas, houses, &c." This is doubtless correct, but has not been confirmed as yet by any of my Southern Indian correspondents, who all talk of finding its nest in holes of trees.

The whole is thinly lined with a few dead leaves, a little grass, and a few feathers, and occasionally with a few small scraps of some other soft material.

They lay from three to five eggs.

From Hansie Mr. W. Blewitt writes :—" During June and the early part of July I found numerous nests of this species in holes of shishum, peepul, neem, and siriss trees situated on the bank of the Hissar Canal. The holes where at heights of from 12 to 15 feet from the ground, and in each a few leaves or feathers were laid under the eggs. Five was the greatest number found in any one hole."

Recording his experience in the Delhi, Jhansi, and Saugor Divisions, Mr. F. R. Blewitt tells us that the Pagoda Myna breeds from May to July, building its nest in holes of trees, selecting where possible those most inaccessible. I have always found the nest in the holes of mango, tamarind, and high-growing jamún trees. Feathers and grass, sometimes an odd piece of rag, are loosely placed at the bottom of the hole, and on these the eggs repose.

" The eggs are pale bluish green, and from four to five form the regular number. I may add that only on one occasion did I obtain five eggs in a nest."

" In Oudh," writes Mr. R. M. Adam, " I took one nest of this species, in a hole in a mango-tree, on the 5th May, containing five eggs."

Major C. T. Bingham remarks :—" All nests I have found at Allahabad and Delhi have been in holes in trees, in the end of May, June, and July. Nest strictly speaking there is none, but the holes are lined with feathers and straw, in which the eggs, four in number, are generally half buried."

Lieut. H. E. Barnes tells us that this Myna breeds in Rajputana in June, and that he found one nest in that month in a hole of a tree with three eggs."

Colonel E. A. Butler records the following notes :—" The Black-headed Myna breeds plentifully in the neighbourhood of Deesa in June, July, and August, but somehow or other I was unlucky this year (1876) in procuring eggs. On the 30th July I found a nest containing four young birds and another containing four eggs about to hatch. On the 2nd of August I found three nests, all containing young birds. On the 20th August I found four more nests; three contained young birds and the fourth four fresh eggs. All of these nests were in holes of trees, in most instances only just large enough at the entrance for the bird to pass through. In some cases there was no lining at all except wood dust, in others a small quantity of dry grass and a few feathers. The average height from the ground was about 8 or 10 feet; some nests were, however, not more than 4 or 5 feet high.

" Belgaum, 21st May, 1879.—A nest in the roof of a house under the tiles; three fresh eggs. Another nest on the same date in a hole of a tree, containing one fresh egg. The hole appeared to be an old nest-hole of a Barbet. Other nests observed later on, in June and July, in the roofs of houses under the tiles. Another nest in the hole of a tree, 27th April, containing four fresh eggs. Three more nests, 4th May, containing three incubated eggs, three fresh eggs, and three young birds respectively. Two of the nests were in the nest-holes of Barbets, from which I had taken eggs the month previous. 7th May, another nest containing four fresh eggs.

"I can confirm Dr. Jerdon's statement, quoted in the Rough Draft of 'Nests and Eggs of Indian Birds,' relative to this species breeding in large buildings, having observed several nests myself this season at Belgaum on the roofs of bungalows. In one bungalow, the mess-house of the 83rd Regt., there were no less than three nests at one time built under the eaves of the roof."

Messrs. Davidson and Wenden, writing of the Deccan, say:— " Not quite so common as *Acridotheres tristis*. Breeds at Satara in May."

Mr. Benjamin Aitken remarks:—" In Nests and Eggs, p. 433, you write:—' Dr. Jerdon says that at Madras it breeds about large buildings, pagodas, houses, &c. This is doubtless correct, but has not been confirmed as yet by any of my Southern Indian correspondents, who all talk of finding its nest in holes of trees.' On the 29th June last year I was at the Anniversary Meeting of the Medical College, and the proceedings were disturbed by the incessant clatter of *two* broods of young of this species. The nests were in holes in the wall near the roof, and the two pairs of old birds, which were feeding their young, kept coming and going the whole time, flying in at the windows and popping into the holes over the peoples' heads. In the following month a nest of young were taken out of a hole in the outer wall of a house I was staying at, and the birds laid again and hatched another brood.

"I very rarely saw the Black-headed Myna in Bombay, Poona, or Berar, but here, in Madras, it is, if anything, commoner than *A. tristis*."

And Mr. J. Davidson, writing from Mysore, also confirms Jerdon's statement; he says:—" *T. pagodarum* breeds here in holes in the roofs of houses as well as in trees."

Of the breeding of this Myna in Ceylon, Colonel Legge says:— " In the northern part of Ceylon this Myna breeds in July and August, and nests, I am informed, in the holes of trees."

Mr. A. G. R. Theobald notes that " early in August I found a nest of *T. pagodarum* at Ahtoor, the hill-station of the Shevaroys. It was down in the inside of a partly hollow nut-tree log, attached to a scaffolding, about 2½ feet down and, say, 35 feet from the ground, and was composed of dry leaves and a few feathers. It contained three fresh eggs."

The eggs of this Myna are, of course, glossy and spotless, and the colour varies from very pale bluish white to pale blue or greenish blue. I have never seen an egg of this species of the full clear sky-blue often exhibited by those of *A. tristis*, *S. contra*, and *A. giuginianus*.

The eggs vary in length from 0·86 to 1·15, and in breadth from 0·66 to 0·8; but the average of fifty-four eggs is 0·97 by 0·75.

546. Graculipica nigricollis (Payk.).   *The Black-necked Myna.*

All that we know of the nidification of this species is contained in the following brief note by Dr. John Anderson :—

"It has much the same habits as *Sturnopastor contra* var. *superciliaris.* I found it breeding in the month of May in one of the few clumps of trees at Muangla."

Muangla lies to the east of Bhamo.

549. Acridotheres tristis (Linn.).   *The Common Myna.*

Acridotheres tristis (*Linn.*), *Jerd. B. Ind.* ii, p. 325 ; *Hume, Rough Draft N. & E.* no. 684.

The Common Myna breeds throughout the Indian Empire, alike in the plains and in the hills.   A pair breed yearly in the roof of my verandah at Simla, at an elevation of 7800 feet.

They are very domestic birds, and greatly affect the habitations of man and their immediate neighbourhood.   They build in roofs of houses, holes in walls, trees, and even old wells, in the earthen chatties that in some parts the natives hang out for their use (as the Americans hang boxes for the Purple Martin), and, though *very* rarely, once in a way *on* the branches of trees.

Captain Hutton says :—"This is a summer visitor in the hills, and arrives at Mussoorie with the *A. fuscus,* Wagl.   It builds in the hole of a tree, which is lined with dry grass and feathers, and on no occasion have I *ever* seen a nest made on the branches of a tree composed of twigs and grass as stated by Captain Tickell."

But in this instance Captain Tickell may have been right, for I have once seen such a nest myself, and Mr. R. M. Adam writes :— "Near Sambhur, on the 7th July, I saw a pair of this species building a large cup-shaped nest in a babool tree ; " while Colonel G. F. L. Marshall affirms that this species "*frequently* lays in cup-shaped nests of sticks placed in trees, like small Crows' nests." And he subsequently writes :—"I can distinctly reaffirm what I said as to this species building a nest in the fork of a tree.   In the compound of Kalunder gari choki, in the Bolundshahr district, I found no less than five of these nests on one day ; the compound is densely planted with sheeshum trees, which were there about twenty feet high, and the nests were near the tops of these trees.   I found several other similar nests on the canal-bank, one with young on the 11th September."

Also writing in this connection from Allahabad, Major C. T. Bingham says :—

"Twice I have found the nest of this bird in trees, but it generally builds in holes, both in trees and walls, and commonly in the thatch of houses.   Once I got a couple of eggs from a nest made amidst a thick-growing creeper."

Neglecting exceptional cases like these, the nest is a shapeless but warm lining to the hole, composed chiefly of straw and feathers, but in which fine twigs, bits of cotton, strips of rags, bits of old rope, and all kinds of odds and ends may at times be found incorporated.

The normal breeding-season lasts from June to August, during which period they rear two broods ; but in Ross Island (Andamans), where they were introduced some years ago, they seem to breed *all. through* the year. Captain Wimberley, who sent me some of their eggs thence, remarks :—" The bird is now very common here. As soon as it has cleared out one young brood, it commences building and laying again. This continues all the year round."

I think this great prolificness may be connected with the uniformly warm temperature of these islands and the great heat of the sun there all through the year rendering much incubation unnecessary. Even in the plains of Northern India in the hot weather when they breed these birds do not sit close, and since at the Andamans the weather is such all the year round that the eggs almost hatch themselves, this may be partly the reason why these birds have so many more broods there than with us, where, for at least half the year, constant incubation would be necessary. I particularly noticed when at Bareilly how very little trouble these Mynas sometimes took in hatching their eggs, and I may quote what I then recorded about the matter :—

" In a nest in the wall of our verandah we found four young ones. This was particularly noteworthy, because from my study-window the pair had been watched for the last month, first courting, then flitting in and out of the hole with straws and feathers, ever and anon clinging to the mouth of the aperture, and laboriously dislodging some projecting point of mortar; then marching up and down on the ground, the male screeching out his harsh love-song, bowing and swelling out his throat all the while, and then rushing after and soundly thrashing any chance Crow (four times his weight at least) that inadvertently passed too near him ; never during the whole time had either bird been long absent, and both had been seen together daily at all hours. I made certain that they had not even begun to sit, and behold there were four fine young ones a full week old chirping in the nest! Clearly these birds are not close sitters down here; but I well remember a pair at Mussoorie, some 6000 feet above the level of the sea, the most exemplary parents, one or other being on the eggs at all hours of the day and night. The morning's sun beats full upon the wall in the inner side of which the entrance to the nest is; the nest itself is within 4 inches of the exterior surface ; at 11 o'clock the thermometer gave 98° as its temperature. I have often observed in the river Terns (*Seena aurantia, Rhynchops albicollis, Sterna javanica*) and Pratincoles (*Glareola lactea*), who lay their eggs in the bare white glittering river-sands, that so long as the sun is high and the sand hot they rarely sit *upon* their eggs, though one or other of the parents constantly remains beside or hovering near and over them, but in the early morning, in somewhat cold and cloudy days, and as the night draws on, they are all close sitters. I suspect that instinct teaches the birds that, when the natural temperature of the nest reaches a certain point, any addition of their body-heat is unnecessary, and this may explain why during the hot days (when we alone noticed them), in this very hot hole, the parent Mynas spent so little of their time in the nest whilst the process of hatching was going on."

They lay. indifferently four or five eggs. I have just as often found the former as the latter number, but I have never yet met with more.

From Lucknow Mr. G. Reid tells us:—" Generally speaking the Common Myna, like the Crow (*Corvus splendens*), commences to breed with the first fall of rain in June—early or late as the case may be—and has done breeding by the middle of September. It nests indiscriminately in old ruins, verandahs, walls of houses, &c., but preferentially, I think, in holes of trees, laying generally four, but sometimes five eggs."

Colonel E. A. Butler writes:—" In Karachi Mynas begin to lay at the end of April. The Common Myna breeds in the neighbourhood of Deesa during the monsoon, principally in the months of July and August, at which season every pair seems to be engaged in nidification. I have taken nests containing fresh eggs during the first week of September; and birds that have had their first nests robbed or young destroyed probably lay even later still."

Lieut. H. E. Barnes informs us that this Myna breeds in Rajputana during June and July.

Mr. Benjamin Aitken has furnished me with the following interesting note :—"A pair of Mynas clung tenaciously for two years, from June 1863 to August 1865, to a hole in some matting in the upper verandah of a house in Bombay. During this period they hatched six broods, one of which I took and another was destroyed, by rats perhaps. I had a strong suspicion that more than one set of eggs were destroyed besides.

"The remarkable thing I wish to note is that every alternate brood of young contained an *albino*, pure white and with pink eyes; being three in all. Every time a new set of eggs was to be laid, a new nest was built on the top of the old one. I once tore down the whole pile, as it was infested with vermin, and found that seven nests had been made, one upon another, showing that the Mynas must have occupied the hole long before I noticed them. Each nest was complete in itself and well lined, and as Mynas are not sparing of their materials, the accumulated heap was nearly two feet deep. Every separate nest contained a piece of a snake's skin, and with reference to your remark on this point I may say that every Myna's nest that I have ever examined has had a piece of snake-skin in it. This may, I think, be simply accounted for by the fact of snake-skin lying about plentifully in those places where Mynas mostly pick up their building-materials. The breeding-season extends into September in Bombay ; and though it usually begins in June, I found a nest of half-fledged young at Khandalla on the 31st May, 1871.

" With reference to your remarks in ' Nests and Eggs,' that you have never met with more than five eggs in a nest, I would mention that I took six eggs from a nest in the roof of a house I occupied at Akola, on the 20th June, 1870.

" At the same station in August 1869 a nest of young Mynas was reared above the hinge of the semaphore signal at the railway-station. One or other arm of the signal must have risen and fallen every time a train passed, but the motion neither alarmed the birds nor disarranged the nest."

Messrs. Davidson and Wenden remark of this Myna in the Deccan :—" Common, and breeds in May and June."

Mr. J. Inglis, writing from Cachar, says :—" The commonest of all birds here. Breeds throughout the summer months. It makes its nest generally in the roofs of houses or in holes in trees. It lays about five eggs of a very pale blue colour."

Finally, Mr. Oates writes from Pegu :—" Commences making nest about 15th March. I have taken eggs as late as 17th July, but in this case the previous brood had been destroyed. Normally no eggs are to be found after June."

The eggs, which are larger than those of either *Sturnopastor contra* or *A. ginginianus*, in other respects resemble these eggs greatly, but when fresh are, I think, on the whole of a slightly darker colour. They are rather long, oval, often pear-shaped, eggs, spotless and brilliantly glossy, varying from very pale blue to pure sky- or greenish blue.

In length they vary from 1·05 to 1·28, and in breadth from 0·8 to 0·95 ; but the average of ninety-seven eggs is 1·19 by 0·86.

550. Acridotheres melanosternus, Legge. *The Common Ceylon Myna.*

Acridotheres melanosternus, *Legge, Hume, Cat. no.* 684 bis.

Colonel Legge tells us, in his ' Birds of Ceylon,' that " this species breeds in Ceylon from February until May, nesting perhaps more in the month of March than in any other. It builds in holes of trees, often choosing a cocoanut-palm which has been hollowed out by a Woodpecker, and in the cavity thus formed makes a nest of grass, fibres, and roots. I once found a nest in the end of a hollow areca-palm which was the cross beam of a swing used by the children of the Orphan School, Bonavista, and the noise of whose play and mirth seemed to be viewed by the birds with the utmost unconcern. The eggs are from three to five in number ; they are broad ovals, somewhat pointed towards the small end, and are uniform, unspotted, pale bluish or ethereal green. They vary in length from 1·07 to 1·2 inch and in breadth from 0·85 to 0·92 inch.

" Layard styles the eggs ' light blue, much resembling those of the European Starling in shape, but rather darker in colour.' "

ACRIDOTHERES

551. **Acridotheres ginginianus** (Lath.). *The Bank Myna.*

Acridotheres ginginianus (*Lath.*), *Jerd. B. Ind.* ii, p.326; *Hume, Rough Draft N. & E.* no. 685.

The Bank Myna breeds throughout the North-West Provinces and Oudh, Behar, and Central Bengal, the greater portion of the Central Provinces, and the Punjab and Sindh. Adams says it does not *occur* in the Punjab; but, as Colonel C. H. T. Marshall correctly pointed out to me years ago, and I have verified the facts, it breeds about Lahore and many other places, and in the high banks of the Beas, the Sutlej, the Jhelum, and the Indus, congregating in large numbers on these rivers just as it does on the Jumna or the Ganges.

It builds exclusively, so far as my experience goes, in earthen banks and cliffs, in holes which it excavates for itself, always, I think, in close proximity to water, and by preference in places overhanging or overlooking running water.

The breeding-season lasts from the middle of April to the middle of July, but I have found more eggs in May than in any other month.

Four is the usual number of the eggs; I have found five, but never more. If Theobald got seven or eight, they belonged to two pairs; and the nests so run into each other that this is a mistake that might easily be made, even where coolies were digging into the bank before one.

There is really no variety in their nesting arrangements, and a note I recorded in regard to one colony that I robbed will, I think, sufficiently illustrate the subject. All that can be said is that very commonly they nest low down in earthy cliffs, where it is next to impossible to explore thoroughly their workings, while in the instance referred to these were very accessible :—

" One morning, driving out near Bareilly, we found that a colony of the Bank Myna had taken possession of some fresh excavations on the banks of a small stream. The excavation was about 10 feet deep, and in its face, in a band of softer and sandier earth than the rest of the bank, about a foot below the surface of the ground, these Mynas had bored innumerable holes. They had taken no notice of the workman who had been continuously employed within a few yards of them, and who informed us that the Mynas had first made their appearance there only a month previously. On digging into the bank we found the holes all connected with each other, in one place or another, so that apparently every Myna could get into or out from its nest by any one of the hundred odd holes in the face of the excavation. The holes averaged about 3 inches in diameter, and twisted and turned up and down, right and left, in a wonderful manner; each hole terminated in a more or less well-marked bulb (if I may use the term), or egg-chamber, situated from 4 to 7 feet from the face of the bank. The egg-chamber was floored with a loose nest of grass, a few feathers, and, in many instances, scraps of snake-skins.

" Are birds superstitious, I wonder? Do they believe in charms? If not, what induces so many birds that build in holes in banks to select out of the infinite variety of things, organic or inorganic, pieces of snake-skin for their nests? They are at best harsh, un-• manageable things, neither so warm as feathers, which are ten times more numerous, nor so soft as cotton or old rags, which lie about broadcast, nor so cleanly as dry twigs and grass. Can it be that snakes have any repugnance to their ' worn out weeds,' that they dislike these mementos of *their* fall *, and that birds which breed in holes into which snakes are likely to come by instinct select these exuviæ as scare-snakes?

" In some of the nests we found three or four callow young ones, but in the majority of the terminal chambers were four, more or less, incubated eggs.

" I noticed that the tops of all the mud-pillars (which had been left standing to measure the work by) had been drilled through and through by the Mynas, obviously not for nesting-purposes, as not one of them contained the vestige of a nest, but either for amusement or to afford pleasant sitting-places for the birds not engaged in incubation. Whilst we were robbing the nests, the whole colony kept screaming and flying in and out of these holes in the various pillar-tops in a very remarkable manner, and it may be that, after the fashion of Lapwings, they thought to lead us away from their eggs and induce a belief that their real homes were in the pillar tops."

Colonel G. F. L. Marshall remarks :—" This species breeds in the Bolundshahr District in June and July. It makes its nest in a hole in a bank, but more often in the side of a kucha or earthen well. A number of birds generally breed in company. The nest is formed by lining the cavity with a little grass and roots and a few feathers. On the 8th July I found a colony breeding in a well near Khoorjah, and took a dozen fresh eggs."

Writing from Lucknow, Mr. G. Reid says :—" During the breeding season it associates in large flocks along the banks of the Goomti, where it nidificates in colonies in holes in the banks of the river. From some of these holes I took a few fresh eggs on the 15th May, and again on the 30th June on revisiting the spot. In the district it breeds in old irrigation-wells and occasionally in ravines with good steep banks."

Major C. T. Bingham, writing from Allahabad, says :—" Breeds in June, July, and August in holes in sandy banks of rivers and nullahs. Eggs, five in number, laid on a lining of straw and feathers."

---

* " When the snake," says an Arabic commentator, " tempted Adam, it was a winged animal. To punish its misdeeds the Almighty deprived it of wings, and condemned it thereafter to creep for ever on its belly, adding, as a perpetual reminder to it of its trespass, a command for it to cast its skin yearly."

Colonel E. A. Butler notes :—"The Bank Myna lays about
Deesa in June and July. On the 26th June I lowered a man down
several wells, finding nests containing eggs and nests containing
young ones, some nearly fledged. The nests are generally in holes
in the brickwork, often further in than a man can reach, and several
-pairs of birds usually occupy the same well. The eggs vary much
in shape and number. In some nests I found as many as five, in
others only two or three. In colour they closely resemble the eggs
of *A. tristis*, but they are slightly smaller, the tint is of a decidedly
deeper shade, and the shell more glossy. July 5th, several nests,
some containing eggs, others young ones. July 13th, numerous
nests in wells and banks, some containing fresh, others incubated
eggs, and others young birds of all sizes. The eggs varied in number from two to five. I took twenty-six fresh eggs and then
discontinued."

Lieut. H. E. Barnes informs us that in Rajputana this Myna
breeds about May.

The eggs are typically, I think, shorter and proportionally broader
than those of other kindred species already described; very pyriform varieties are, however, common. They are as usual spotless,
very glossy, and of different shades of very pale sky- and greenish
blue. Although, when a large series of the eggs of this and each
of the preceding species are grouped together, a certain difference
is observable, individual eggs can by no means be discriminated,
and it is only by taking the eggs with one's own hand that one can
feel certain of their authenticity.

In length they vary from 0·95 to 1·16, and in breadth from
0·72 to 0·87; but the average of forty-seven eggs is 1·05 by 0·82.

552. Æthiopsar fuscus (Wagl.). *The Jungle Myna.*

Acridotheres fuscus (*Wagl.*), *Jerd. B. Ind.* ii, p. 327; *Hume, Rough
Draft N. & E.* no. 686.

The Jungle Myna eschews the open cultivated plains of Upper,
Central, and Western India. It breeds throughout the Himalayas,
at any elevations up to 7000 feet, where the hills are not bare, and
in some places in the sub-Himalayan jungles. It breeds in the
plains country of Lower Bengal, and in both plains and hills of
Assam, Cachar, and Burma, and also in great numbers in the
Nilgiris and all the wooded ranges and hilly country of the
Peninsula. The breeding-season lasts from March to July, but the
majority lay everywhere, I think, in April, except in the extreme
north-west, where they are later.

Normally, they build in holes of trees, and are more or less
social in their nidification. As a rule, if you find one nest you
will find a dozen within a radius of 100 yards, and not unfrequently
within one of ten yards. But, besides trees, they readily build in
holes in temples and old ruins, in any large stone wall, in the
thatch of old houses, and even in their chimneys.

The nest is a mere lining for the hole they select, and varies in

size, and shape with this latter; fine twigs, dry grass, and feathers
are the materials most commonly used, the feathers being chiefly
gathered together to form a bed for the eggs; but moss, moss and
fern roots, flocks of wool, lichen, and down may often be found in
greater or less quantities intermingled with the grass and straw
which forms the main body, or with the feathers that constitute
the lining, of the nest. I have never found more than five eggs,
but Miss Cockburn says that they sometimes lay six.

From Murree, Colonel C. H. T. Marshall writes:—"This Myna,
which takes the place of _A. tristis_ in the higher hills, breeds always
in holes in trees. We found five or six nests in June and early in
July."

They breed near Solan, below Kussowlee, and close to Jerripani,
Captain Hutton's place below Mussoorie, in both which localities I
have taken their nests myself.

Captain Hutton remarks:—"This is a summer visitant in the
hills, and is common at Mussoorie during that season; but it does
not appear to visit Simla, although it is to be found in some of the
valleys below it to the south. It breeds at Mussoorie in May and
June, selecting holes in the forest trees, generally large oaks, which
it lines with dry grass and feathers. The eggs are from three to
five, of a pale greenish blue, shape ordinary, but somewhat inclined
to taper to the smaller end. This species usually arrives from the
valleys of the Dhoon about the middle of March; and, until they
begin to sit on their eggs, they congregate every morning and
evening into small flocks, and roost together in trees near houses;
in the morning they separate for the day into pairs, and proceed
with the building of nests or laying of eggs. After the young are
hatched and well able to fly, all betake themselves to the Dhoon in
July."

In Kumaon I found them breeding near the Ramghur Iron-
works, and, writing from Nynee Tal, Colonel G. F. L. Marshall
says that they " breed very commonly at Bheem Tal (4000 feet),
but I have not noticed them at Nynee Tal. I took a great many
eggs; they were all laid in holes in rotten trees at a height of 2 to
8 feet from the ground; they average much smaller than the eggs
of _A. tristis_, but are similar in colour."

Writing from Nepal, Dr. Scully says:—" This species is common
and a permanent resident in the Valley of Nepal, but does not occur
in such great numbers as _A. tristis_. It is also found in tolerable
abundance in the Nawakot district and the Hetoura Dun in winter.
It breeds in the Valley in May and June, laying in holes in trees
or walls; the eggs are very like those of _A. tristis_, but smaller—
not so broad. I noticed on two or three occasions an albino of
this species, which was greatly persecuted by the Crows."

Mr. G. Vidal remarks of this bird in the South Konkan:—
" Exceedingly common. Breeds in May. The irides of all I have
seen were pale slate-blue."

" In the Nilgiris," writes Mr. Wait, " the Jungle Myna's eggs
may be found at any time from the end of February to the beginning

of July. They nest in chimneys, hollow trees, holes in stone walls, &c., filling in the hole with hay, straw, moss, and twigs, and lining the cavity with feathers. They lay from three to five long, oval, greenish-blue eggs, a shade darker than those of the English Starling."

From Kotagherry Miss Cockburn tells us that " these Mynas breed in the months of March and April, and construct their nests (which consist of a few straws, sticks, and feathers put carelessly together) in the holes of trees and old thatched houses. They lay five or six eggs of a beautiful light blue, and are extremely careful of their young. The nests of these birds are so common in the months above mentioned that herd-boys have brought me more than fifty eggs at a time.

" About a year ago a pair took up their abode in my pigeon-cot, and although the eggs were often destroyed they would not leave the place, but continued to lay in the same nest. At last one of them was caught ; the other went away, but returned the next day accompanied by a new mate. At length the hole was shut up, as they committed great depredations in the garden, and were useful only in giving a sudden sharp cry of alarm when the Mhorunghee Hawk-Eagle, a terrible enemy to Pigeons, made its appearance, thus enabling the gardeners to balk him of his intended victim."

Dr. Jerdon states that " it is most abundant on the Nilgiris, where it is a permanent resident, breeding in holes in trees, making a large nest of moss and feathers, and laying three to five eggs of a pale greenish-blue colour."

Mr. C. J. W. Taylor informs us that at Manzeerabad, in Mysore, this Myna is common everywhere, and breeds in April and May.

Captain Horace Terry notes that in the Pulney hills the Jungle Myna nests in April.

Mr. Rhodes W. Morgan, writing from South India, says in ' The Ibis ' :—" It breeds on the Neilgherries in holes of trees. The hole is filled up with sticks to within about a foot of the entrance, and a smooth lining of paper, rags, feathers, &c. laid down, on which are deposited from two to six light blue eggs. The young are fed on small frogs, grasshoppers, and fruit. An egg measured 1·2 inch by ·88. Breeds in May."

At Dacca Colonel Tytler found them nesting in temples and houses about the sepoy lines.

Mr. J. R. Cripps tells us that at Furreedpore, in Bengal, this species is " pretty common, and a permanent resident. This species associates with *A. tristis*, but is seen on trees away from villages, which the latter never is. Prefers well-wooded country, whereas *A. tristis* never goes into jungle. On the 29th of June, 1877, I found a nest in a hole of a tree, about 12 feet off the ground. The diameter of the entrance-hole was two and a half inches, and inside it widened to six inches and about twenty inches in depth. The nest was a mere pad of grass and feathers, and contained four very slightly incubated eggs. And again on the 17th July, seeing the hole occupied, I again sent up a boy, who found

another four fresh eggs. The tree formed one of an avenue leading
from the house to the vats, and as men were always going along
the road it surprised me to find these birds laying there; the hole
had been caused by the heart of the tree rotting."

Mr. Oates remarks of this Myna in Pegu:—" This bird does not
appear to lay till about the 15th April. I have taken the eggs,
and I have seen numerous nests with young ones of various ages
in the middle of May. They breed by preference in holes of
trees and occasionally in the high roofs of monastic buildings."

The eggs of this species, which I have from Mussoorie, Dacca,
Kumaon, and the Nilgiris, approximate closer to those of *Acridotheres tristis* than to those of *A. ginginianus*. They are rather long
ovals, somewhat pointed usually, but often pyriform. They are
perhaps, as a rule, somewhat paler than those of either of the
above-named species, and are of the usual spotless glossy type,
varying in colour from that of skimmed milk to pale blue or
greenish blue. Typically, I think, they are proportionally more
elongated and attenuated than those either of *A. tristis*, *A.
ginginianus*, or *S. contra*.

In length they vary from 1·03 to 1·31, and in breadth from 0·78
to 0·9 ; but the average of forty eggs is 1·19 by 0·83.

555. Sturnopastor contra (Linn.).   *The Pied Myna.*

Sturnopastor contra (*Linn.*), *Jerd. B. Ind.* ii, p. 323; *Hume, Rough
Draft N. & E.* no. 683.

The Pied Pastor, or Myna, breeds throughout the North-Western
Provinces and Oudh, Bengal, the eastern portions of the Punjab and
Rajpootana (it does not extend to the western portions nor to
Sindh), the Central Provinces, and Central India.

The breeding-season lasts from May to August, but the majority
of the birds lay in June and July. It builds in trees, at heights of
from 10 to 30 feet, usually towards the extremities of lateral
branches, constructing a huge clumsy nest of straw, grass, twigs,
roots, and rags, with a deep cavity lined as a rule with quantities
of feathers. Occasionally, but very rarely, it places its nest in
some huge hole in a great arm of a mango-tree. I have seen
many hundreds of their nests, but only two thus situated.

As a rule these birds do not build in society, but at times,
especially in Lower Bengal, I have seen a dozen of their nests on
a single tree.

The nest is usually a shapeless mass of rubbish loosely put together, rough and ragged.

A note I recorded on one taken at Bareilly will illustrate sufficiently the kind of thing:—

" At the extremity of one of the branches of these same mango-
trees, a small truss of hay, as it seemed, at once caught every eye.
This was one of the huge nests of the Pied Pastor, and proved to be
some 2 feet in length and 18 inches in diameter, composed chiefly

of dry grass, but with a few twigs, many feathers, and a strip or two of rags intermingled in the mass. The materials were loosely put together, and the nest was placed high up in a fork near the extremity of a branch. In the centre was a well-like cavity some 9 inches deep by 3½ inches in diameter, at the bottom of which, amongst many feathers, lay four fresh eggs."

Five is the full complement of eggs, but they very often lay only four, and once in a hundred times six are met with.

From Hansie Mr. W. Blewitt writes that he "found numerous nests during May and June. They were all placed on keekur-trees, at heights of from 10 to 15 feet from the ground, the trees for the most part being situated on the banks of a canal or in the Dhana Beerh, a sort of jungle preserve.

"The nests were densely built of keekur and zizyphus twigs, and thickly lined with rags, leaves, and straw. Five was the greatest number of eggs that I found in any one nest."

Writing of his experience in the Delhi and Jhansi Divisions, Mr. F. R. Blewitt remarks that "the Pied Pastor breeds from June to August, making its nests between the outer branchlets of the larger lateral branches of trees, without special choice for any one kind. The nest is altogether roughly made, though some ingenuity is evinced in putting all the material of which it is composed together. Twigs, grasses, rags, feathers, &c. are all brought into requisition to form the large-made structure, which I have found, though less commonly, at a higher altitude from the ground than the 8 or 10 feet Jerdon speaks of."

Major C. T. Bingham writes:—"Breeds in Allahabad in June, July, and August; and at Delhi in May, June, and July. The nest is a large shapeless mass of straw, feathers, and rags, having a deep cavity for the eggs, which are generally five in number. The nest is almost always placed at the extreme tip of some slender branch, and there is no attempt at concealment."

Mr. J. R. Cripps tells us that at Furreedpore, in Bengal, this Myna is "very common, and a permanent resident. They eat fruit as well as insects. Lay in May and June, building their huge nests at various heights from the ground, and in any tree that comes in handy. I have generally found the nests lined with the white feathers of the paddy-birds; some of the feathers being as much as six and seven inches in length. The nests were composed principally of doob-grass; three to four eggs in each nest."

From Cachar Mr. J. Inglis writes:—"The Pied Pastor is very common all the year. It breeds during March, April, May, and June, making its nest on any sort of tree about 15 feet or more from the ground; about 100 nests may often be seen together. It prefers nesting on trees on the open fields. I do not know the number of its eggs."

The eggs are typically moderately broad ovals, a good deal pointed towards one end, but pyriform and elongated examples occur; in fact, a great number of the eggs are more or less pear-shaped. Like those of all the members of this subfamily, the eggs are blue, spotless, and commonly brilliantly glossy. In shade they vary from a delicate bluish white to a pure, though somewhat pale, sky-blue, and not uncommonly are more or less tinged with green.

They vary in length from 0·95 to 1·25, and in breadth from 0·75 to 0·9 ; but the average of one hundred eggs is 1·11 by 0·82 nearly.

556. Sturnopastor superciliaris, Blyth. *The Burmese Pied Myna.*

Sturnopastor superciliaris, *Bl., Hume, Rough Draft N. & E.* no. 683 bis.

Of the Burmese Pied Pastor, or Myna, Mr. Eugene Oates says that it is common and resident throughout the plains of Pegu. Writing from Wau he says :—

" On the 28th of April, having a spare morning, I took a very large number of nests and eggs. The eggs were in various stages of incubation, but the majority were freshly laid. On May 7th I took another nest with two eggs. These were quite fresh.

" The nest is a huge cylindrical structure, about 18 inches long and a foot in diameter, composed of straw, leaves, and feathers. It is placed at a height of from 10 to 25 feet from the ground, in a most conspicuous situation, generally at the end of a branch which has been broken off and where a few leaves are struggling to come out. A bamboo-bush is also a favourite site. This Myna will, by preference, build near houses, but in no case *in* a house; it must have a tree."

The eggs, which I owe to Mr. Oates, are, as might be expected, very similar indeed to those of our Common Pied Pastor, but they seem to average somewhat smaller.

They are moderately broad ovals, a good deal pointed towards one end, and in some cases more or less compressed there, and slightly pyriform.

The specimens sent are only moderately glossy. In colour they vary from *very* pale bluish green to a moderately dark greenish blue, but the great majority are pale.

In length they vary from 1·0 to 1·1, and in breadth from 0·73 to 0·82 ; but the average of fifteen eggs is 1·04 by 0·77.

INDEX.

INDEX.

PRINTED BY TAYLOR AND FRANCIS, RED LION COURT, FLEET STREET.